Light Vision Color

Light Vision Color

Arne Valberg

The Norwegian University of Science and Technology

John Wiley & Sons, Ltd

Other Wiley Editorial Offices

John Wiley & Sons Inc., 111 River Street, Hoboken, NJ 07030, USA

Jossey-Bass, 989 Market Street, San Francisco, CA 94103-1741, USA

Wiley-VCH Verlag GmbH, Boschstr. 12, D-69469 Weinheim, Germany

John Wiley & Sons Australia Ltd, 33 Park Road, Milton, Queensland 4064, Australia

John Wiley & Sons (Asia) Pte Ltd, 2 Clementi Loop #02-01, Jin Xing Distripark, Singapore 129809

John Wiley & Sons Canada Ltd, 22 Worcester Road, Etobicoke, Ontario, Canada M9W 1L1

Wiley also publishes its books in a variety of electronic formats. Some content that appears in
print may not be available in electronic books.

Library of Congress Cataloging-in-Publication Data

Valberg, Arne.
  [Lys, syn, farge. English]
  Light, vision, color / Arne Valberg.
      p.   cm.
  Includes bibliographical references.
  ISBN 0-470-84902-9 (cased) – ISBN 0-470-84903-7 (pbk.)
1. Color vision. I. Title.

QP483.V35   2005
152.14' 5–dc22   2004021735

British Library Cataloguing in Publication Data

A catalogue record for this book is available from the British Library

ISBN 0 470 84902 9 Hardback
      0 470 84903 7 Paperback

Typeset in $10\frac{1}{2}$/13 pt Times by Thomson Press (India) Ltd, New Delhi
Printed and bound in Great Britain by Antony Rowe Ltd., Chippenham, Wiltshire
This book is printed on acid-free paper responsibly manufactured from sustainable forestry
in which at least two trees are planted for each one used for paper production.

"...sometimes you look inside and find something outside and sometimes you look outside and find something inside."
Anna in Fynn (1979)

With eyes and nerve cells on the outside, most of vision is inside.

Contents

Preface

This book is based on courses given at the University of Oslo and at the Norwegian University of Science and Technology in Trondheim. The lectures were held for interdisciplinary groups of students, for biophysicists, biologists, psychologists, architects and engineers. While studying vision and its limitations is valuable as an applied science directly related to visual function, it has also been important in furthering our understanding of general principles underlying higher brain processes. The rapid expansion in brain research has led to an increasing interest in the visual system and its extraordinary performance, and vision science continues to be an active branch of neuroscience. Vision research is often cited as an example of what can be achieved in this area through interdisciplinary collaboration.

It is a risky venture to write a book in a field that is in such rapid development. I have not tried to cover the whole field of vision sciences, but rather to give an introduction to some basic themes. Since my own background is in the natural sciences, the book will have this bias, with emphasis on basic experiments and quantitative data. I hope it will convey some of my own fascination with combining knowledge from many different areas.

Arne Valberg

Acknowledgments

After completing the book, I wish to express my gratitude to many colleges and friends. Some of them contributed with professional knowledge, and others provided valuable perspectives and personal views. In alphabetical order I want to mention Otto D. Creutzfeldt, Torger Holtsmark, Anders Johnsson, Aart Kooijman, Jan Kremers, Barry B. Lee, Richard Jung, Karl Miescher, Borgar T. Olsen, Baingio Pinna, Klaus Richter, Inger Rudvin, Thorstein Seim and Lothar Spillmann. All have been an inspiration, along with the students in my classes, either by asking the right questions or by giving me the challenge and the support I needed.

Special thanks go to Inger Rudvin for language advice and for assistance in clarifying the text, and to Rune Kjær Valberg for his endurance in preparing the many figures. I also want to thank Heidi Arnesen, Per Fosse, Tor Gjerde, Jo Tryti and Jan Henrik Wold for valuable assistance during this work and for allowing me to use some of their data.

I also want to thank Rachael Ballard, Robert Hambrook and Andrew Slade at Wiley & Sons for their advice, great patience and courtesy.

This work has been supported by Thonning Owesen's Foundation, The Norwegian Non-fiction Writers and Translators Association, and Norwegian University of Science and Technology.

Arne Valberg
Trondheim, January 2005

1 Introduction

Vision and experience

In his book *An Anthropologist on Mars* (1995), Oliver Sacks relates the story of Virgil, a 50-year-old man who had been blind from early childhood, but who regained his sight after an operation to remove cataracts from both eyes. It is natural to ask if his vision was normal from this moment on, or if something was missing.

It appears that Virgil's vision was not that of a normally sighted person, but rather like that of somebody who was mentally blind – meaning that he was able to see, but that he did not understand what he saw. He could not judge distances; spatial relations and perspective was meaningless to him. He could decipher letters, but not connect them to form words. Images were without meaning. He could not make out the form of an object using his eyes, but had to rely on the familiar method of touch. It might have been possible to train a younger person to achieve a meaningful visual capability, but Virgil gave up after a while, accepted a functional blindness, and felt great relief.

Similar problems with orientation in a visual world after an eye operation have been described in other books. For instance, the psychologist Richard L. Gregory (1990) wrote about a similar case of a 50-year-old man who was expected to use his eyes after an operation. In a collection of cases in which vision was regained in individuals who had been born with congenital cataract, M. von Senden (1960) called attention to the remarkable difficulties these individuals had in relating to visual space, and in dealing with concepts like distance and size. Such deviations are hard to understand for people with normal vision, which comes so effortlessly. However, when you think of it, even understanding normal vision is not straightforward. For instance, how can the visual world be so stable despite the fact that we are constantly moving our eyes, heads and body? Images of borders, colors and objects shift rapidly

Light Vision Color. Arne Valberg
© 2005 John Wiley & Sons Ltd

back and forth across the retina without in any way undermining our ability to orient ourselves and to move around.

In this introduction I shall call attention to some fundamental elements pertaining to the 'act of seeing' and construct a mosaic where, at first, the parts may seem only loosely connected. Nevertheless, all elements are important at one level or another of our visual encounters with the external world. Many aspects of vision are enigmatic, and we shall, for a moment, dwell on some philosophical questions. I hope that this broad background may cause some reflection about the active role we play when engaged in 'seeing with the mind' as opposed to 'seeing with the eyes', and that these considerations may help us to better understand Virgil's destiny and similar stories.

The electromagnetic spectrum elicits an orderly sensation of colors, and vibration frequencies of strings give rise to a sound spectrum and to music. These are examples of how physical stimulation is transformed by sense organs to give the physical dimension a quality and a structure specific for each sense. Sensory organs and neural processes exhibit many physical and logical structures. However, in order to comprehend vision, we need to take a broader perspective. We need to go beyond the idea of objects eliciting sensory processes and to also ask what are the important perceived qualities of each stimulus in the image. As subjects, we experience qualitative properties, like for instance colors, in a direct and immediate way, but are they a property of the external world, of 'reality', or constructs of our brains? Maybe they are some intriguing attribute of the signals that the brain receives from the eyes? Is it at all possible to find an answer to such questions by studying, for instance, the responses of visual neurons to optical stimulation of the eye?

The eye and the visual sense have always interested philosophers and scientists. One reason for this interest, and especially that of many physicists, lies close at hand. Vision is our most important tool for understanding the organization of objects and events in the external world in which we are moving, thus it is an important means of obtaining experience and knowledge. Relying on their senses, physicists in earlier times had a direct approach to the natural phenomena they studied. The impression of order mediated by the senses was crucial in experimental investigations. The famous Scottish physicist James Clerk Maxwell (1831–1879) said that

> 'if the sensation which we call color has any laws, it must be something in our own nature which determines the form of these laws' (Maxwell, 1872).

Today, neuroscientists and biophysicists are attempting to elucidate these laws at the level of sensory cells in the retina, in rods and cones, and in nerve cells in the brain. In addition to studying the chemical and electrical phenomena that follow the stimulation of photoreceptors with light, and the neural processing thereafter, modern science also focuses on the structuring of perceptual attributes which, according to earlier beliefs, follows from the functioning of the peripheral sense organs. However, as we shall see later, these relationships are not that simple.

In order to arrive at an understanding of the external world, we rely on a variety of sensory experiences, of light, space, gravity, sound, taste, smell, etc. Scientists from many different fields study sensory processes, and they are often integrated in an interdisciplinary environment. Vision science has become an advanced area of research, and it is a key example of how progress in science depends on collaboration of many disciplines.

Visual perception

A naive realist would perhaps tell us that 'the way the world appears to us is the way it really is'. However, you need not have studied vision to discover that this statement is problematic. Our conscious perception of, for instance, a place or a situation, is different from a simple copy, like a photograph. Euclid (Greek philosopher; *ca* 300 BC), for instance, thought that visual experience was mediated by 'rays of vision' projected from the eyes to touch the surfaces of objects like a million invisible fingers (Zajonc, 1993). It is not clear to what extent these rays of vision were considered dependent on learning and interpretation, being 'rays of thought and comprehension' leading to mental images, or if they were regarded as carriers of the immediate and subjective elements, the qualitative sensory attributes of external objects (like color, sound, etc.).

The poetic idea of 'rays of vision' may appear strange to us, but according to more modern views it is also too simple to think that the visual scenes one perceive originate solely from images, i.e. from light reflected off objects and imaged optically on the retina, like in a camera. Between these opposing views one finds other theories that take neural mechanisms into account to a variable degree. Such theories allow for different levels of processing, from the lowest retinal neurosensory level to the higher cognitive level(s) of the brain. There is no clear dividing line between a 'bottom-up' explanation and a cognitive interpretation ('top-down' view) of visual perception. It is a border that will be moved in this or that direction as our knowledge about the brain increases. Maybe some day a synthesis will be reached where the various approaches are not exclusive, but complement each other (Spillmann and Dresp, 1995). From Oliver Sack's account of Virgil's story, and from modern neuroscience, we know that vision is not a passive process – it has an active mental component.

In the seventeenth century Rene Descartes (French mathematician and philosopher; 1596–1650) was already wondering why we see the objects so vividly 'out there', in three-dimensional space, when the brain only has two-dimensional images to start with, which are relatively diffuse and distorted. In addition, the images on the retinas are inverted (Figure 1.1). One might explain this apparent paradox by saying that our perception of the things around us is not congruent with the things themselves, but more like a hypothesis, or a model. According to Richard L. Gregory and the German physiologist Hermann von Helmholtz, this model is a mental construct that goes beyond what is mediated directly by the retinal image and the sensory organs, building on earlier experience and on our memory. Consider how we, as children, establish a relationship with our environment. We crawl around touching things, weighing them in our hands and putting them in our mouth. As children we are curious and explore

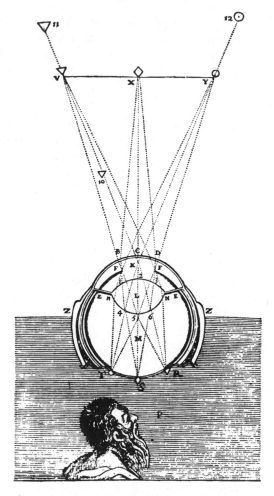

Figure 1.1   Rene Descartes and the retinal image. Light rays from the object points VXY form an inverted image RST on the retina. The drawing describes an experiment where Descartes made a little hole in the back of a bull's eye and covered it with a thin paper. When the front of the eye was directed towards a well-lit scene, he could see a distorted image of this scene on the paper. The drawing is from *La Dioptrique* (1637).

the environment with all our senses. In this way we build a relationship between the outer world and ourselves, giving the objects an identity.

It is tempting, for a moment, to interpret Euclid's rays of vision as being the qualitative attributes that the individual projects upon the things 'out there' in front of us, for example the form and the red color of an apple. Red is a qualitative sensory experience that is normally associated with long-wavelength light being reflected from the apple's surface. However, it does not require much thinking to realize that our experience of the redness of an apple cannot be deduced from this physical radiation alone. Physical concepts like energy and spectral distributions are different from experiences of 'red'. Colors are elements of our subjective, qualitative experience, and thus they are different from physical radiation and processes evoking impressions of color.

'Red' and 'green' can be associated with wavelengths in the electromagnetic spectrum. Likewise, 'coldness' and 'warmth' can be related to low and high readings on a one-dimensional temperature scale, although in both cases the sensory impressions are qualitative and thus quite different from the physical entities. Depth can similarly be treated as a perceived quality. Its 'projected property' as subjective experience (as 'rays of vision') becomes clear whenever you perceive depth as an 'illusion' in an autostereogram (see Figure 1.2).

It is probably even more counterintuitive to view our experience of motion as belonging to the projected sensory qualities. This may be easier to accept when recalling how one can see moving images when a series of pictures are rotated on a cylinder or pass quickly in front of a film projector. Moreover, perception of movement without changes in position is a common phenomenon. It is not unusual to have a strange experience of 'aftermovement' when the train stops at a station. You have the impression that the trees and buildings are moving although you know that

Figure 1.2 An autostereogram. Look at the picture at reading distance but try to avoid fixating on the plane of the paper. The perception of depth occurs when one focuses in front or beyond the picture. If this is difficult, you can hold the book up to your nose and try focusing on it. If you then move the book slowly away from your face without changing your focus, you may succeed in seeing mountains and valleys. However, some people are unable to experience depth in autostereograms because they lack the necessary stereo vision capabilities.

they are not, or you feel that the train moves backwards. The phenomenon is a physiological consequence of having been 'adapted to' movement in one direction. One interpretation of the motion aftereffect in terms of a neural correlate is that your movement-sensitive neurons need a re-calibration after having been exposed and adapted to movement in one direction for some time. Another example of apparent movement is shown in Figure 1.17.

> The motion after-effect is a phenomenon of which we can say: 'that which moves, does not move at the position where it is, nor at the position where it is not', in analogy with the Greek philosopher Zenon's (about 460 BC) statement about physical movement.

Like the perception of color and depth, perceiving motion is thus also a qualitative experience. If you are still in doubt, you may think about the phenomenon of 'motion blindness', a strange disability that has been shown to occur in persons with certain brain injuries. Such a person sees the moving object, like a moving car, in different positions, but he or she cannot observe the movement from one position to the next (Zihl *et al.*, 1983; see also p. 400).

This discussion may have made clear why we suggest it is a good idea to reinterpret and revitalize Euclid's 'rays of vision'. We can choose to view them as the immediate, qualitative attributes of our visual experience (like color, lightness, depth, movement, etc.) that we so effortless project out onto the objects of the outside world. These and other acquired properties – collection of elementary *perceptions* that together make up our model of the world – contribute, together with our interpretation and our understanding, to the mental image in front of us. The importance of interpretation and meaning is strikingly demonstrated in so-called cognitive, reversible 'visual illusions', where we can choose among alternative meanings. Figure 1.3 shows some examples. Figure 1.3(a) is called 'young woman/ old woman', and in Figure 1.3(b) we can more or less decide if we want to see a vase or two faces in profile, facing each other. Figure 1.3(c) and (d) are also ambiguous; that is, our perception is multistable and alters between two perspectives, one from above and the other from below.

Experience and qualia

The physical entity that 'elicits' a perception of, for instance, 'red', or the chemical agent eliciting a subjective taste-experience, is called a *stimulus* (one *stimulus*, many *stimuli*). The qualitative experience itself may be called *qualia*. One of the most difficult problems to account for scientifically is the subjective experience of sensory qualities (qualia) and their relationship to the physical and chemical stimulation of a sense organ. Even if the perception of 'red' may arise when the retina is stimulated by long wavelength light, this is neither a sufficient nor a necessary condition.

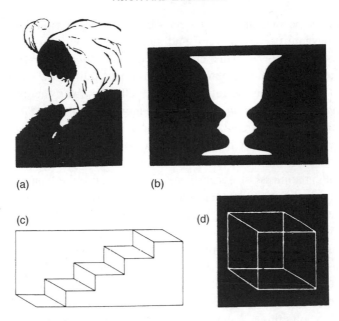

Figure 1.3 Visual illusions in which you can choose what you want to see. You see either a young girl or an old lady (a); a vase or the profiles of two people facing each other (b); a staircase and a cube that can be seen in different perspectives and from different directions – from above or from beneath (c, d). We see only one of these interpretations, never both at the same time. With some practice you can choose which one you want to see.

The perceptions resulting from sensory stimulation – the pain from a prick of a needle, the sour taste of a lemon or the coldness of ice – are subjective and private experiences, even though the feeling that something is 'cold' usually requires the stimulation of specific cold receptors in the skin. Although the receptors' reactions upon stimulation are measurable, my experiences of qualities like pain, smell, taste, sound, color, etc. are not *directly* available for observation by another person. However, under certain circumstances they can be described *indirectly* by means of observing the subject's reaction to stimulation. For instance, it is not possible to know if my experience of 'redness' or 'greenness' is the same as yours, but this has not prevented us from agreeing on ways to arrange our color perceptions in a color space. In color vision, phenomenological geometrical systems, like color circles and color scales, have been devised to organize our qualitative experiences (see Figures 5.1–5.3).

The response of sensory cells to stimulation, in the form of electrical potentials and chemical processes, is observable and measurable in an objective and scientific manner. Yet how can our conscious subjective feelings of 'coldness', or of a bitter taste, or of the redness of an object, arise from changes in electrical potentials of nerve cells? Are these realms of subjective and objective realities compatible? A common answer to this question is to point to a chain of stages and reactions, beginning with the peripheral event (the physical or chemical stimulation), continuing with the intermediate nerve reactions, and ending with conscious awareness, or the whole organism's response. This chain consists of *transducer mechanisms* (for instance the rods and cone receptors in the primate retina), *signal pathways* (the nerve cells and channels in the retina and the pathways from the eye to the brain), *specialized receiver stations in the brain* (which are different, for example, for lightness contrast, for movement, for color, etc.), and *binding mechanisms* that join appropriate elements together in conscious experience. Strictly speaking, however, nobody has yet identified or even localized the brain processes that lead to the 'redness' or the 'greenness' of, for instance, a pepper, or to the 'sourness' of a lemon.

Although it has been made plausible that qualities and feelings are mediated by, or originate in, the brain, they are, as we have seen, often related to – or projected upon – objects and events out there in the physical world. Following the amputation of a limb, for example, people are left with perceptions associated with the non-existing (phantom) body part. We see the redness of the pepper lying on the table, and not inside our heads, and we consider the sourness of the lemon as a property of its pulp. The red percept is projected onto the pepper and sourness is also something that obviously belongs to the external object. One is tempted, like Hermann von Helmholtz, to regard such qualitative attributes as natural symbols – as potential carriers of meaning – open to fast, direct and non-verbal interpretations of the state of a fruit, for example, to determine if it is ripe, fresh or foul.

Experience and language

Every scientific endeavor requires a specialized language. It is necessary to identify the subject of scientific inquiry and to find a line of demarcation relative to related fields. In vision it is necessary to distinguish stimuli from percepts, i.e. to separate the abstract physical description of measurable quantities from a symbolic representation of subjective phenomena. To meet this end, nomenclatures are continuously being developed and refined. Let us take a look at a few misconceptions that originate in a less precise use of language.

We have already pointed out that the word 'color' belongs to the realm of subjective experience. Related terms, denoting correlated physical entities, are *spectral distribution, wavelength, chromaticity,* or something else, depending on the topic of discussion. *Lightness* and *brightness* refer to the subjective impressions of intensity of a surface (of being more or less bright), whereas the related physical terms may be *luminance, luminance ratio or luminance contrast. Light* also refers to our experience, whereas *electromagnetic radiation* refers to a physical phenomenon. In many other cases, our language does not (yet) have separate words for the subjective and the physical. This is still the case when referring to size, movement, texture, contours, depth, etc.

However, even with the most refined language, there will always be something left out when talking about subjective experiences. Some of our perceptions and emotions, and some of our thinking, can be expressed symbolically by words, but not all. Language can build a bridge between 'subjective' inner experiences and 'objective' outer events, but of course the immediacy and the quality of the real experience is lost in the symbolic, verbal descriptions – although something else is gained. The written word, for example 'red', or the sound of the word when we say it aloud, are substitutes for 'the real thing'. Language uses written or spoken symbols to express or point to feelings and qualities. Language symbols point to something meaningful for those who, through their own experience, know what is meant by, for instance, a 'yellow banana' and a 'green banana'. Communication presupposes that one has such abilities.

The limitations of language also become clear when you try to describe or explain the experience of 'red' or 'green' to a color-defective person who cannot distinguish between them. This task is utterly frustrating, but it is, in fact, not easy either to use common language to explain a color that you saw yesterday for a person with normal color vision, or even to ourselves. Usually, therefore, we are content with using words only as pointers to an implicit meaning.

For the especially interested, a thorough discussion of this philosophical topic can be found in the article by B. A. Farrell, in particular, in *The Philosophy of Mind* (1962), and in Ludwig Wittgenstein's *Remarks on Color* (1977).

The concept of consciousness normally includes our ability to perceive – to be aware of – and to reflect on what we experience, and to eventually give verbal expression to it. If one is unable to distinguish colors like red and green, it has some consequences for one's awareness and view of the world. This applies to about 8 percent of all men and 0.45 percent of all women whose color vision deviate from the normal. The most common form of deficiency is a reduced ability to distinguish between red and green. It is assumed that people with normal color vision have the same qualitative impressions as the color-defective, but that their color palette is richer.

When a verbal account of a common experience is missing, i.e. when a person has problems with putting certain relations into words in his or her mother tongue, one may conclude that there is a conceptual deficiency, or some kind of 'concept blindness'. When important concepts are missing, one often gets the impression that conscious experience is also suffering; we have a tendency to see what we expect to see, or what we are used to. What then is the more fundamental process, sensation or reflection? Is Aristotle's statement still true that nothing can exist in consciousness without first passing through the senses, or is this like asking what comes first, 'the chicken or the egg?'

This question about the role of sensation and reflection, from early on considered to be of great importance, has divided philosophers into two groups: the empiricists against the rationalists. The empiricist holds that everything we can know about the world is based on sensory experience, whereas the rationalist thinks that we can make true statements about the world independently of our experience, and uses logic and mathematics to strengthen his case.

To the child, the phenomena are very much present, and seeing and hearing come before words. The intellectual mind will find it hard to separate what it sees from what it knows or believes. This lack of 'direct access' can sometimes be an obstacle to understanding, such as when confronted with another culture or with art. A painting, for instance, may represent a content and a meaning that cannot be fully described in words. This other reality – if it is impressionistic, like that of Claude Monet, or expressionistic, like that of Edward Munch – can in fact be overshadowed by the use of common language (Gombrich, 1972; Grenness and Magnussen, 1989).

We have said that our sensations, thinking and emotions are private and not directly accessible for others. Yet this does not prevent us from trying over and over again 'to point to' our feelings, by gestures and signs, or by expressing their contents in a verbal language. This language can, of course, be developed and refined to create new symbols of experiences and knowledge. Neuroscience has reminded us that learning, including visual learning and recognition, depends on the active use of our sensory organs, and that use and experience bring about a certain structuring of the nervous system. As described below, Thorsten Wiesel and David Hubel (Nobel Prize Laureates, 1981) have shown, for example, that adult cats are only able to distinguish types of visual stimuli they have been exposed to in their first weeks with open eyes. These stimuli leave lasting traces in the nervous system (Hubel and Wiesel, 1963).

It appears that normal sensitivity to a visual stimulus depends on the organism receiving adequate stimulation in an early phase of development. If such stimulation is wanting, or if it is too restricted, this can influence the nervous system's structure and functioning later in life.

A classical example from Thorsten Wiesel and David Hubel's research is the effect of isolating kittens in a cage and allowing them to see only vertical black and white stripes in the first few weeks after they have opened their eyes. These kittens will later be unable to distinguish horizontal black and white stripes. Hubel and Wiesel (1963) found that the cells in the cat's visual cortex were particularly selective for and could only discriminate between stimuli that had been presented to them during the first few weeks of vision. Primate cells also possess this property, but in higher vertebrates the critical period lasts longer. If cells are deprived of adequate visual stimulation during the critical period, these cells will not die, but will be recruited for other functions.

This may throw some light on the case of Virgil, the blind man who regained his eye sight. It may be that that neurons in his brain which were originally designed to deal with and interpret visual information had been allocated for other purposes. Super-normal tactile capabilities may have originated through frequent use. The reports of von Senden (1960) seem to support this view.

This conditioning or habituation of the nervous system seems like a 'bottom-up' process of adjusting and adapting the sensitivity of the nervous system to certain features of the external stimulus. This activity leads the nervous system to acquire a structure and hence 'knowledge' that can be used later to interpret sensory impressions in a 'top-down' process where new sensory information is compared with previous information. This combination of 'bottom-up' and 'top-down' processes may appear as a closed loop with no beginning or end, except that it is limited to a particularly plastic period of the animal's life. It may be difficult to decide where one process begins and the other one ends during the plastic learning period. Perhaps it is easier to assume that both processes are active at the same time.

In this context it would be of great interest to identify the predispositions (heredity, instinct, etc.) that newborns carry with them and which allow them to develop more complicated perceptions. Studies of sensory experience, and particularly sensory deprivations in animals (when stimulation is too weak or absent), can contribute to a better understanding of the complex interplay between heredity and environmental influences.

Through modern non-invasive techniques for monitoring brain activities, like functional magnetic resonance imaging (fMRI), it has become possible to study some of the problems mentioned above experimentally. Modern neuroscience has opened new gates to the understanding of cognitive functions, and visual science is one of the most important sources for such knowledge.

'The sweet smell of purple'

There are many unanswered questions related to how 'the hypothesis of the world' – i.e. our understanding of reality – comes into being. For example, how are sensory data treated by the nervous system before they become conscious perceptions? We shall repeatedly come back to this difficult problem throughout this book, but let us here just take a brief look at an interesting phenomenon that might be of relevance.

Normally, it is assumed that the dimensions of what we might call the sensory spaces of different sensory modalities are separated at birth, and that they gradually become related to each other as a result of experience. For instance, vision is separated from hearing, and both are separated from touch. Nevertheless, one may ask whether sensory structures emerge during development (this is the standpoint of the empiricists), or if they already exist as abstract representations in the newborn. In the latter case (which resembles the standpoint of the rationalist), what is left for the child is to further develop, modify and refine an already existing program through active engagement and experience.

One may wonder if our understanding of the world around us depends more on abstract properties that are extracted from sensory data than on the sensory data themselves. One theory claims that babies are born with an undifferentiated sensory system. A newborn does not 'know' if it sees, hears or touches something. Everything belongs to the same sensory space. A baby a few months old can apparently distinguish the forms of objects equally well by tactile or visual manipulation. Eye-hand coordination may already exist at birth, but it is not stable and can be influenced. Since vision develops rapidly, cross-modal organizations changes during the first half-year of life, and afterwards the hand becomes a fascinating and important tool for grasping things and an instrument for the eye (Streri, 1993).

According to this theory there is a kind of primitive unity of the senses at birth, and it is the repeated interaction with the environment that leads to differentiation and autonomy of separate sensory systems. The more specific representations of object properties, such as color, form, size, etc., may develop gradually. As a consequence of differentiation, the coordination of information about an object from different sensory systems requires integration at a higher brain level.

This theory suggests the possibility that we enter this world with a form of synesthesia, and that we grow away from it. Synesthesia refers to a strange phenomenon that is sometimes encountered in adults, namely that words or letters evoke an impression of color, or that sounds have a taste. In an adult light activates the visual center, and this activity can be measured by non-invasive methods like PET (positron emission tomography) or fMRI. A flash of light on the eye of a newborn does not cause a reaction in the part of the brain where the visual center is later located. According to this theory, the specific contacts between the eye and the visual areas in the brain are formed gradually as the brain matures. When some adults with synesthesia feel, for example, that 'purple smells sweet', this may be because something went wrong in establishing the normal connections and some of the original cross-connections between modalities have been maintained. In this way the senses 'mix' to produce the strangest phenomena, which are said to be of great pleasure to the persons who have them.

Observations also suggest that some connections between sensory modalities develop during adulthood. The purpose here may be that information that belongs together, for instance in the auditory and visual systems, leads to amplification when both occur simultaneously (e.g. looking at the mouth while listening to somebody talking). Such processes may also be crucial for learning.

Vision and natural science

Psychophysics is the name of a discipline that studies the responses of organisms to stimulation. The safe mechanistic approach is to study small responses at or near threshold. In humans one can, for instance, measure how much light energy is necessary to produce an impression of light and color. Such psychophysical measurements are important for our understanding of the functioning of the nervous system, and they consequently supplement neurophysiological studies.

A comparison of the information capacity of the auditory and visual systems shows that vision has been allocated the most resources. This dominance of the visual system is anatomically reflected in that about 60 percent of all nerve fibers from a sensory organ to the brain come from the eyes. From the ear about 30 000 nerve fibers transmit acoustic information to the brain, but from each eye it is between 1 and 2 million. In the cortex there are about 800 000 nerve cells in the auditory center compared with about 500 million in the visual centers.

Below is a simplified 'bottom-up' characterization of the main physical, neural and mental processes that lead to vision:

1. *Imaging* – this process accounts for the spatial light distribution on the retina by the optical eye media (cornea, pupil, lens, vitreous, etc.). In recent years, the development of *adaptive optics* has allowed the quality of the images projected on the retina, as well as those taken of the retina, to be improved.

2. *Detection and discrimination* – detection refers to the transformation of the energy in the light quanta absorbed by the photoreceptors (rods and cones) to electrical potentials (*transduction*) and to the neural activity that follows. This includes the chemistry and genetics of visual pigments. A comparison of the weakest stimulus (the minimum energy) required to elicit a sensory impression – the threshold value – with that required to elicit a response in nerve cells makes it possible to correlate psychophysics with molecular biology and neurophysiological processes. There may be a difference between detection ('seeing something') and discrimination of what that something is.

3. *Neural encoding and signal transmission* – the input from about 100 million receptors in the retina converges on somewhere between 1 and 2 million retinal ganglion cells. After reception there are four functional cell types that organize the information from a retinal image, decomposing it and encoding it before the signals are transmitted to higher brain centers. Already at the retinal level it is possible to suggest hypotheses that connect neurophysiology and psychophysics (e.g. about spectral sensitivity and receptors).

4. *Adaptation* – the eye can adjust to changing light levels, from moonlight to the brightest sunlight, as well as to the changing colors of light. Light adaptation covers an intensity range greater than 1×10^{12} and thus cannot be due to regulation by the pupil (which accounts for a factor of only about 10 in intensity).

5. *Differentiation and structure* – we can imagine a cooperation of diverging, converging and parallel pathways in the retina and in the cortex that all receive input from a common set of receptors. Processing of this information may take place in a hierarchy of functional units, or in cell types that treat different components of the image separately. Today, about 40 areas in the cortex that deal with different aspects of vision have been identified.

6. *Identification, recognition and interpretation* – this represent the central brain processes which, in addition to processing in the visual cortex, include the other senses and higher mental activities, like memory, context, will, etc. Although it has been claimed that mental activity is 'nothing more' than complex and highly organized neural activity, throughout this book we shall use the concept of mental activity (*qualia*) to mean what we intuitively know as such, through personal experience and introspection. The gap in our understanding of mental and neural activities is still too large for us to use a common language for both.

The emotions and the actions of an individual (e.g. facial expressions, speech, gestures, movement, etc.), that follow a chain of physical and neural processes, reflect the personality and the cultural background of the individual. They are organized at a level that cannot, for reasons mentioned earlier, currently be explained by natural science, within a detailed scheme of cause and effect.

As we have seen, the visual process can be studied at many different levels and by many disciplines. The list above, which is not complete, shows a progression from physical stimulation towards 'higher' functions that are governed by complex mental activities. The imaging under point (1) is satisfactorily explained by physical and physiological optics. Biophysics, molecular biology and neuroscience deal primarily with points (2)–(5), investigating processes in single cells, the interactions between individual cells and the interplay of many cells in neural networks.

Signals from about 100 million rods and 6 million cones converge on between 1 and 2 million ganglion cells and an equal number of nerve fibers. In the brain we see divergence of these inputs as these ganglion cells successively establish contact with roughly 500 million nerve cells. The perception of contrasts, contours, colors, textures, three-dimensional space, movement and orientation, to mention a few elements, depends on connections and interaction between many cells in a network. Later, we will look into these processes in more detail.

Interpretation, understanding and behavior are subject to scientific investigation within the fields of neuropsychology and cognitive psychology. The visual process is, due to its complexity, studied by many overlapping disciplines.

Decomposition of 'the optical image'

When an organism is adapted to the prevailing lighting conditions, the most important features of the visual environment are differences in lightness and color – what we normally call contrast. Contrast relative to a background is a prerequisite for seeing anything at all. Without contrast, neither the form of an object nor its movement can be detected (identification of the object usually requires even larger contrasts than detection).

Contrary to what is often believed, the projection of the optical image on the retina to the brain is not a simple transmission – such as, for instance, a TV broadcast.

Rather, different types of retinal cells seem to split up an image into a set of different, fundamental components. The pattern of light and shadow in a scene imaged on the retina is transformed in complicated ways by the interaction of retinal cells, and the resulting signals that are transmitted to the higher brain areas are subject to further decomposition. This process involves several representations or abstract 'maps' of the optical image.

What do these 'maps' look like? What is the structure and meaning to the seemingly chaotic activity evoked in millions of nerve cells by even the simplest stimulation? First of all, we know that already in the retina cells respond only to stimuli that are situated within a small, limited area of the visual field. In the central part of the eye, the fovea, this territory of a particular cell – called its receptive field – comprises a limited number of photoreceptors. According to the traditional view, the retinal cell is marginally affected by what happens outside its receptive field. Recent research, however, has demonstrated that more remote areas of the retina may affect the cell's response. Taken together, a million *ganglion cells* (each of which has direct contact with a nerve fiber) cover the whole visual field, each cell being dedicated to processing input from a small, fixed position on the retina. Some transient ganglion cell types are specialized for *signaling temporal* or *spatial changes* at their particular location in the optical image, whereas others respond in a sustained fashion to *stationary*, colored surfaces.

Higher-level brain cells are more specialized. Some cells respond only to a narrow range of wavelengths while other cells are activated only when their receptive fields are exposed to a contour of a certain angular orientation, e.g. to an oblique line drawn on a piece of paper. If this line were rotated 10° away from its orientation, these cells would stop responding, and others would take over. Some cells respond in a selective fashion to size and contrast. While some respond to broad stripes, others prefer narrow stripes of the proper orientation. Still other cells signal movement, but usually only if it occurs in one particular direction, for instance from left to right within their territory of visual space.

From the earliest retinal levels to the visual cortex there thus seems to be a parametric decomposition of physical properties into different components, or dimensions. The extraction of visual form, for instance, may be achieved in many ways and from several visual inputs. It may be constructed by means of light, color, movement, contours or even depth. Information derived about these attributes at an early level is sent to the higher brain for further processing by way of diverging and converging visual pathways, in the *magno-, parvo-* and *koniocellular* cells. We shall return to these important cell systems later.

Different aspects of an image seem to be distributed to different areas of the brain for further structuring before the information is integrated to prepare for a course of action. One can imagine a letter, for instance the letter 'B', which to the cells is composed of many contour segments of many different orientations – short quasi-straight contours that appear curved when taken together, with filled-in areas of even black contrast in between. Bits and pieces of the letter are detected separately before they are brought together and integrated to reappear as a unity – the letter 'B'. The letter typically appears in a word, together with many other letters that are subject to the same filtering and integration processes. The integration of many such letters in a word, and finally of words to a meaningful sentence, requires the involvement and interplay of an increasing number of neural units, and of mental functions.

It is tempting to ask whether there is a connection between these neural processes and the decompositions that are familiar to us. Are the different attributes of objects all reflected in corresponding neural processes? If so, how is this achieved? It is not yet possible to give a complete answer to this question, but in order to appreciate possible answers, we must first identify which are the different perceptive elements that make up a visual image of the world. Here is a (incomplete) list of some important dimensions and properties:

Light – shadow – lightness – darkness – glare – contrast – color – texture – line – contour – form – orientation – depth – distance – size – spatial position – foreground – background – movement – direction – speed – temporal changes, etc.

Some of these attributes can be broken down to finer elements, such as several million colors of different hue, saturation and lightness, contours of more than 20 orientations, and so on. This list of visual attributes, which includes those which the skilled artist exploits, reminds us of the problems that have been extensively discussed within the field of artificial intelligence (AI), in connection with the development of artificial vision systems (e.g. computer vision). The aim of AI is to make artificial vision systems that respond adequately to these and to other attributes of a visual scene. These attributes are often, and without the necessary caution, ascribed to the physical nature of the external world. In the way they are presented here, they are the psychophysical entities of visual observations by individuals, and they do not represent a pure physical description. For instance, the corresponding physical concept of 'color' would be 'spectral distribution of electromagnetic radiation'.

The physiological basis for observing these properties, and all different aspects of an image, is the activation of retinal receptors – the same ones in all cases. Yet these receptors are very limited in their responses and can only signal that they have absorbed more or less light. Therefore the rather complex properties listed above must be constructed at later stages of processing, beyond the receptors, based not only on all the information available to the receptors, but also on several different transformations of this information in successive layers of nerve cells.

Form vision

The perception of an object requires that it is segregated from its background, and the process of figure–ground separation is therefore of the greatest importance. For instance, contours like the ambiguous images of Figure 1.3 delineate a figure and not the background. In the animal world, it is of equally importance to disguise and to hide by means of camouflage or other means (e.g. by 'freezing' movement). It is therefore challenging to search for the factors that are the most important in object perception, a field that has been intensively studied by the Gestalt psychologists (Max Wertheimer, Wolfgang Köhler and Kurt Koffka) and Edgar J. Rubin. They claim that definition of a figure as a visual entity that stands out against a background is dominated by four factors: *symmetry, smooth continuation, closure* and *parallelness*. In addition, *proximity, similarity* and *common fate* are important for the grouping of elements that belong to the same object, such as for instance when a running rabbit is occluded by trees and other objects, or when a line drawing of a bird is moved relative to the hatched background (Figure 1.4).

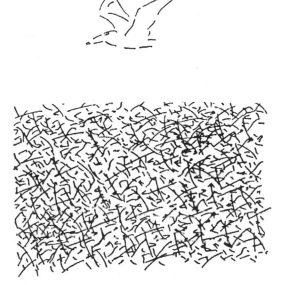

Figure 1.4 Make an overhead of the line drawing of the bird and superimpose it on the hatched area below. The form of the bird will first disappear, but then the form reappears as the overhead is moved relative to the hatched background. (Reproduced from Regan, 1986, *Spatial Vision* **1**, 305–318 by permission of USP.)

The significance of Gestalt factors for visual processing may be shared throughout the animal world. When an animal is adapting to its environment, for example when a fish changes its color to become less visible against the water surface, Gestalt factors are at work. *Bothus ocellatus*, a bottom-dwelling flatfish, can mimic on its skin a checkerboard pattern on which it is placed by the experimenter, and it can even follow changes in check size (Ramachandran *et al.*, 1996). These are examples of how Gestalt factors, seemingly similar to animals and man, are counteracted.

Several studies of the monkey visual cortex have revealed neurons that respond to 'Gestalt factors', like those mentioned above, and neurophysiological research is now taking Gestalt rules seriously (for further discussion see Spillmann and Ehrenstein, 2003).

Modeling vision

There has been an increasing need to model primitive visual processes for use in practical detection tasks and pattern recognition by optical devices. Such devices are used in the control of production processes, in surveillance and security, etc. In some cases, such as in the monitoring of products on an assembly line, matching with standard templates is sufficient. For more general optical tasks, many efforts have been made to develop adequate models. While algorithms for pattern recognition by template matching need not be related to how biological vision works, the more elaborate the task becomes, the more interested model builders become in knowing how similar problems are solved by nature. Model builders have, over the years, also become interested in size and color constancy, and in odd perceptual illusions and their neurophysiological correlates. Illusions may serve as tests of the biological relevance of a model. Following the studies of how such problems are solved in nature, the next step may be to implement similar, physiologically plausible mathematical models in machines. Such higher level models clearly have as their goal the mimicry or simulation of brain functions, and, since knowledge of these functions is still rudimentary, many different analytical and mathematical approaches are competing. In recent years, models have begun to apply psychological principles to object perception, such as some of the Gestalt factors mentioned earlier.

Models are also important in testing out ideas. In this respect they represent 'quantitative thinking', and they produce predictions that can be tested. The construction of models for artificial vision has occured in parallel with the rapid development of computing power. This is currently a very dynamic field, and in the future we are likely to see many more models and fascinating new applications of them. Some simple and well-tested models, often parts of more complicated and general ones, will be mentioned throughout this book. We shall present mathematical formula for neural color processing, for combined cone contrasts, scales of perceptive differences in lightness and color, the sensitivity of receptive fields, and lateral inhibition.

Visual illusions

According to Plato (Greek philosopher, 427–347 BC), our perceptions of the world are all illusions. We are like inhabitants of a cave who cannot experience the outside world directly, but try to make sense of it from the shadows that fall on the cave walls. Here we take a more restricted view and consider those phenomena that arise when our perception of a visual scene conflicts with its physical characterization or our expectation. The illusory phenomena that are of principal interest here are those that 'happen in the visual organ' and where our perceptual 'mistake' reflects normal visual function and its limitations. The popular belief that the distortions and phantoms that we usually call illusions are, in one way or another, 'errors of judgement' needs qualification. What at a first glance appears as a disturbing and useless distortion or misinterpretation may on closer inspection be appreciated as representing a feature of great significance. The underlying processes may be advantageous both for the construction of a neural image and for its interpretation. Thus, what appears as an illusion when a particular function or phenomenon is viewed isolated from its natural context may offer insights into normal neural and cognitive processing. Consequently, illusions are not simply subjective distortions of the contents of objective perception. On the contrary, since many of the processes leading to illusions are important for vision to function properly, the study of illusions advances our understanding of the normal behavior of the visual system. Today, illusions are actively used as tools to understand how the brain represents reality. For instance, does the activity of cells in the primary visual cortex represent the mental image or the physical stimulus input, or does it mirror something of both? Is it possible to find a brain region where stimuli turn into perceptions? Studies of the perception of illusions help us answer such questions. In fact, the study of visual brain processes needs the guiding principles offered by perception.

Experience has taught us not to pay much attention to some common, everyday illusory phenomena. When an oar in water looks broken, we disregard it because we have learned that it is an optical distortion. It is, like a broken spoon in a tea-cup, explained by the refraction of light at the transition from water to air. Mirror images of people who look at themselves in convex and concave mirrors give rise to much amusement, and may also be called optical illusions. The same applies to stroboscopic movement, where movement is suggested by a row of small lights that are turned on and off in rapid succession.

Other distortions are better hidden, or we have become so used to them that they are difficult to detect, such as the enhancement of lightness differences. Normally we are not aware that the perceived lightness of surfaces is due to active neural processes that introduce gray levels and enhance the contrasts between adjacent surfaces. Lightness is not a property of the luminance of an isolated surface, i.e. of how much light it absorbs or reflects, it is something which is *induced* laterally into a target surface by its neighbors (also called *simultaneous contrast*). A piece of black coal in the sunlight outdoors reflects more light than a white sheet of paper in a dark corner of a room indoor.

In Figure 1.5 the gray stripes have the same reflectance in both images, even though their appearance is different. The significant condition is that the same stripe is embedded in surrounds of different luminance, and that these surrounds affect its

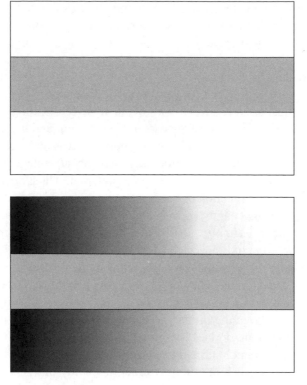

Figure 1.5   Simultaneous lightness contrast. The inner gray stripe is homogeneous in that it reflects equal amounts of light along its entire length (upper picture). The lightness appears to change only because light and dark surrounds induce different gray levels (see also Figure 4.32).

apparent lightness and the contrast. A light surround makes the stripe look darker, and a darker surround results in a lighter stripe. This is not a product of our imagination but is an induced effect, brought about by a lateral neural process in the visual system. It is not difficult to appreciate that this particular alteration of the physical condition – that the stripes are equal – is useful for discrimination. The visual system works on the physical difference of reflectance between an object and its background and exaggerates it in order to make the object more visible. To achieve this, the neural system uses methods that sometimes, when isolated, make the illusion stand out. There is no immediate or direct physical correlate in the object itself. This may be even more compelling in Figure 1.6. In this celebrated Logvinenko (1999) illusion the diamonds marked with a circle have the same reflection factor and are thus physically equal (you can convince yourself by masking out these areas). Note that even the small circles are subject to the same illusion. The circles have equal reflectance, and therefore the one on the darker diamond would be expected to appear brighter than that on the lighter diamond. However, in this context of an *illusory* darker diamond one is surprised to see that the opposite happens and that the one on the dark diamond appears darker.

Figure 1.6 The marked diamonds are physically equal in that they reflect equal amounts of light. This applies to all diamonds in the same two rows. The illusion of difference is brought about by a sinusoidal luminance distribution in the background. This modulation of the background usually escapes the untrained eye. You can convince yourself of the illusion by cutting two small holes in a paper and viewing the fields isolated from their surrounds (Logvinenko, 1999). Note that the small circles used to mark the two diamonds are also physically equal, although the lower one appears to reflect more light than the other. Thus, it is the physical surround rather than the induced gray level of the diamonds that determines the perceived lightness of the two spots. For an observer unaware of the sinusoidal luminance distribution of the background, the perceived lightness of the two spots seem to contradict the demonstration of simultaneous contrast in the previous figure.

Simultaneous contrast effects are active also in color vision. A gray patch surrounded by green will take on a reddish tint. This is called *simultaneous color contrast* or *color induction*. Simultaneous color contrast is particularly striking when demonstrated as *colored shadows* (see Figure 5.38). Color contrast enhances the difference between adjacent surfaces, and it works to maintain *color constancy*, i.e. to preserve the colors of objects when the color of the light source changes (for instance from daylight to different kinds of artificial indoor lighting). Constancy and enhancement of differences, as in simultaneous color contrast, seem to be two sides of the same coin. For instance, an illusion can arise when the color of the illumination is concealed. A white or gray surface, illuminated by a hidden colored light source, may mistakenly be thought to have that color. If the color of the illuminant were recognized, this hypothesis would be discarded and color constancy would be a more likely outcome.

The opposite phenomenon – contrast reduction or assimilation – is also common, particularly for smaller objects. The color and lightness of a surround seem to float or to melt into a narrow pattern of another color (see Figures 1.7 and 5.37), thus reducing contrast. We shall return to this and other color contrast phenomena later.

Figure 1.7   Assimilation. (See also Color Plate Section).

Area lightness contrast spreads over large distances, and we have already men-tioned its biological value. Another lightness contrast of short range is seen as a demarcation of edges in the so-called 'Mach bands' (Figures 1.8 and 1.9). If you make a hole in a piece of paper and move it across the two diffuse triangles of Figure 1.8, you see no sudden change in lightness as you would expect if the effect was physical. This phenomenon led early on to hypotheses about lateral interactions between nerve cells. Mach bands were named after the Austrian physicist and philosopher Ernst Mach (1838–1916), who devoted much of his life to psychophysical studies of vision. Mach bands are strongest and give rise to contour enhancement at a border with a smooth luminance transition between light and dark areas (Mach, 1965). This phenomenon is different from the enhancement of lightness and color contrasts that is effective over large distances (Figure 1.5), often called *Hering area contrast* after the German physiologist Ewald Hering (1834–1918). These effects were observed and exploited by artists long before they were made the subject of analytical studies (Ratliff, 1965). Figure 1.8 shows an example of particularly strong Mach bands. The effect in Figure 1.8(a) can be obtained when a triangular piece of cartoon is placed in front of a triangular light source casting a shadow on a screen. Figure 1.8(b) comes from using a triangular aperture in front of the same light source (Wold, 1998). Figure 1.9 shows Mach bands produced by a rotating disk.

Mach bands have been described in Floyd Ratliff's (1965) book *Mach Bands.* See also *Outlines of a Theory of the Light Sense* (Hering, 1964).

Figure 1.8 Mach bands. A smooth transition of luminance between, for example, a white triangle and a black background, and a black triangle on a white background, gives rise to the perception of white and dark bands not present in the reflection patterns. (Courtesy of J.H. Wold. See also http://www.phys.ntnu.no/~arneval/illusjoner/Mach.doc).

Other examples of simultaneous contrast, in other domains of visual perception, are shown in Figure 1.10. In Figure 1.10(a) parallel lines appear to be converging and diverging. In Figure 1.10(b) the horizontal stripes in the disks appear to have different orientations because of the orientations of the stripes in their surrounds, and in Figure 1.10(c) perspective induces a difference in apparent size. Figure 1.10(d) shows the Ebbinghaus illusion of size differences.

These enhancements of contrasts are very robust phenomena. Some less robust, but equally simple and fascinating contrast phenomena, are those that change their appearance depending on context, or according to the interpretation of the viewer. The Wertheimer illusion in Figure 1.11(a) is one such example, the appearance depending to a large extent on the interpretation (see, for instance, Schober and Rentschler, 1979, p. 73). If you decide to see two Vs, the gray appearance of the two parts is different from when you interpret it as a W. Such illusions can be explained in a top-down fashion, as psychological or cognitive in nature.

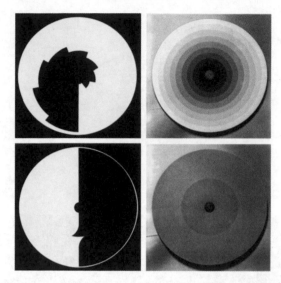

Figure 1.9   Border contrast phenomena demonstrated by rotating disks. (a) When the disk on the left is rotated, narrow circular stripes arise that are physically homogeneous, but each of which appears lighter towards the center of the disk than towards the periphery. This is the Mach-band phenomenon. (b) Although the inner disk and the outer ring are physically equal, a narrow local luminance perturbation produces a contrast that spreads over half the disk. This effect resembles that of Figure 1.13.

In the illusions of Figure 1.3, our perception tends to flip back and forth between two alternative interpretations in a less controlled manner, and we experience occasional reversals of form. It is as if the brain cannot decide what is figure and what is background, and therefore it engages actively in trying out the most likely possibility. Such ambiguous or multistable stimuli have become invaluable tools in the study of the neural basis of visual awareness (Eagleman, 2001; Logothetis, 2002). They allow us to distinguish between neural processes that correlate with basic sensory inputs ('bottom-up' processes) and those that correlate with interpretation ('top-down'). This distinction is also possible for the phenomenon of binocular rivalry that manifests itself when different images are presented to the two eyes. In order to avoid conflict, usually one eye will dominate over the other one for some time, and this dominance is correlated with neural responses at higher stages of processing, probably representing the brain's interpretation of what is observed.

Other illusions make use of contrasts and perspective to create a paradox, as in many of M.C. Escher's drawings (Ernst, 1988). In one of them, the eye is fooled to see water flowing upward, against gravity. After-impressions of illusory movement without actual displacement, such as when you look out of the window of a train that has just stopped at the station, is another example of paradoxical experiences (cf. Zenon's paradox). From this illusion, and from Figure 1.17, we learn that there is dissociation between 'perception of movement' and the actual 'physical movement'.

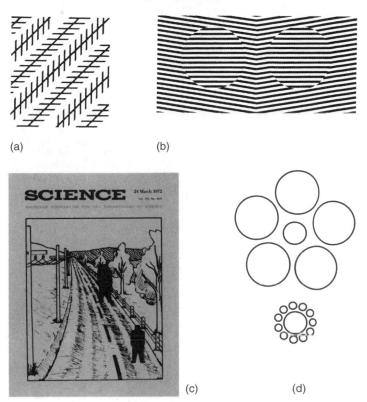

Figure 1.10 (a) The Zöllner illusion. The long diagonal lines are all parallel, but seem to converge and diverge. (b) The stripes inside the disks are all horizontal, but appear tilted in a direction opposite to that of the surrounding stripes. (c) We are fooled by perspective. The two black figures are equal in size. (d) The Ebbinghouse illusion (also called Titchener's illusion). The two inner disks have the same diameter, but appear different in size.

A consequence of this can be seen in some rare cases of brain injury that lead to the odd defect of 'motion blindness'. In a famous case, a woman was reported not to see movement, not even of a car approaching in the street (Zihl *et al.*, 1983). She could see it at a distance and then again close to her, but she was unable to see its movement towards her. When the same woman poured water into a glass, she could neither see the water running from the jug, nor the glass being filled up – overfilling the glass and spilling water as a result.

Sketchy images and incomplete pictures, like caricatures, may be completed 'top-down', leading to the perception of mental content that is not physically present in the image. Other examples of extrapolations and interpolations are shown in Figure 1.12(a)–(c). Being able to read the word FEET in Figure 1.12(b) is surprising because it requires extensive, and probably knowledgeable, extrapolations of contours. Try not to complete the illusory contours (turn the image upside-down), and you will be surprised by the difference. Kanisza's triangle is another celebrated

(a)

(b)

Figure 1.11   (a) Examples of contrasts that change according to interpretation. If you decide to see two Vs, the one embedded in black looks lighter than the one on the white background. If you instead choose to interpret the two Vs as a single letter W, its lightness is more evenly distributed. (b) The colors of the two semicircular rings change with separation and interpretation. The difference between the two halves of a continuous circle is perceived as smaller than that of the two separate halves. (Reproduced from Schober and Rentschler, 1979, *Das Bild als Schein der Wirklichkeit* by permission of Moos.) See also Color Plate Section.

example, as is the picture of the Dalmatian dog. Forms are constructed from a combination of real and illusory contours, by what may be called 'subconscious interpolations'. The brain seems to be trying, as far as possible, to make sense of what is seen by interpreting it as something familiar. Such phenomena have been taken to demonstrate how cognitive processes and experience contribute to perception (see below). However, recent findings that cells in primary visual cortex respond to the triangle as if real contours were present (von der Heydt *et al.*, 1984; Peterhans and von der Heydt, 1991) indicate a strong neural component to the illusion.

Cartoonists are experts at exploiting our fine-tuned ability to recognize faces and interpret facial expressions, and it is disturbing to find that this capacity is lost in some people with a brain injury called *prosopagnosia*. A person with prosopagnosia may see the details of a face, like a nose, ears, eyes, mouth, and so on, but cannot put it together into a familiar face. A prosopagnostic parent, for instance, cannot even recognize his or her own child by the face alone, but must rely on recognition of the voice or of the clothes.

A stroboscope that illuminates people dancing in a disco cuts time into intervals that are too short to allow perception of physical movement, although an impression

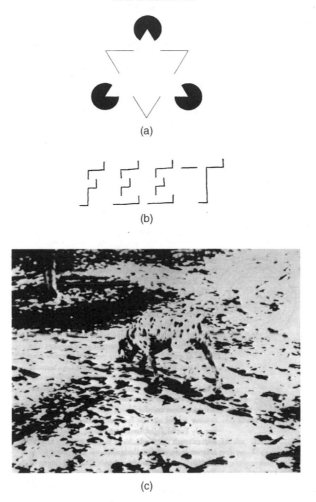

(a)

(b)

(c)

Figure 1.12 (a) The Kanisza triangle demonstrates that we often see more than what is present in the image. We complete the figure by drawing the 'illusory' contours of a triangle. The triangle even appears brighter than its white background. (b) This is another example of completion. Once you have seen the word FEET, the illusion becomes very strong, although it can be weakened by turning the book upside down. (c) If you see a Dalmatian dog, it is as a result of the completion of partial contours.

of (illusory) movement is conveyed. Likewise, in the cinema we are not only attending to a story, but we are watching a rapid succession of two-dimensional still images that appear rich with depth and motion. Two-dimensional images of three-dimensional objects do not give a sense of depth in the way stereograms do, but perspective reintroduces the third dimension that may, in certain circumstances, lead to an illusion, as in Figure 1.10(c). In the popular autostereograms, real depth is seen by combining two flat images of the same scene (Figures 1.2 and 3.8). Here again, we are confronted by a perceived dimension that is not physically present.

It may be surprising to learn that properties that we normally attribute to the external world, like the physical properties of contrast, movement and depth, rely so heavily on perception, i.e. on a subjective process common to most people. Having realized this, we can perhaps appreciate the difficulty in defining illusions, and we are better able to understand philosophers who hold that all vision is, in fact, an illusion. This insight might bring us to reconsider Galileo Galilei's (Italian philosopher, 1564–1642) influential division between primary and secondary sensory qualities. In his view, and in that of John Locke (English philosopher, 1632–1704), 'primary characteristics' such as hardness, mass and extension in space and time are properties of objects, whereas 'secondary characteristics' such as color, texture and shape are created by the mind.

It is tempting to categorize visual illusions into main groups, depending on what one believes to be the underlying mechanisms:

- physical (optical);

- physiological (neural);

- psychological (cognitive).

However, as we have seen, it is not easy to arrange all illusions in such distinct categories. Whereas optical distortions are relative easy to separate from the rest, it is not always straightforward to distinguish between 'neural' and 'cognitive' levels of illusory effects. Contrast enhancements, as in Figures 1.5 and 1.6, and the construction of form by illusory contours in Figure 1.12 may serve as examples of different levels of processing. However, one may take the view that this distinction relates more to the extent of higher level neural involvement than to which of two distinct levels are engaged. It seems difficult to deny that perception, at least in part, is influenced and determined by knowledge and concepts that we have developed through experience. For instance, subjective, illusory contours like those in the word FEET [Figure 1.12(b)] seem to result from the nervous system's anticipation of a known word. It is as if the mature nervous system has acquired a hypothesis about 'what is out there'. This may either stem from a genetic predisposition for recognizing known forms, or a neurocognitive process occurring during development and active experience – if not both. The boundaries between inheritance and environmental influences are probably not fixed. They are subject to changes as we go along and learn more, and it is important to note that some neurons in the brain respond, although weakly, to physically absent illusory contours, such as those of Figure 1.12(a, c).

It is likely that neurons responding to the same basic elements of an image, such as color or orientation of contours, are grouped together in the brain. Think about a triangle; it is composed of three lines or contours and three corners of different orientations. These lines, which excite different groups of cells, must be joined together before we can perceive the figure as a triangle. The question of which object properties cells respond to is also a question about which physical parameters are important for detection, discrimination, recognition and interpretation, i.e. for our understanding of a visual image. Properties such as contrast, orientation and direction may be regarded as autonomous elements in the perception of an object. These are among the primitive properties that must be integrated to form a unitary perception. Other more composite and complex properties may be dealt with in separate, dedicated locations in the brain. When attributes such as form, color and movement are joined together in the perception of an object, for example in a yellow car that drives by, this requires the combined activity of several cortical modules. All the attributes are perceived as belonging to the car; they do not hover around freely. Thus the early splitting up (analysis) and subsequent integration of elements (synthesis) by the nervous system raises a new question: how is integration accomplished? This is often referred to as the *binding problem*. Partial integration takes place in a hierarchy of converging pathways. Integration seems to be as general a principle as the early segregation of object properties, and it is likely to occur at several levels of the visual pathway.

In the retina and the first stages of processing thereafter, neurons are functionally multidimensional. This means that the information they transmit to other neurons is largely unspecific. A change in contrast, size, orientation, spectral distribution, etc. of a stimulus may affect a cell's response. Only information gathered at a later stage of processing, from other cells types that are more specialized or respond to another combination of parameters (or to another weighted sum of stimulus dimensions), can help sort out the proper stimulus. In principle, this can be achieved by combining activation and inhibition of neurons belonging to converging and diverging pathways.

Some illusions are even more intriguing and subtle than the above examples, and it is, at least at first glance, harder to assign them to a single category. In Figure 1.13, the contrast is spread over large areas, much larger than can be shown in this picture. It can, for instance, be seen on a normal projection screen a few meters away. In Figure 1.13, the surprise comes on masking the two oblique borders with two long thin paper strips. Then the contrast effect suddenly disappears, and the top and bottom triangles become equal in lightness. This contrast effect is thus a border phenomenon. An impressive and convincing demonstration can be made by projecting lights on a screen, where it can easily be demonstrated by photometric measurements that the upper and lower triangle are, indeed, of identical luminance. This phenomenon bears some resemblance to the Craik–O'Brian–Cornsweet illusion. A similar contrast spreading from borders also occurs with colors (Figure 1.14; Pinna *et al.*, 2001). In this example, the chromatic edges are likely to induce long-range processes in neural networks. Vision in everyday situations is, of course, affected by these phenomena, and we can assume that the 'filling in' of contrast and color shown in Figures 1.13 and 1.14 is regularly at work without us noticing it.

Other illusions are more complex. In Figure 1.15, the disks can be interpreted as hollow or convex, depending on the shadows formed by an imagined light source. However, this is not the whole explanation. The figure may appear as randomly

Figure 1.13 The curtain illusion. If you mask the oblique borders with two pencils, the illusion will disappear. An optical arrangement for demonstrating this effect is shown in the chapter on shadows. (Courtesy of J.H. Wold. See also http://www.phys.ntnu.no/~arneval/illusjoner/Mach.doc).

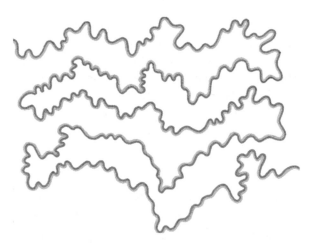

Figure 1.14 The watercolor effect. Color spreading from chromatic borders (Pinna *et al.*, 2001) gives the appearance of a slight difference in tint on the two sides of the line. (Reproduced from Pinna *et al.*, 2001, *Vision Res.* **41**, 2669–2676 by permission of Elsevier Science Ltd.). See also Color Plate Section.

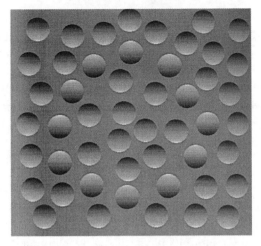

Figure 1.15   Do you see convex and concave spheres that form a circular pattern? If not, turn the book 90° and the convex and concave forms may appear. The spheres will invert if you then rotate the book 180°. An attempt at an explanation is given in the text. (Reproduced from Ramachandran, 1995, *The Artful Eye* by permission of Oxford University Press.)

shaded disks, but if you bend your head towards your shoulder, or rotate the book, you will see 'hills' and 'valleys' and that some of the disks form a large circle. The form of the disks (if they are convex or concave) seems to depend both on the most probable direction of illumination and how the shadows fall relative to what is 'up' on the retina. That is why the impression changes when you turn your head (or rotate the book) 90°. This illusion of three-dimensional form seems to depend on some unconscious hypothesis, including the following: (a) there is only one light source; (b) the position of the light source corresponds to what is 'up' on the retina (and not to what is up according to gravity); and (c) objects are usually convex (Ramachandran, 1995).

Recently, an interesting piece of evidence for the dissociation between vision and motor action has been reported. In a task arranged like the Ebbinghaus illusion, shown in Figure 1.10(d), the subject was asked to grab the central target with his hand. The visual illusion of size differences had no influence on the size of the grip. The illusion was apparently ignored by the motor actions, and the size difference which was clearly seen with the eyes was not what guided the hand (Goodale, 2000). This is counterintuitive, so one may ask if, at some stage in the visual process, the visual information required for motor action takes another neural route than that which provides for conscious perception, or if the size illusion is somehow compensated for after it has manifested itself in vision?

Another example of object constancy is when an object does not change its apparent size in accordance with the size of its retinal image. Try, for instance, looking at your hands. Hold one hand close to your face and move the other one as far away as you can. You see hardly any difference in size between them, despite the retinal image of the closest hand being about 10 times larger than that of the distant hand. In other cases size constancy may give rise to the opposite effect. In Figure 1.10(c), for instance, the size illusion is generated by perspective. Another famous example, already mentioned by Aristotle, is the 'moon illusion'. Like the sun and the stars, the moon looks bigger at the horizon than high up in the sky, even though measurements have shown that the angular size is the same. This illusion was first analyzed as a problem of physics, then of physiology, and finally of psychology, and today it is considered to be a size illusion caused by the receding horizon (Wade, 1996, 1998; Kaufmann and Kaufmann, 2000). Various cues in the landscape place the horizon moon at an effectively greater distance than the elevated moon. Although both moons have the same angular size, near the horizon, the full moon is perceived to be further away and therefore its physical size is interpreted to belong to a larger object. A similar effect may occur when you misjudge the distance to an object, for instance if you have impaired binocular depth perception.

Finally, cross-modality is known to influence perception. For instance, facial expressions influence the perception of emotion in the voice of the speaker, and the direction of gaze, or your attention alone, will influence the perception of sound. In a crowd, for example, you can pick out which conversation you want to listen to (the cocktail party effect) with or without shifting your gaze.

Is there a real difference between neural and cognitive explanations of visual illusions? If so, where does the neural interpretation end and the cognitive start? In the view of some prominent neuroscientists, what we call cognition is simply a reflection of the activity of nerve cells (Crick, 1995). Over recent years, and with the increasing influence of neuroscience on our thinking about perception and the brain, we have witnessed a change in views on this matter. The prevailing view from some years ago of vision as a model, or as a mental construct, is about to give way to explanations in terms of neural processes. One example is the explanation of the Kanizsa triangle (Figure 1.12(a)]. Kanizsa himself appears to have believed in a cognitive interpretation, but he did not hesitate to revise his opinion once it became clear that cells in the primate brain reacted to his illusion as if there were, indeed, physical contours present in the image.

Neural correlates of such a perceptual completion, in the absence of a physical luminance difference, have been described as occurring in different species. The state of the animal, whether it is alert and how it is behaving, is important for the neural activity in the early brain areas, indicating a cognitive amplification of the response through feedback from higher brain levels. Illusions, like the Kanzsa triangle and the more cognitive ones of Figure 1.12, are well suited for studying how cognitive activities influence and modulate neural brain processes.

Here, we have compromised in that, when an illusory effect is influenced by our interpretation, i.e. by our way of thinking about it, then it shall be regarded as mainly cognitive, such as the effect of seeing a W or two Vs in Figure 1.11(a) or the color

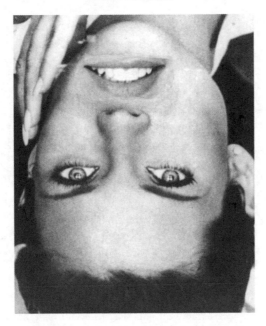

Figure 1.16 Turn the book upside down. What do you see? Can you suggest an explanation?

contrasts of the demi-circles in Figure 1.11(b). Other examples are caricatures, and the images where we are voluntarily able to flip figure and background, from seeing one figure or another, as in the reversing or multistable images in Figure 1.3. What we see in these examples is dependent on our interpretation and our conscious thinking. Neural correlates should in these cases be understood as correlates of consciousness, i.e. correlates that would depend more on the state of the conscious mind (a 'top-down' process) than on the physical entity. This definition would not necessarily include the Gestalt factors as being cognitive. Gestalt might as well be viewed as the ways neural networks of living organisms organize experience, depending on biological predisposition and genetic programming.

It is unclear how the illusion of Figure 1.16 relates to cross-modality or to any of the other phenomena described above. The illusion is demonstrated by rotating the book 180°. A key to an explanation may lie in separate coding of a face as a unitary form from its different parts (eyes, nose, mouth, etc.).

While illusions are surprising and usually considered a lot of fun, they also give rise to confusion and bewilderment. Knowing about them makes us less sure of our judgements and maybe a little uncomfortable because they seem to question our rationality. This is also what makes them so fascinating. Every now and then new, surprising effects are discovered, and in Figure 1.17 an intriguing illusion of movement is shown, discovered by the Italian psychologist Baingio Pinna (Pinna and Brelstaff, 2000). Here one would expect movement-sensitive neurons to play an important part (Gurnsey et al., 2002).

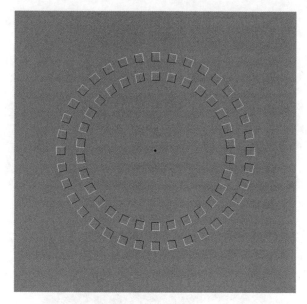

Figure 1.17   A rotating wheel? Pinna's illusion. (Reproduced from Pinna *et al.*, 2001, *Vision Res.* **41**, 2669–2676 by permission of Elsevier Science Ltd.)

When a blind person becomes seeing

We started with the story of Virgil's lack of understanding of the visual world. His experiences, together with the several illusions experienced by the normally sighted, clearly demonstrate the role we play in our encounter with the external world. Our adult world-views depend on inherited predispositions of the brain together with acquired conditioning. The latter depends on our ability, in younger years, to interact with the environment using all our senses. The dimensions of a visual world, which are taken for granted by the normally sighted, do not hold a similar reality for the blind. People who have been blind for a long time, with stories similar to Virgil's, support the idea that the act of seeing is not a simple reflex – it is not automatic (von Senden, 1960). If, in the early years of life, we are deprived of the opportunity for normal, active engagement of the visual sense, irreversible impairments will result.

2 Optics

Light

Physiological optics treats topics that are relevant for understanding how images are formed in the eye. *Geometric optics* treats light as bundles of straight lines or beams, like when the light from a pinhole passes through a smoke cloud or, in a laboratory demonstration, through a bowl of glass filled with a blend of water and skimmed milk. It was Euclid (about 300 BC) who introduced the concept of light rays (Grini, 1997). In modern physics light is described dualistically as a wave and as a stream of particles. The concept of the wave nature of light is attributed to Christiaan Huygens (1629–1695). Isaac Newton (1642–1727) preferred the particle theory, and these two theories were considered to be in conflict with each other. Even today we do not have good unifying alternatives to the dualistic concept, and this is a dilemma to anyone who wishes to 'explain' light. Modern theories combine the wave concept and the concept of photons, and we use either the particle or the wave analogy depending on which one best suits the phenomenon being dealt with.

Figure 2.1 shows light as a part of the spectrum of *electromagnetic radiation*. Light – or visible radiation – is the small 'window' in this spectrum that allows us to see with our eyes, as a small part of a much greater spectrum that consists of everything from radioactive radiation to radio waves. Normal eyes can detect radiation with wavelengths, λ, between 380 (violet) and 760 nm (red).

$$1 \text{ nanometer} = 1 \text{ nm} = 0.000000001 \text{ m} = 10^{-9} \text{ m}$$

For wavelengths below 380 nm, and down to about 100 nm, the radiation is called *ultraviolet light* (*UV*), although this radiation does not lead to a visual impression, which is normally associated with light. Above 780 nm and up to a wavelength of 1 mm (1 000 000 nm), the radiation is called *infrared light* (*IR*).

Light Vision Color. Arne Valberg
© 2005 John Wiley & Sons Ltd

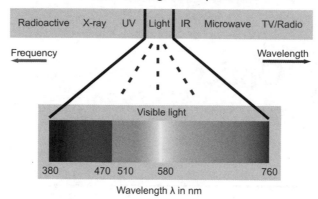

The electromagnetic spectrum

Radioactive   X-ray   UV   Light   IR   Microwave   TV/Radio

Frequency ←

Wavelength →

Visible light

380   470 510   580   760

Wavelength λ in nm

Figure 2.1 The visible range of the electromagnetic spectrum stretches from 380 to 760 nm. In addition to visible light, the electromagnetic spectrum consists of shorter-wavelength ultraviolet radiation, radioactive X-rays and longer-wavelength, infrared radiation, microwaves and TV and radio waves. See also Color Plate Section.

The description of light as a stream of particles, called *quanta* or *photons*, goes back to the beginning of the twentieth century. Quanta are small 'energy-packets', each with an energy, E, that is proportional to the light's frequency, ν:

$$E = h\nu$$

h is a universal constant, called Planck's constant, and the frequency, ν, represents the number of cycles of oscillation the wave completes per second. This equation can also be written as

$$E = hc/\lambda$$

where c is the speed of light. As for any wave, the speed of light is a product of its wavelength and frequency, $c = \nu\lambda$. Albert Einstein's (1879–1955) theory of relativity was developed on the assumption that the speed of light in vacuum is a universal constant.

The higher the frequency of electromagnetic radiation (and the shorter the wavelength), the more energy there is in each quantum. Radioactive γ-radiation and X-rays have higher quantum energy than visible light. TV and radio stations broadcast at frequencies where the wavelengths range from 1 m to 30 km. The quantum energy for this radiation and also for microwaves is lower than that of visible light. In vision, the particle nature of light is of practical importance only at very low intensities, close to the absolute threshold for night vision.

Regardless of wavelength, light and all other electromagnetic radiation propagate with a speed $c = 299\,792\,458$ m/s in vacuum (or about $300\,000$ km/s $= 0.3$ m/ns;

1 ns = 10^{-9} s). In an optically more dense material the speed of light will be lower. Since the speed of light in vacuum is constant, the length of a standard meter is now defined as the distance that light travels in vacuum in a time interval of 1/299 792 458 of a second. The meter-stick, which was previously used as a length standard, and which is kept in Paris, is therefore now obsolete.

Ultraviolet radiation

Most of the sun's *ultraviolet radiation* (UV) below 380 nm is absorbed by the ozone layer in the Earth's atmosphere. The long-wavelength ultraviolet radiation closest to visible light (from 315 to 400 nm) has been given the name UV-A. The next, somewhat more energy-rich region of the spectrum, is called UV-B (280–315 nm), while the highest-energy ultraviolet region (100–280 nm) is referred to as UV-C. While the cornea of the eye is transparent for both UV-A and UV-B, all ultraviolet light (UV-A, -B, -C) is absorbed by glass. Ultraviolet radiation can cause absorption of light in a substance and excitation of electrons from lower energies to higher energy levels in atoms and molecules. When returning to lower energies, longer-wavelength visible light is emitted, a process called *fluorescence*. Materials such as teeth, bones and minerals fluoresce on exposure to UV, as do some optical whiteners used with paper and textiles.

With respect to its effect on humans, ultraviolet light has 'two faces'; a positive effect initiating the synthesis of vitamin D, which is important for calcium to be absorbed in bone and tissue, and the harmful effect of provoking skin cancer and *cataract*. Sunburns (of the skin) and inflammation of the cornea are also well known effects of exposure to intense sunlight. UV-C kills germs and bacteria and is therefore often used to sterilize instruments.

Infrared radiation

Infrared radiation (IR) is heat radiation with wavelengths longer than 780 nm. About half of the energy of the sun reaching the surface of the earth is in the visible spectrum range, and only about 3 percent of the total is in the ultraviolet. The rest is life-giving infrared radiation.

Infrared cameras register temperature differences, and can be used to 'see' in the dark. They can also be used to see through opaque films. In a painting, for instance, infrared radiation is less effectively scattered by small particles in varnish films than visible light, and it is able to overcome the opacity of the upper layers of the painting. It therefore becomes possible to observe detail in the paint layer which has become obscured by old varnish. Forgeries can sometimes be detected in this way.

Infrared radiation is absorbed by water and by all the media in front of the retina. Only a small percentage of IR radiation reaches the retina. Overexposure to infrared radiation may cause cataract, as indicated by the expression 'a glassblower's cataract'.

Light sources

Light and its effect on the organ of vision is the subject of all chapters in this book. Here, we want to introduce some fundamental physical concepts that will be used later. Let us start with the *black body radiator*. A black body thermal radiator emits light with intensity and spectral distribution depending on the temperature of the material (usually a metal). A tungsten filament lamp comes close to being such a light source. The higher the temperature is, the shorter the wavelength of maximum energy. At low temperatures, most of the energy is in the infrared region. Even around room temperature, the black body radiates energy. Visible light begins to appear around 1000 K (1 K = −273 °C), and ultraviolet does not appear before the temperature has reached about 2000 K. At about 3000 K, the source looks bright yellow, and at 5000 K, approximately the temperature of the sun's surface, energy is about evenly distributed over the visible spectrum, and the body appears white. At still higher temperatures, the radiating body becomes bluish.

The color of natural and artificial illumination can be compared with that of a black radiating body radiator of a certain temperature. The temperature of the body that gives the closest color match is called the correlated color temperature of the light source. This characterizes the color of a wide range of illuminants, including fluorescent light, although the actual temperature of the fluorescent light may be rather different from that of the black body radiator (it depends more on the source's spectral distribution). The color of different daylight phases also comes close to that of a black body radiator of different temperatures. Color temperature is more meaningful when comparing sources having about the same spectral distribution. Incandescent light has a spectral distribution that comes close to that of the black body. The color temperature of the light from a common tungsten filament is in the range 2700–3200 K, depending somewhat on wattage. However, such lamps are not very efficient light sources. Most of the energy is released as heat, and only about 15 percent is released as light. Color rendering is a problem. Because the light is yellowish, it is difficult to see yellow lines on a white paper, and it is also hard to distinguish slightly different blue shades.

Fluorescent lamps are gas-filled tubes with a material coating, called phosphor, on the inside of the tube. When the ultraviolet radiation emitted by the mercury gas excites the phosphor, it fluoresces and radiates visible light. The result is a broad spectral distribution of emitted light with superimposed emission lines from the mercury gas.

Light from all kinds of light sources affects the color appearance of surfaces, for instance a blue surface looks different in daylight and in incandescent light, but it is usually hard to predict such differences (see chapter on Color rendering, p. 268).

Light and the diurnal rhythm

Changes in the light/dark cycle can lead to disturbances in the diurnal rhythm and to mental problems (Noell, 1995; Singer and Hughes, 1995). Shift workers are typical examples of groups that might be affected. The seasonal changes in light in the northern and southern parts of the globe may cause sleeplessness in the bright time of the year and depression in the winter. On the other hand, in institutions for the elderly, the variations of brightness between day and night may be too small to maintain a normal diurnal rhythm (Holsten and Bjorvatn, 1997). The circadian rhythm, the sleep–awake cycle of about 24 h (circadian means 'almost a day'), is determined by a hormone, *melatonin*, which is produced in the dark. The production of melatonin is inhibited by strong light; artificial light can therefore be used to control the diurnal rhythm. A special kind of light sensitive ganglion cells (with their own photopigment) that project to the hypothalamus are probably responsible for synchronization of circadian rythms with the solar day.

In many cases light therapy, the exposure to intense light for a period every day, solves problems related to diurnal rhythm and depression in the polar night. In the literature, one can find reports that positive results have been achieved with exposure to light levels of 2000 lx for a period in the morning. White light has been reported to be more effective than colored light.

Geometrical optics

Shadows

The fact that light travels as straight rays results in the formation of shadows, some examples of which are shown in Figure 2.2. With the light radiating from a single, small, point-shaped light source, objects will give rise to homogeneous dark shadows with relatively sharp edges. With two point sources of light side by side, a central dark core will result (often called the umbra, U), surrounded by two half-shadows (the penumbra, P) each of which is illuminated by only one light source.

With an extended light source, for example a conventional fluorescent tube, a long shadow will be seen with an orientation parallel to the tube. Thus, when there is a certain distance between the object and the shadow, we reach the conclusion that the form of a shadow depends more on the geometry of the light source than on the shape of the light-obstructing object. The shadow of a round object (such as a coin) at some distance from a triangular light source will be triangular, not round. If we pass light from a triangular light source through a mask with a circular aperture, it will form a triangular bright spot, rather than a circular disk, on the opposite wall.

Figure 2.3 shows a photograph of the shadow of some foliage taken during a partial eclipse of the sun. The moon masked about half of the sun, and one can clearly see

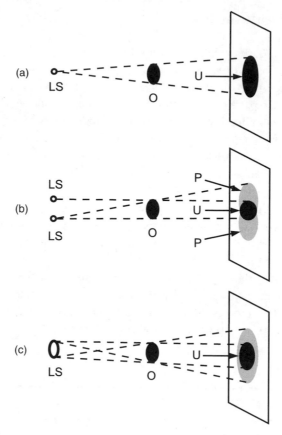

Figure 2.2 Shadows created by a point light source (top), by two point sources (middle), and by an extended light tube (bottom). LS, light source; O, object; U, umbra; P, penumbra.

this in the pattern formed on the wall by the leaves. These bright crescents demonstrate that the shape of the bright spots that you normally see reflects the disk-shape of the full sun, rather than the shape of the leaves.

Exercise

Make a cross-formed light source (for example, fixing a cardboard sheet with a cross-formed aperture in front of a table lamp). What is the shape of the shadow formed by a round object, for example a coin, positioned at some distance from the light source?

We are not used to thinking that shadows have much importance in recognizing objects. However, the shadows we are accustomed to seeing in our everyday lives also influence our interpretation of visual images. For example, in Figure 1.15 we have the impression that some circles are concave, or curved inwards, while others are convex, or curved outwards. Those seen as coming out of the paper are usually white in the

Figure 2.3 Shadows of leaves on a wall photographed during the partial eclipse of the sun in the summer of 1999. Note the sickle-shaped form of the light spots. How does this pattern differ from the normal ones in full sunlight? (Courtesy of Harald M. Anthonsen.)

upper halves and dark gray on the lower parts, whereas those which have the opposite distribution of lightness are hollow. These three-dimensional illusions are usually explained by the fact that shapes and shadows are normally illuminated from above. However, as we have already mentioned in the section on illusions, this is only a part of the explanation (see Figure 1.15).

Reflection – mirrors

When standing in front of a mirror, the image is seen behind the mirror. This image is *virtual* (opposite of *real*), since the light is not reflected from the positions in space corresponding to the image.

One can construct the virtual image that is created in a mirror by using the angles of incidence and reflection of light rays entering and leaving the mirror's surface. Figure 2.4(a) shows two of the many light rays one can imagine coming from an object *O*, at a distance *d* from the mirror. A ray that enters the mirror perpendicular (normal) to the surface, is reflected backwards in the same direction, and a ray that

OPTICS

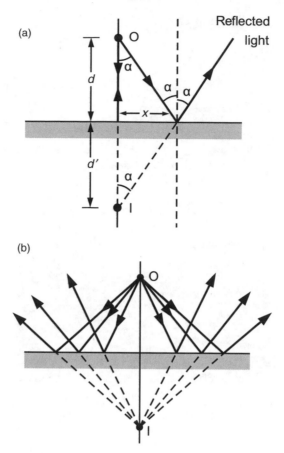

Figure 2.4 Reflections in a mirror. (a) The lengths of d and d' are the same, and the angles of incidence and of reflection are the same. O, Object; I, image. (b) The reflected light appears to come from the image point I.

hits the surface with an angle α relative to the normal is reflected back on the other side of the normal with the same angle. It thus appears as if the light is coming from a point behind the mirror, in the extension of the reflected ray. The backward projections of all reflected rays cross each other at the image point I. The image of every point on the object is directly in front of the respective object point, with the image distance d' being the same as the object distance d. Ordinary mirrors have a certain thickness, reflecting from both the front and back surfaces, thus forming double images.

The face that you see in a mirror is not how others see you. Compared with a photograph, mirror images are flipped horizontally but not vertically, switching your left and right. However, when we lay down horizontally in front of the mirror, the image is reversed vertically, but not horizontally. A better description of how your mirror image differs from a photograph would be that your mirror image is reversed back to front without changing the relative positions of your head and your feet or your left and right hands. If you were standing on a mirror, the relative position of your feet and head would be reversed in the image, without reversing left/right or front/back.

The appearance of a surface is closely related to the way it reflects light. Specular reflection characterizes light coming from a smooth, polished surface, like a mirror, and diffuse reflection occurs from a roughened, matte surface. The same paint may look different on surfaces of metal, paper, glass and textiles. The appearance of gloss corresponds to a high degree of specular reflection.

Refraction

The propagation of light changes direction when it enters a transparent medium of different optical density, for example in passing from air to glass. This refraction is important for the image representation of the object. Refraction is said to follow Willebrord Snellius's law (called Snell's law of refraction; Snellius was a mathematician und physicist from the Netherlands, 1591–1626):

$$n_2/n_1 = \sin\alpha_1 / \sin\alpha_2$$

Here n_2 and n_2 are the *indices of refraction* in mediums 1 and 2 (e.g. air and water). The angle of incidence, α_1, and the angle of refraction, α_2, are the angles which a ray of light makes with the normal to the plane between the two media (see Figure 2.5).

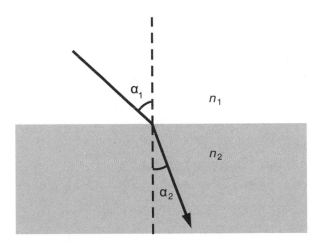

Figure 2.5 Refraction at a plane surface between air and an optically denser medium. Air has an index of refraction $n_1 = 1.0$, whereas for normal glass $n_2 = 1.50$, and for water it is 1.33. When entering a denser medium, light is refracted towards the normal to the surface. α_1 is the angle of incidence and α_2 the refraction angle.

The refraction index, n, is larger in the optically denser medium (here medium 2). The index of refraction for a light of a given frequency is the ratio, $n = c/v$, of the speed of the light, c, in vacuum and the speed of light, v, in the medium, the latter always being slower than the speed of light in vacuum. The ratio n_2/n_1 then becomes equal to the inverse ratio of the speeds of light, v_1/v_2, in the same medium.

Snell's law implies that, if n increases, the sine to the angle of refraction decreases, and therefore the angle itself decreases. When a light beam enters an optically denser medium it is deflected towards the normal, while the ray is deflected away from the normal when passing from a more dense to a less dense medium.

As far back as in 984, an Arab, Ibn Sahl, formulated the law of refraction (Grini, 1997). There are indications that in Europe too the law was known before Snell. Studies carried out by a Norwegian, Johannes Lohne (1959), suggest that Thomas Harriot in England knew about the law of refraction in 1602. However, its formal expression was first published by Descartes in *La Dioptrique* in 1637 (see Descartes, 1953).

The refraction index in air is usually set at $n = 1.0$. More accurately it is 1.00028 at a wavelength of 589 nm and 15 °C. In water, n is 1.333 ($= 4/3$), in light crown glass $n = 1.517$ (3/2), and in heavy flint glass it is approximately 1.65.

In a flat sheet of glass the light ray is displaced parallel to its original path by a distance proportional to the thickness of the glass (Figure 2.6). Some light is also reflected at the border surfaces (air/glass and glass/air).

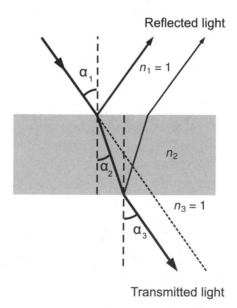

Figure 2.6 Refraction and reflection at parallel plane borders between air and glass. When the transmitted ray leaves the glass and enters air again, it is displaced parallel to the entry ray ($a_3 = a_1$). Some light is reflected at all borders.

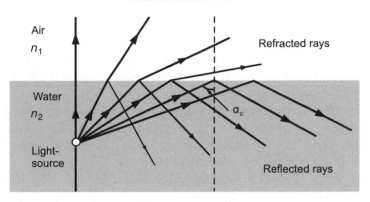

Figure 2.7 Total inner reflection occurs when the angle of incidence is greater than the critical angle α_C. This angle depends upon the refraction indices of the two media.

Figure 2.7 shows what happens when light rays originate from a source under water. When a ray meets the surface, some of the light will be reflected back into the water and some will be refracted into the air. The intensity of the reflected light will increase as the angle of incidence increases (towards the right in the figure), and at the same time the intensity of the refracted light will decrease. When the angle of refraction reaches 90° relative to the normal (at the position of the vertical dashed line), the intensity of the refracted ray has dropped to zero. When this happens, the angle of incidence has reached the critical angle, α_c. When the angle of incidence increases beyond this, there will be no refracted light; all light will be reflected back into the water at the border between the two media. This is called *total inner reflection*. Total inner reflection will only take place at the border to a material with a lower index of refraction.

In mirror reflections the reflected light will always be slightly less intense than the incident light. The intensity of the reflected light depends on the refraction indexes n_1 and n_2 in the two media (air and glass, for example). Light falling normally on the surface obeys the following relation for the ratio, R, between the intensity I_r of the reflected light and the intensity I_o of the incident light:

$$ R = I_r/I_o = (n_2 - n_1)^2/(n_2 + n_1)^2 $$

From this expression we find that, at the border between air and glass ($n = 1.5$; for example in spectacles), 4 percent of the intensity is lost at each surface reflection and about 8 percent is reflected in total. The greater the difference, $\Delta n = n_2 - n_1$, the more light is reflected. Unvarnished paint surfaces appear matte, and a varnished surface is glossy. More light is reflected from the varnished surface, leaving less light to be transmitted to and reflected by the paint layer. This effect causes a darkening of a freshly varnished surface or of a wet sidewalk after rain (covered by a smooth, glossy coating of water; Brill, 1980).

A material with a high index of refraction tends to appear more opaque. A diamond owes its brilliance to a large difference, Δn, relative to air, as well as to a high degree of color-producing dispersion. For total inner reflection there is no intensity loss. This is why prisms with total inner reflections are often used instead of mirrors in binoculars, periscopes and cameras.

The fact that there is no energy loss in total inner reflection is exploited in fiber optics, where light is sent through long, thin strands of glass. Glass fibers consist of two types of glass with different refraction indices; the glass with the smaller refractive index surrounding the other. Such fibers can transmit light over a distance of many kilometers without significant attenuation. This technology has become increasingly important for the transmission of data over long distances and for telecommunications. Similar fibers are also used in medical procedures to image inner organs before and during surgery.

Dispersion

The refraction index of a material changes with the wavelength of the incident light. The lens of the eye has a refractive index that varies between 1.451 for short-wavelength light around 430 nm and 1.437 for 690 nm wavelength. Short-wavelength light, which appears blue or violet in the laboratory, is thus refracted more than long-wavelength orange and red lights. This phenomenon is called *dispersion*.

Whitish light from a light source consists of many different wavelengths. These wavelengths will be separated at the border between air and glass, except when the direction of the incident light is normal to this surface. In the prism in Figure 2.8 this happens twice, once when light enters the prism at the surface between air and glass and again between glass and air when the light leaves the prism. The resulting separation of wavelengths results in a spectrum of different colors that can be projected on to a screen.

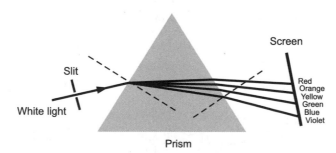

Figure 2.8 The index of refraction depends on the wavelength of the light (dispersion). When a narrow beam of white light is refracted by a glass prism, it is spread out as a color spectrum on the opposite wall. Short-wavelength light (that has a violet color in darkness) is refracted more than long-wavelength (red) light.

If the beam of light is sufficiently narrow, for example after passing through a narrow slit, dispersion may cause light of different wavelengths to separate enough to give rise to a multicolored spectrum on the screen. When viewed in darkness at moderate intensity, these wavelengths will appear as colors in a sequence going from red (760–600 nm), orange (600–580 nm), yellow (580–560 nm), green (540–490 nm), turquoise (490–480 nm), blue (480–460 nm) to violet (450–380 nm). When the slit and the light beam are wide we see the so-called border colors; an orange and red border opposing a turquoise, blue and violet border with white in between. Unless it is corrected for, dispersion will cause black/white borders to be rendered with colors when imaged by an ordinary lens. Later we shall see that it is possible to compensate for this dispersion in lenses by combining materials with slightly different refractive indices.

Polarization

Radiation from most light sources is unpolarized. This means that the oscillations of the electric and magnetic vectors of the electromagnetic waves are orientated randomly in all directions in a plane perpendicular to the direction of propagation. If the amplitude of this oscillation is diminished in some specific directions, either as a result of reflection or transmission, we say that the light is polarized. This happens, for instance, when light is scattered by small particles in the atmosphere, when it is reflected by water and glass, and by transmission through a polarizing filter, such as polaroid sunglasses. Blue sky light scattered by small particles in the atmosphere has a high degree of polarization. This polarized light is not registered by humans, but is of great importance to bees, ants and flies that navigate by it. Bees, for instance, can take into account the changing polarization of sunlight due to the sun's altitude above the horizon.

In linearly polarized light, the electric vector of the waves oscillates in one direction only, like a cork on water. A polarizing filter absorbs light of all directions of oscillation except for one, where it transmits light. If another polarizing filter is held before the first one, the intensity of the light which is let through both filters will depend on the relative orientation of the two filters. If the light entering the filters is unpolarized, and both filters have the same orientation, the light that is polarized by the first filter will also pass through the next. However, if the second filter is rotated 90° from this direction, all light that is let through the first filter will be absorbed by the second. At all angles in between, more or less light will be transmitted, depending on the angle between them. In this way the intensity of light can be adjusted gradually, with the same effect as adjusting an iris diaphragm in a camera.

The orientation of polaroid filters in sunglasses is such that they optimally absorb the polarized light that is reflected from horizontal surfaces, such as water and wet roads. They have little effect on light reflected from vertical surfaces, such as windows, etc.

Strain regions in glass can be made visible by placing it between two polarizing filters.

Scattering

The light scattered from an object's surface is important for how the surface appears. A matte appearance is the result of diffuse reflection, when light is scattered in all directions as a result of a randomly irregular surface. The matte appearance does not depend on the angle of viewing, because light is uniformly scattered in all directions.

If the difference in the index of refraction between air and object is large, the surface will be a good scatterer of light and its transparency reduced. If there is no difference, no light will be scattered at the interface, and all the light will pass into the material. Such considerations are important in painting. For instance, the transparency of a paint will be high if the index of refraction is small. 'Whitewash' (chalk mixed with water and a little glue) is not entirely opaque when first painted on a wall. However, as it dries, its opacity increases and so does its whiteness. The explanation is that the wet chalk–water interface has a difference in refractive index of about 0.27, but when dry the chalk–air difference is 0.60. The same applies to watercolors. They tend to be more transparent when first applied than when dry – their hiding power increases as they dry (Brill, 1980).

If a particle is much smaller than the wavelength of light, the light will bend around the particle and be scattered in a manner different from larger particles. This is known as Rayleigh scattering (after the English physicist J. W. Rayleigh, 1842–1919) and is responsible for the blue appearance of the sky. This scattered light is proportional to $1/\lambda^4$, which means that light at 380 nm is scattered 16 times more than light at 760 nm, and that the violet and blue of the spectrum are more effectively scattered towards the observer than are the longer wavelengths.

Diffraction

Light waves can bend around corners and propagate into the shadow of an object, like waves on water when they are bent into the 'shadow' of an obstacle. Waves that originate on opposite sides of an object can interfere with each other when they meet again; they can either amplify or cancel, dependent on their relative position (their relative phase). This is called diffraction and occurs for all kinds of waves.

This phenomenon can easily be observed when the light is monochromatic, i.e. when it consists of only one wavelength, for example from a laser. When this light passes a small, circular aperture, an intensity pattern will arise on the opposite wall that is produced by the interference of waves from different positions of the aperture. The diffraction pattern from a distant point source is an intensely illuminated center area surrounded by concentric dark and light rings (Figure 2.9).

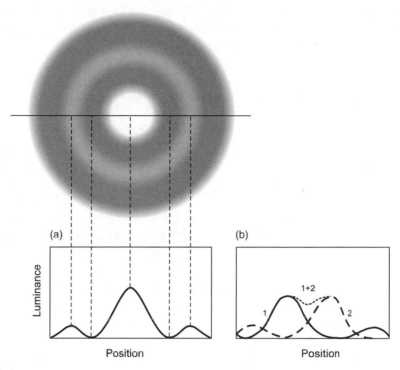

Figure 2.9 Diffraction and the Rayleigh criterion for visual resolution. (a) A small circular aperture leads to a diffuse image because of a diffraction pattern in the form of light and dark rings. (b) The limit of resolution of two close stars. They can be distinguished if the intensity maximum in the diffraction pattern of one of them falls on the intensity minimum of the diffraction pattern from the other.

The Rayleigh criterion for resolution, a rule often used in astronomy, states that two points of light, e.g. two nearby stars, can be separated if their angular distance is larger than the distance from the bright center to the first dark intensity minimum in the diffraction pattern. For stars that are closer together, their images cannot be separated as two points, but are smeared out to a somewhat elongated diffuse area.

For imaging through a circular aperture, such as for the pupil, the smallest angular distance, θ, for the resolution of two images is:

$$\theta = 1.22 \, \lambda/D \, (\text{rad})$$

Here, λ is the wavelength of light, and D is the diameter of the pupil, both measured in meters. The angle is given in radians (rad; $360° = 2\pi$ rad, and 1 rad = $57.296°$).

For a wavelength of 560 nm and a pupil diameter of 3 mm, the angle θ and the limit of the eye's resolution due to diffraction is:

$$\theta = 2.3 \times 10^{-4} \, \text{rad}$$

This corresponds to an angle of 0.013° or 0.79 min arc (60 min arc = 1°). We get the same value for the angle if we use the following combination of wavelength and pupil size: 380 nm and 2 mm pupil, or 760 nm and 4 mm pupil.

The Nyquist criterion for visual resolution

The Rayleigh criterion relates to diffraction, but there are other limits to resolution as well: the two stars mentioned above must also be resolved by the retinal mosaic of photoreceptors. Let us assume that the anatomical criterion for resolution of two point sources is that there is at least one photoreceptor between the two light maxima that can register the first minimum in their diffraction patterns (Figure 2.9(B)]. This is often referred to as the *Nyquist criterion* for resolution. In the fovea this doube distance between cones is about 5 μm. When using an eye length of 2.4 cm, this gives a minimum angle for resolution of 2.1×10^{-4} rad, or 0.012° (0.72 min arc).

This number is close to the diffraction limit that we calculated from the Raileigh criterion (in comparison, an angle of 2.3×10^{-4} rad as calculated above would imply a distance on the retina of 5.5 μm). This means that the density of the cone packing in the fovea is well adapted to the physical limit set by diffraction. As we see from the Rayleigh criterion, for a 3 mm pupil diameter, wavelengths shorter than 560 nm give an angle to the first diffraction minimum that is too small to resolve. The diameter and the spacing of the cones in the retina will then be the limiting factor. Thus we may say that, for a constant pupil diameter of 3 mm and wavelengths shorter than 560 nm, the distance between the foveal cones determines resolution (acuity), and for wavelengths longer than 560 nm, inter-cone distance is smaller than needed, and diffraction will set the resolution limit.

Let us also consider the case with a constant wavelength of 560 nm, but a variable pupil size. A larger pupil will remove the limitations set by diffraction and leave the cone spacing as the main limiting factor. However, at the same time image quality will deteriorate because a larger corneal surface introduces more irregularities and amplifies lens aberrations.

The above calculations lead to the conclusion that diffraction problems are irrelevant for all wavelengths in the visible spectrum when the pupil diameter is larger than 4 mm. Theoretically, diffraction will reduce the quality of the image for smaller pupil sizes, increasingly so for longer wavelengths. However, in practice, optimal image quality depends on minimizing several adverse factors such as diffraction, spherical aberration, chromatic aberration, stray light from the eye media, etc. Therefore, a pupil size somewhat smaller than that derived from considering diffraction alone is usually recommended.

Figure 2.10 shows a diagram of normal pupil sizes for light- and dark-adapted eyes as a function of age. Maximum pupil size is reduced with age, this reduction alone lowering maximum retinal illuminance in the dark by about 50 percent.

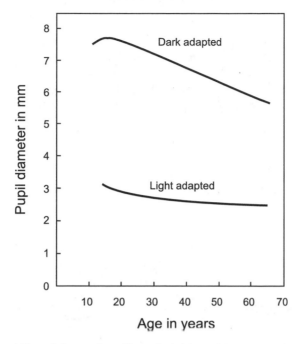

Figure 2.10 The ability of the pupil to dilate diminishes with age. The figure shows the pupil diameter as a function of age for two different adaptation levels, one corresponding to light adaptation and the other to dark-adapted eyes.

Vergence

Lenses are made from transparent materials, usually glass or plastic, that have a geometric form that will focus light and make images. Here we shall mainly deal with thin, spherical lenses, lenses that are thin in relation to the radii of the curves of their surfaces. However, before dealing with lenses that have two refracting surfaces, we shall take a look at what happens when light passes through a single curved surface.

When light from a point source enters a surface it has a certain *vergence*. Vergence is a useful concept in geometric optics, and therefore we shall devote it a little space. Light that spreads out in space around a point source is called *divergent*; it has a negative vergence. The wavefront of the light is concave towards the light source, with the center of curvature being the light source itself (see Figure 2.11(a)).

Focused light is called 'convergent'. To the right in Figure 2.11(a), the wavefront is concave towards the center of curvature, i.e. towards the point of focus. It is defined as having a *positive vergence*. Vergence, both divergence and convergence, has the unit m^{-1}, called a *diopter*. For example, light coming from a point 2 m away has a wavefront with a radius of curavture, $r = -2\,m$, and therefore a vergence, V:

$$V = 1/r = -0.5\,m^{-1}$$

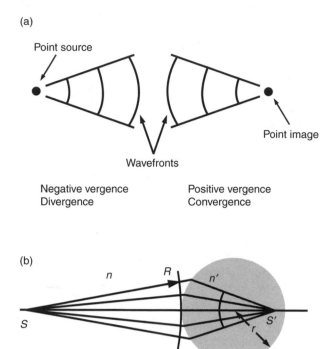

(a)

Point source

Wavefronts

Negative vergence
Divergence

Positive vergence
Convergence

Point image

(b)

Figure 2.11 (a) We use the term vergence when light spreads out from a point source (left) or when it is focused in a point image (right). (b) Divergence is followed by convergence when light leaves a point source and subsequently enters the curved surface of a medium with higher positive lens power, $P = n'/r$, than the light's negative vergence, $V = n/R$.

Let us analyze what happens when light meets the curved surface of a medium with a refractive index n, say glass with a *refractive power*, P. As with vergence, we define a surface that is convex in a direction opposite to the direction of light propagation to have positive power (in the case of light coming from the left, a convex surface towards the left has a positive power). Another surface that is concave towards the left has a negative power. If a surface has a positive power greater than the negative vergence of the light, the refracted light will have positive vergence, V'. This means that the light will converge and form a real image of the source in a point S' inside the medium, at a certain distance from its surface [see Figures 2.11(b) and 2.12(a)].

When the source is brought closer to the refracting surface, as in Figure 2.12(b), the radius of curvature of the wave front which enters the surface will be smaller, and hence, according to the definition, the negative vergence will be more strongly negative. At a certain distance, which we will call f, the negative vergence will just cancel the positive vergence of the surface and the net result will be zero vergence.

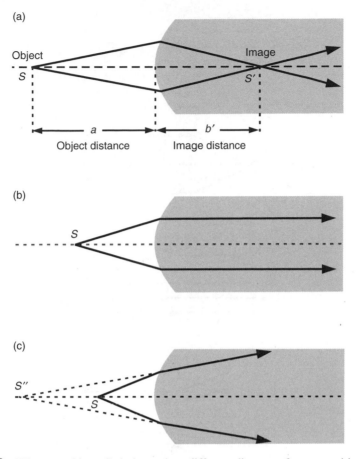

Figure 2.12 When an object, S, is located at different distances from a positive, refracting surface the image may be real, as S' in (a), at infinity as in (b), or virtual as in (c).

This means that inside the medium the light rays will be parallel and that the image will be projected at infinity [Figure 2.12(b)].

When the light source is closer than that, the surface does not have enough converging power to compensate for the larger negative vergence of the light, and the light beams will remain divergent inside the second medium. No real image can then be formed inside the second medium, but a so-called *virtual image* can be constructed outside the medium. In this case [Figure 2.12(c) to the left], the inner rays would appear to come from the point S''.

When parallel light rays are directed towards a spherical surface with positive vergence, as in Figure 2.13, they will all intersect at *the second focal point, F'*, at a distance, f', from the surface. In the next example [Figure 2.13(b)], the distance to the light source and to the curvature of the surface has the combined effect of allowing

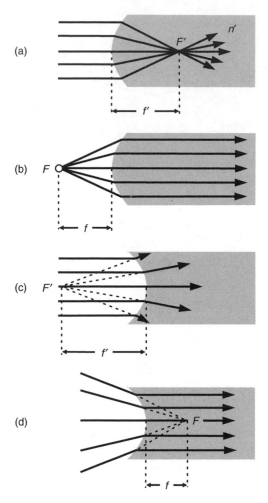

Figure 2.13 The focal points, F and F′, and the corresponding focal distances, f and f'.

the rays to propagate inside the medium in a direction parallel to the optic axis. The light source is located at the *first focal point*, F, at a distance, f, from the surface. f and f' are both *focal lengths*. The same convention applies to a surface with negative curvature [Figure 2.13(c,d)]. The refractive power, P, of a surface with *refractive index n′*, is:

$$P = n'/f'(\mathrm{m}^{-1})$$

An advantage of concepts like vergence, V, and of refractive power, P, is that the distance to the image, b', inside a medium, with refractive index n', can easily be calculated. The calculation only requires knowledge of the distance to the object, a, the refractive indices, n and n', of the two media, and the refractive power, P, of the imaging medium. The following equation applies:

$$V + P = V'$$

which is the same as

$$n/a + n'/f' = n'/b'$$

This simple summation allows us to make computations on images inside refractive media, including the eye. When the light enters the refracting surface from air, $n = 1$.

Example

We have a point source at a distance of 0.5 m to the left of a glass rod. The glass has a refractive power $P = 6$ diopters. The refractive index, n', of the glass is 1.6. Where will the source be imaged inside the rod?

In this case the vergence in air is $V = -1/0.5 = -2$ diopters. The vergence inside the rod, $V' = V + P$, giving $V = -2 + 6 = 4$ diopters. Since $V' = n'/b'$, we can calculate the distance b' to the image as $b' = n'/V' = 1.6/4 = 0.40$ m.

After doing this exercise, you will be able to calculate where the image would be if $P = 60$ diopters and $n' = 1.40$, values closely approximating those of the human eye.

We will return to images formed on the retina of the eye and in other optical media later on, but first some words about thin lenses. From the view of geometrical optics, ordinary spectacles can be treated as thin lenses.

Thin lenses

Lenses can collect light (converging lenses) or they can spread light (diverging lenses). Normally, when a lens is in a medium with a smaller refraction index than that of its own material (e.g. glass in air or water), converging lenses are convex and diverging lenses are concave (see Figure 2.14). A convex lens refracts a beam of

Converging lenses   Diverging lenses

Figure 2.14   Converging and diverging lenses.

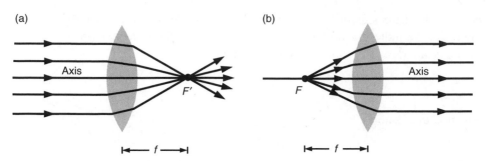

Figure 2.15 Focal points and focal lengths for a converging lens. The focal length, f, is positive for a converging lens.

light that is parallel to the optical axis towards the focal point F' behind the lens. Light coming from the focal point F in front of the lens will leave the lens as parallel rays [Figure 2.15(a, b)]. A concave lens will refract parallel light away from the axis, making it appear as if the light is coming from the focal point F' in front of the lens. Rays initially converging towards the focal point F behind the lens will leave the lens as parallel beams (Figure 2.16). The focal length f is defined as positive

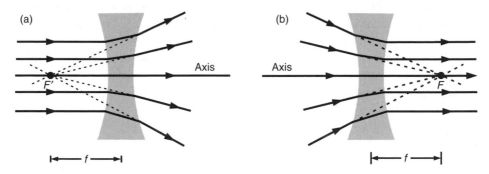

Figure 2.16 Focal points and focal lengths for a diverging lens. The focal length, f, is negative for a diverging lens.

for collecting lenses and negative for diverging lenses. Thus, thin lenses have two focal points, F and F', one on either side, the focal lengths being the same on both sides.

The lens maker's formula

The power of a lens is defined by its focal length: shorter focal lengths mean stronger refraction and higher lens power. The focal length is determined by the material of the lens and the curvature of its surface. Convex lenses have positive radius of curvature, whereas diverging lenses have negative radius of curvature. A plane surface has an infinitely large radius of curvature. These conventions are described in Figure 2.17.

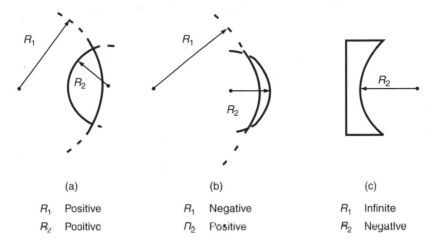

(a)   (b)   (c)

R_1.   Positive R_1 Negative R_1 Infinite
R_2 Positive n_2 Positive R_2 Negative

Figure 2.17   Curvatures and radii of curvatures for lens surfaces. In (a) the curvatures are convex, in (b) they are concave and convex, and in (c) plane and concave.

Examples of refraction and imaging using thin lenses are shown in Figure 2.18(a–c). The focal length, f, of a thin lens with refractive index n and two radii of curvature, r_1 and r_2, is given by the lens-maker's formula (n is given relative to air for which $n = 1.00$):

$$1/f = (n - 1)(1/r_1 + 1/r_2)$$

Since the radius of curvature is positive for a convex lens surface and negative for a concave surface (see Figure 2.17), the formula above gives a positive value of f for converging lenses, whereas diverging lenses have a negative value. It is important to note that, in the formula above, the index of refraction, n, is a relative index. This is of no significance in air with a refraction index equal to 1.00, but if the medium is not air, n must be interpreted as the ratio $n(\text{lens})/n(\text{medium})$.

A consequence of the lens-maker's formula is that a lens will have a longer focal length when used underwater. This explains why we see so poorly underwater. The refraction indices of the eye media are only a little higher than that for water ($n = 1.33$) and therefore there is less refraction in the transition from water to eye than between air and eye. When using diving goggles, light is refracted as normal at the air/eye boundary, and vision will be normal close to the optical axis.

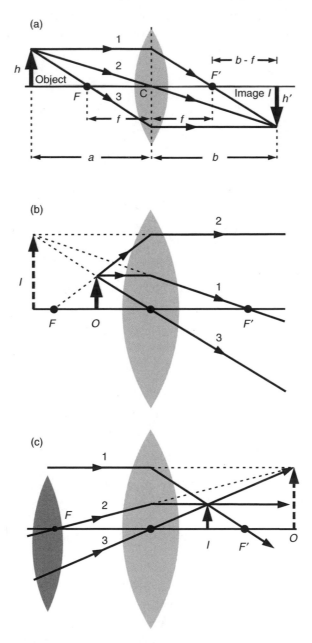

Figure 2.18 Three light rays can be used to construct the image. In (a) a real object and a real image are illustrated. In (b) there is a real object, O; and a virtual image, I. In (c) we have a virtual object, O, and a real image, I. A real object and a real image are drawn as a solid arrow, while the virtual ones are dashed.

An air-filled biconvex lens made from plastic and placed in water would be a diverging lens. This is easy to see from the lens maker's formula when the expression $(n-1)$ becomes negative, when, as in this case, the ratio n is less than 1; $n = n(\text{air})/n(\text{water}) = 1/1.33 = 0.75$).

Lens power

Lens power is measured in diopters

$$P = (1/f)\,(\text{m}^{-1})$$

For a biconvex lens with two equal positive radii of curvature, and a biconcave lens with two equal negative radii of curvature, the lens power P will be

$$P = 1/f = 2(n-1)/r$$

The lens power P (and the focal length f) then becomes positive for a collecting lens and negative for diverging lenses. The advantage of expressing the lens power in diopters is that diopters can be added together straightforwardly.

$$P = P_1 + P_2 + P_3 + \cdots$$

Two thin lenses, with diopters 2 and 3, will, when placed one right behind the other, have a total lens power of 5 diopters.

Imaging with thin lenses: the lens formula

We have already mentioned that images can be defined as real or virtual. A real image is one that can be formed on a screen, like in Figure 2.18(a). The figure demonstrates the formation of an image with the help of a few light rays. A virtual image, on the other hand, seems to be formed somewhere in front of the lens. This image cannot be captured on a screen [Figure 2.18(b)].

Virtual images can be formed in a system of many lenses, for example, when converging light rays from one lens pass through another. Images that would be formed by the first lens alone serve as objects for the next lens, as in Figure 2.18(c). The imaging performed by this lens system can be explained by the lens on the left making a virtual object (the dashed arrow to the right in the figure) that is imaged as a real image (short solid arrow). Three rays from an object-point off the axis can localize the image. Two are sufficient; the third ray can serve as a control. This way of constructing an image is shown in Figure 2.18 (a–c). In Figure 2.18(a) the arrow on the left serves as a real object at a distance a from the lens. The lens forms a real, but upside-down image in the image distance b. The three rays can be drawn like this: ray no. 1 leaves the arrow head parallel to the axis and is refracted towards the focal point

F'; ray no. 3 passes through the focal point F and leaves the lens parallel to the axis; and ray no. 2 passes through the center of the lens without refraction since the two lens surfaces are parallel. The ray travels as if it passes through a flat sheet of glass, and because this is an (ideal) thin lens, the displacement is so small that it can be neglected.

Figure 2.18(b) shows how a larger virtual image is formed by a real object, as in a magnifier (see below). Figure 2.18(c) shows how a virtual object can form a real image. The three rays crossing at the tip of the imaged arrow in I were drawn through the focal points F and F', and through the center of the lens. It is also possible to find the distance to the image by simple calculations. In order to do this, we must introduce some rules for the use of positive and negative signs:

1. The object distance a is positive for a real object and negative for a virtual object.

2. The image distance b is positive for a real image and negative for a virtual image.

3. The height of the object h is positive if the object is above the axis and negative if it is below the axis.

4. The image height h' is positive for an image that is above the axis and negative for one that is below.

It is possible to derive a mathematical expression for the relationship between the distances a, b and f from the geometry of Figure 2.18(a). A consideration of congruent triangles determines that:

$$-h'/h = b/a$$

and the linear magnification becomes

$$m = h'/h = -b/a$$

The linear magnification m is the ratio between image size and object size. This ratio is negative when the image is upside-down and positive when the object is right-side up. Geometrical considerations lead to the relationship of $b/a = (b-f)/f$, and from this we get the lens formula (the lens-user's equation) for thin lenses:

$$1/f = 1/a + 1/b$$

For far-away objects, the ratio $1/a$ approximates zero, and $1/b = 1/f$. Thus, in this case $b = f$, which is in accordance with the definition of the focal length. Likewise, when the image distance is large, b approximates infinity, and $a = f$. Note that all object points that have the same distance from the lens also will have the same image distance from the lens and will be imaged in the same plane.

Figure 2.18(a) shows that, when the distance of a real object to a collecting lens is larger than the focal length, a real image will result. Placing the object between the focal point and the lens will result in a virtual image.

Angular magnification

A distant object projects a small image on the retina; by bringing it closer, the image is enlarged. While the limit for how close an object can be brought and still be imaged sharply on the retina (the near point) depends on a person's age (see later), normal reading distance is taken to be 25 cm. By using a magnifying glass, the object (e.g. a text) can be brought closer than the near point and still be sharply focused. When the object is placed at the focal point of the magnifying glass, the object will appear to have been moved to infinity, spanning a visual angle u'. If, when seen at a distance of 25 cm without the magnifying glass, the object is found to span an angle u, then the angular magnification, M, of the magnifying glass is defined as:

$$M = u'/u$$

M, then, is the ratio of the angular size of an object in the focal point of a magnifying glass, divided by its angular size at 25 cm from the eye – without any magnifying glass. For small angles we can replace the angles with the ratio between the object hight and the object distance (i.e. $u = \tan u = y/0.25$ m), where y is the hight of the object in meters:

$$M = (y/f)/(y/0.25) = 0.25/f = P/4$$

Angular magnification is thus one-quarter of the power of the magnifying glass. This magnification applies for an object in the focal point of the magnifier and for an image seen at infinity, e.g. without accommodation. A magnifying glass with focal length of 5 cm has a power of $P = 20$ diopters, and it will magnify an object $5\times$. Magnification greater than $3\times$–$4\times$ for a simple bi-convex lens is not practical because lens aberrations will be conspicuous. If the image is seen with accommodation, it will be slightly more magnified than when seen without accommodation.

Imaging in the eye

The eye consists of many different parts that contribute to imaging. Most of the refraction occurs at the cornea, with its refraction index n close to 1.38. The lens, with $n = 1.44$, would have a greater collecting power in air, but since it is embedded in fluids which also have a high index of refraction, its refracting power is reduced (aqueous humor and vitreous humor both have $n = 1.34$). However, the lens is able to increase its curvature and thus increase the refracting power. This allows focus on objects close to the eye (to accommodate). Accommodation means changing the lens' focus from infinity to a shorter distance. This adjustment can be expressed in diopters in the same way as the power P of the lenses in a pair of ordinary glasses. The ability of the lens to change its curvature, and hence its accommodation power, deteriorates with age. For persons older than 25 years the accommodative power of the eye is typically less than 10 diopters.

When swimming underwater, the refractive power of the eye is dramatically reduced. The lens is unable to accommodate sufficiently to fully compensate for the reduced refraction at the cornea. The refractive power of the cornea drops from 60 m^{-1} in air to about 1.8 diopters under water. With an additional 10 diopters available by accommodation, the total refractive power underwater is 11.8 diopters. This renders the eye farsighted or hyperope. Goggles trapping a layer of air between the glass and the eye will restore normal refraction.

In a model eye, where we need not consider small differences in refractive indices, we can consider the eye to be a homogeneous optic medium with only one surface of refraction and one refractive index, n'. The following general formula can be derived from Snell's law of refraction:

$$n/a + (n' - n)/r = n'/b'$$

Here a is the object distance, and b' is the image distance within the medium, while n and n' are the refraction indices of the medium containing the object and of the medium containing the image, respectively. If the eye is surrounded by air, we have:

$$1/a + (n' - 1)/r = n'/b'$$

By definition, the power of the refracting surface relative to air is

$$P = n'/f' = (n' - 1)/r$$

Using the concept of vergence, we can rewrite Snell's law of refraction for one refracting surface:

$$V + P = V'$$

If the surrounding medium is not air with an index of refraction equal to 1.00, but another medium with refractive index n, we have:

$$P = n'/f' = (n' - n)/r$$

Let us see what implications this has for seeing underwater. We write $n = 1.33$ for water and $n' = 1.34$ for the average eye. The radius of curvature of the cornea is $r = 0.0057$ m. The equation above for P gives us a power of 1.75 diopters. Underwater this gives a focal distance of about 76 cm inside the eye, whereas the focal distance would be 2.25 cm in air.

This reduced spherical eye simplifies the calculations by assuming only one refracting surface and only one curvature (Meyer-Arendt, 1995). In reality there are many transitions of refracting indices (cornea/aqueous humor, aqueous humor/lens, lens/vitreous humor). In addition, the lens accommodates by increasing its curvature. If we insert the following values for the reduced model eye, $n = 1.00$, $n' = 1.34$, $r = 0.0057$ m, we obtain for the focal length inside the model eye,

$$f' = 22.5 \, \text{mm}$$

Going in the opposite direction, from eye to air, by setting $b' = \infty$, we get

$$f = 16.8 \, \text{mm}$$

The two focal lengths are related to each other by the expression $nf' = n'f$, but the reduced eye has only one magnitude of the power:

$$P = (n' - n)/r = (1.34 - 1)/0.0057 = 60 \, \text{diopters}.$$

Models of the eye

Instead of the values for a reduced spherical eye used above, one might have used a model with somewhat different values (Schober, 1957): only one refracting surface with the same refraction index as water ($n = 4/3$), a diameter of the eye equal to the focal length on the image side (i.e. inside the eye) of 2.4 cm, and a radius of curvature of 0.006 m. These values give a power of 56 diopters, not far from the diopters of real eyes.

The schematic eye of the Swedish ophthlmologist A. Gullstrand (1862–1930) is yet another simplified model eye with the following values:

Radius of curvature

Cornea	7.80 mm
Lens (frontal surface)	10.00 mm
(back surface)	6.00 mm

Refraction index

Lens	1.413
Vitreous	1.336

Since the cornea has a refraction index of about 1.38 in air and a radius of curvature of 7.80 mm at the first refracting surface, it alone accounts for 49 of the 56 diopters.

Farsightedness and nearsightedness

In the normal eye, objects are imaged on the retina, about 2.4 cm from the cornea. In nearsighted eyes (myopia), the eyeball is usually longer than this or the refraction is too strong, so that the image of distant objects falls on a plane in front of the retina [see Figure 2.19]. This nearsightedness can be corrected by concave (negative) lenses which move the image plane farther away from the lens.

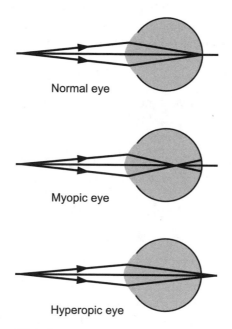

Figure 2.19 Refraction of light in a normal eye, a nearsighted (myopic) eye, and a farsighted (hyperopic) eye. In a myopic eye, the image is in front of the retina and can be brought into place by a diverging lens. In a hyperopic eye, the image is behind the retina, and this refraction error is corrected by a converging lens.

An eyeball that is too short causes farsightedness (hyperopia) by bringing the image plane behind the retina. Such conditions are corrected by convex lenses that collect the light and shorten the distance to the image plane, so that it coincides with the retina.

Mild forms of myopia, hyperopia and a lens aberration called astigmatism (see below) can also be corrected by changing the curvature of the cornea by using a laser (LASIK). In cases of myopia, the cornea is flattened; by increasing the radius of curvature the focal length is increased, reducing the eye's refracting power. In hyperopia one does the opposite, increasing the refractive power. Every pulse of the laser removes an extremely thin sheet of corneal tissue, only about 0.25 μm.

(a)   (b)   (c)

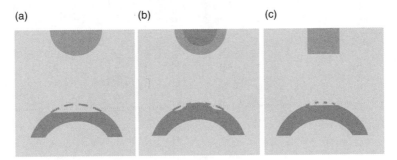

Figure 2.20 The curvature of the cornea can be altered by laser treatment. A thin sheet of the cornea is evaporated by means of a computer-guided laser. About 10 µm is removed per diopter. A myopic person who needs a correction of −5 diopters must have about 50 µm of tissue removed from the central part of the cornea, corresponding to a reduction of the thickness of the cornea by about 10%. The figure shows correction of myopia (a), hyperopia (b) and astigmatism (c). The upper parts of the figures show the correction as seen from above.

Ideally, this operation provides the patient with optimal refraction without glasses or contact lenses (see Figure 2.20).

Children can focus and see sharp images at distances shorter than 10 cm, but since the ability of the lens to accommodate declines with age, it is increasingly difficult to achieve the lens curvature needed in order to focus on close objects. The near point moves farther away from the eye, and after a certain age books must be held at arm's length. This condition is called *presbyopia* and can be corrected with convex lenses.

As the ability to accommodate deteriorates with age, individual differences in accommodative power increases. Typical accommodation range as a function of age is shown in Figure 2.21(b). In early years, the accommodation range can be 12–16 diopters, whereas after 50 years of age it is typically less than 2 diopters.

Using the lens formula and the concept of vergence

Example 1

A nearsighted woman has her far point at a distance $a = 0.2$ m. She has a 4 diopter accommodation range. What would be the power of glasses that would allow her to see distant objects?

If we assume that the image distance of her eye is 2.4 cm, then from the simple lens formula, $1/f = 1/a + 1/b$, we get

$$P = 1/f = 1/0.2 + 1/0.024 = 46.7 \text{ diopters}$$

Because this simple formula is valid only for thin lenses, the calculated power is wrong, but we can live with that since, in this example, we only care about changes of powers, the absolute power of the eye being irrelevant. To move the woman's far point, a, to

(a)

(b)

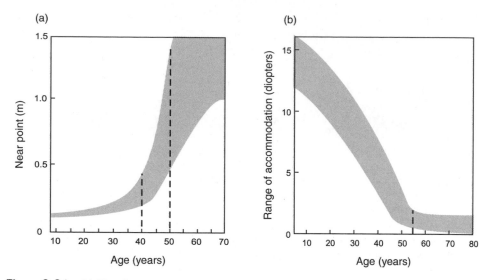

Figure 2.21 (a) The distance to the near point (the shortest distance of focus) increases with age. (b) The width of accommodation in diopters between the far point and the near point decreases with age. The hatched areas illustrate the natural variation within the population.

infinity, $P = 1/0.024 = 41.7$ diopters. This means she needs glasses with -5 diopters $(= 41.7 - 46.7)$.

With an accommodation range of 4 diopters, maximum refractive power without glasses will be $46.7 + 4 = 50.7$ diopters. Using the lens formula again and substituting the near point x for a, we get $x = 0.11$ m, her near point without glasses. When using eyeglasses of -5 diopters, and maximum accommodation of 4 diopters, then maximum total power is $41.7 + 4 = 45.7$ diopters, and the near point x will now be

$$45.7 = 1/x + 1/0.024; \qquad x = 0.25 \, \text{m}$$

Example 2

A farsighted person has his near point at a distance of 1 m. What power must his eyeglasses have to move the near point to 0.25 cm?

When focusing at 1 m $(a = 1$ m$)$, the power of the eye is $P = 1 + 41.7 = 42.7$ diopters. From example 1, we know that focusing at 0.25 cm requires a power of 45.7 diopters. The difference, 3 diopters, is what he needs to focus at 0.25 cm.

We might have used the concept of vergence, $V + P = V'$, to calculate these values. In example 1, using the numerical values for the reduced eye $(n' = 1.34;$ $b' = 0.024$ m$)$, we obtain $V' = 55.8$ diopters and $V = -1/0.2 = -5$ diopters. This corresponds to a power of $P = V' - V = 55.8 - (-5) = 60.8$ diopters at a distance of 0.2 m. For an object at infinity, $V = 0$, and $P = 55.8$ diopters. Thus, in order to move the far point to infinity, she would require −5 diopter eye glasses, just like we found before.

The result obtained for the required corrections is the same with the two methods, although the numerical values for the power of the eye are different. In prescribing glasses, only differences in refracting power are of interest. The numerical value of the power of the eye itself is a constant that is cancelled out by subtraction.

Where is the image plane in the retina?

The image plane of the fovea is normally close to the receptor pedicles (Williams *et al.*, 1994). The cones are about 80 μm long (0.08 mm), corresponding to 0.2 diopters. It is believed that imaging can take place along the whole length of the receptors without compromising the sharpness of the image, so determining refraction at a resolution finer than 0.2 diopters is therefore not necessary.

Lens aberrations

It may come as a surprise to learn that even the normal eye suffers from a series of optical defects. In fact, the defects are so severe that an optician trying to sell lenses of a similar poor quality would ruin his reputation (a comment usually referred back to Hermann von Helmholtz). We shall take a closer look at some of these defects, or aberrations. Aberrations seen for white light, as a result of the refractive index changing with the wavelength of the light (*dispersion*, see Figure 2.8), are called *chromatic aberrations*. Other deviations that occur even when the light is monochromatic are called *monochromatic aberrations*.

Chromatic aberrations

As a result of dispersion, short- and long-wavelength light is not focused together in the same plane on the retina. Normally, wavelengths around 580 nm are in focus, rendering the eye nearsighted for short-wavelength light (blue and violet objects) and farsighted for long-wavelength, red light. In other words, in normal white illumination, the short-wavelength components are imaged somewhat in front of the retina, whereas the long-wavelength components are imaged a little behind the retina. This difference in focus is called *longitudinal chromatic aberration* [Figure 2.22(a)]. The images formed by wavelengths other than 580 nm are therefore not rendered sharply on the retina. The images formed by short and long wavelengths are also of different sizes, since the blue image closer to the cornea is seen as larger than the red image. This is due to *transversal (or lateral) chromatic aberration*.

The dependence of refraction on wavelength leaves the eye with about 2 diopters higher refracting power for short wavelengths around 400 nm than for long wavelengths around 700 nm. Figure 2.22(b) shows the difference in diopters for all

(a)

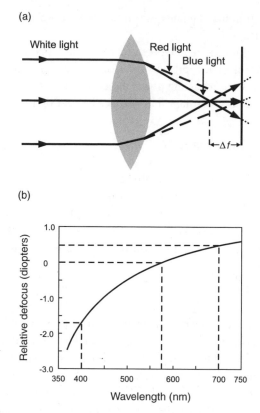

(b)

Figure 2.22 (a) Longitudinal chromatic aberration is caused by different wavelengths having different focal lengths. (b) Lens powers necessary to bring monochromatic lights between 370 and 750 nm in focus on the retina, relative to an initial focus at 575 nm.

wavelengths in the visible spectrum when 575 nm is in focus on the retina. From this diagram one can see that 700 nm needs about 0.4 diopter more than 575 nm to be in focus, and 400 nm needs a −1.7 diopter correction. This corresponds to a difference in refractive index of about 0.02.

When an object is illuminated by white light, the image on the retina will be sharp for one wavelength range and out of focus for another. The image will have colored borders – similar to that of a cheap lens – but we are usually not aware of chromatic aberration in our own eyes. One reason may be that the central foveola (where we have the sharpest vision) is free from the blue-sensitive S-cones, making this part of the retina insensitive to short-wavelength light. Another possibility is that the nervous system, in some unknown way, corrects for chromatic aberrations.

It is not possible to remove these color defects from a lens, but it is possible to combine two or more lenses in such a way that the aberrations of one lens are to some

Refraction index			
	656 mm (Red)	589 nm (Yellow)	486 nm (Blue)
Crown glass	1.517	1.520	1.527
Flint glass	1.644	1.650	1.664
Power (diopters)			
$P1$ (crown)	10.34	10.40	10.54
$P2$ (flint)	-6.44	-6.50	-6.64
$P=P1+P2$	3.90	3.90	3.90

Figure 2.23 An example of an achromatic lens composed of crown glass and flint glass. The lens power of the combination is the same for all wavelengths.

extent cancelled by the aberrations of the other(s). Figure 2.23 shows an example of a relative simple achromatic lens. Here, the color aberrations are neutralized by combining a convex lens, made of flint glass, with a concave lens made of crown glass. It can be seen from the table in Figure 2.23 how the index of refraction and the lens power varies for each lens, and that in the combination the effective power, $P = P1 + P2$, is constant for all wavelengths. This lens combination is therefore free from chromatic aberrations.

Correcting for the difference in refraction for short and long wavelengths does not lead to a significant improvement of visual acuity. However, it has been claimed that yellow sunglasses (that absorb short wavelength light and transmit long wavelength light) lead to higher acuity. One likely explanation is that, with removal of the short-wavelength, blue light, there are fewer chromatic aberrations and therefore sharper border contours in the image.

Spherical aberration

The relatively simple lens formulae used in this chapter are, strictly speaking, only valid when the light rays form a small angle with the optical axis. In reality, light at a distance from the optical axis will also contribute to image formation, and all light rays from one object point are generally not going to meet at one point in the image, even if the light is monochromatic. This affects the sharpness of the image. Monochromatic light passing near the periphery of a lens is refracted more than that passing close to its center. This can be viewed as a prism effect and is called *spherical aberration* (see Figure 2.24). The same thing happens in the eye, and its focal length decreases with distance from the optical axis. Consequently, this aberration is more disturbing the larger the pupil diameter. A small pupil size yields a greater depth of focus and requires less accommodation.

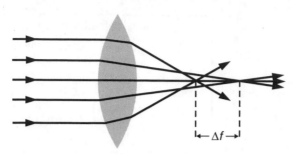

Figure 2.24 Longitudinal spherical aberration can be characterized by the difference in focal length, Δf, between central and peripheral light beams.

Astigmatism

In the eye, astigmatism is caused by irregularities in the cornea that lead to a variable curvature of its surface, i.e. the radius of curvature depends on orientation. Sometimes, the difference resembles that of a cylinder lens, with one orientation having a greater curvature than the other. In the presence of astigmatism, the sharpness of contours depend on their angle relative to the horizontal; horizontal and vertical stripes will not be imaged in the same plane and their contrast on the retina will be different. This defect can be compensated for by a cylinder-lens correction in the eyeglasses.

When optical and other distortions of vision occur in children, they should be corrected for as early as possible in order to give the neural system a chance to develop normally. When severe distortions prevail after pre-school age, it might be difficult to achieve the desired improvements. Such conditions are known as *amblyopia*.

Adaptive optics

The correction of the eye's main optical aberrations, such as defocus and astigmatism, by spectacles and contact lenses has become routine. Spectacles have been used for more than 700 years, and astigmatism has been corrected since shortly after 1801, when Thomas Young discovered this defect in the human eye. A pupil diameter of 3 mm is considered to be optimal for optical performance for a wide range of visual tasks. While increasing pupil size would remove the deleterious effect of diffraction, dilating the pupil would increase the blur due to several kinds of optical aberrations. If all aberrations could be compensated for, one would not only achieve improved spatial resolution in vision, but one would also increase the resolution of images taken of the retina. This goal appears closer after the recent developments in adaptive optics. The development of new and rapid methods for measuring the lower-order as well as the higher-order aberrations of the eye has provided the necessary background to develop technologies that correct for them. Such adaptive corrections have been achieved by deformable mirrors, and a retinal camera has been constructed that can take images of the retina, with a resolution that allows discrimination of single receptors (Miller, 2000). The first image of the distribution of cone receptors in the retina was taken with a camera using adaptive optics (Roorda and Williams, 1999; see also Figure 3.15). Such images could not be obtained earlier because the imperfections of the eye's optics caused too much blur. It is conceivable that this technology for imaging the finer details of the retina, down to the size of receptor cross-sections, will allow for the monitoring of physiological and pathological changes during retinal diseases.

Adaptive optics, which also compensate for higher-order aberrations such as *coma* and spherical deviations, bears promise of improving vision as well. The largest benefits can be expected for large pupils where diffraction plays no role. The effect of adaptive optics would be to render scenes and objects in finer detail; they would appear crisper and with higher contrast. In an eye provided with adaptive optics, the spacing of the retinal receptors would set the final, neural limit of visual resolution. For large improvements of the eye's optics, the cone mosaic would be too coarse in comparison to the optical resolution, and the finer details of the optical image would pass unnoticed by the neural system. This supernormal vision may heighten contrast sensitivity and visual resolution, but it is not clear how these improvements, achievable in the laboratory, would transfer to everyday use. How could these adaptive corrections be applied? Traditional spectacles are not the best means because the eyes rotate relative to a fixed spectacle lens. The most likely choice is to design special contact lenses. Another option might be refractive surgery (see also Figure 2.20).

Spectral power distributions

Every light source has its characteristic spectral energy or effect distribution. Figure 2.25 shows a few examples of radiant energy distributions of light as a function of wavelength. The distribution, P_λ, of power (expressed as Joules per second or Watts) in the spectrum characterizes energy per second and nanometer:

$$P_\lambda = \Delta P / \Delta \lambda (\text{W/nm})$$

This quotient describes the concentration of power within the small spectral range $\Delta \lambda$ centered on λ, with λ, ΔP and $\Delta \lambda$ being magnitudes that can be measured. P_λ can change smoothly throughout the spectrum, as for light from an incandescent lamp [Figure 2.25(a)], or in a discontinuous manner, as for a fluorescent tube with emission lines due to gas discharges [Figure 2.25(b)].

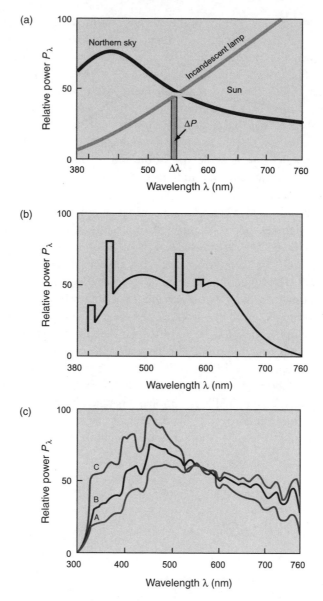

Figure 2.25 Relative spectral distribution of light from an idealized northern sky, the sun and an incandescent lamp (a); fluorescent tubes with overlaid spectral lines from the gas in the tube (b), and different phases of daylight (c): direct sunlight (A) northern sky (C), and a mixture of sunlight and skylight (B). (See also color plate section.)

Daylight has a much more irregular spectral distribution of power than does light from an incandescent lamp [Figure 2.25(c)]. Fluorescent light has a continuous distribution overlaid with intense spectral lines, generated by radiation from gas discharges. A pure gas discharging lamp (a so-called spectral lamp) has a spectrum consisting only of spectral lines. The pattern of lines in the spectrum is characteristic of the particular gas contained in the lamp (e.g. gaseous mercury or sodium). Each type of gas has its own line spectrum, and the gas can be identified on the basis of these 'fingerprints'. By studying the spectral lines in the light emitted from stars, one can determine the materials these stars are composed of.

Spectral reflection and transmission

Newly fallen dry snow is an example of a spectrally non-selective surface with diffuse reflection of the incoming light equally in all directions. This is called a Lambertian surface. Its reflection factor β tells us how much of the incoming light is reflected. If a spectral non-selective surface has a large β, higher than 0.7, the surface looks white, and if β is below 0.1 the surface will look dark gray or black. Paper reflecting about 20 percent of the incoming light, i.e. that has a value of $\beta = 0.2$, looks middle gray (not $\beta = 0.5$, because lightness is nonlinearly related to β).

Mathematically, the reflection factor is expressed as the ratio between the light flux that is reflected from the surface, Φ, and the flux, P, that is falling upon it:

$$\beta = \Phi/P$$

Colored surfaces are spectrally selective and reflect more light in one part of the spectrum than in another. Such surfaces can be characterized by their spectral reflection factor

$$\beta_\lambda = \Phi_\lambda/P_\lambda$$

These relations are shown in Figure 2.26. Spectral reflection curves show how much of the incoming light is reflected for light in different, narrow wavelength intervals, $\Delta\lambda$. Some examples are shown in Figure 2.27. For most natural surfaces, β_λ changes slowly over the spectrum.

A spectral reflection factor $\beta_\lambda = 1.0$ means that all light is reflected for the given wavelength, λ, whereas a value of 0 means that all the light of this wavelength is absorbed in the material and none of it is reflected back. For values in between, some of the light is reflected and the rest is transmitted, scattered or absorbed in the material and converted to heat. The intensity of the light incident on an object is divided into the categories: specular reflection and diffuse scattering of light from a surface, transmission, and absorption of light by the object. Conservation of energy requires that the sum of these intensities must equal the intensity of the incident light.

An experiment

Put some pieces of metal painted with different colors (for example red, yellow, green, blue, white and black) on top of the snow on a sunny winter day. After the sun

(a)

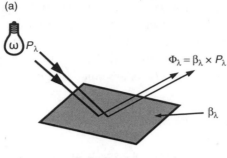

Reflection from a surface

(b)

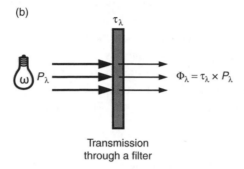

Transmission
through a filter

Figure 2.26 (a) Reflection of light from a surface. (b) Transmission of light through a filter. P_λ denotes the incident spectral power distribution from the light source, and Φ_λ that of the transmitted or reflected light. β_λ represents the spectral reflection factor and τ_λ the spectral transmission factor. See also Color Plate Section.

has shone on them, which color sinks deepest into the snow and which one is the highest?

Pressed magnesium oxide (MaO_4) and snow reflect light diffusely and have spectral reflection factors that are close to 1.0 for all wavelengths (in practice this means about 0.95). Black velvet is blacker than most materials; here nearly all light is absorbed.

Transparent materials, such as colored glass and color filters, have uneven spectral transmission curves telling us how much light passes through the filter for each wavelength [Figure 2.26(b)]. The spectral transmission factor is defined as follows:

$$\tau_\lambda = \Phi_\lambda / P_\lambda$$

The curves in Figure 2.27 are examples of spectral reflection or transmission curves. The optical density, D, of a filter is defined as $D = \log(1/\tau)$ (a filter with a transmission factor of 10 percent has an optical density $D = 1.0$).

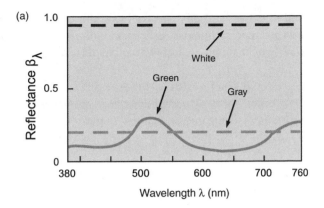

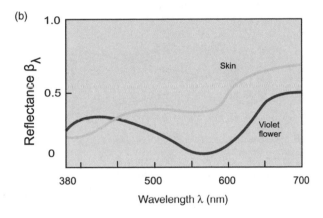

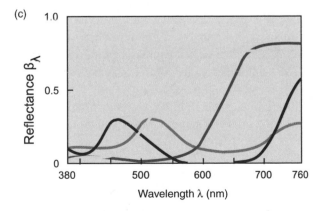

Figure 2.27 Typical spectral reflection curves for (a) a green, a gray and a white surface; (b) for Caucasian skin color and a violet flower; and (c) for some other colored surfaces (yellow, green, blue and red) corresponding to the colors used for drawing the curves. See also Color Plate Section.

Some surfaces and filters are strongly spectrally selective, meaning that they reflect or transmit only narrow spectral bands. *Interference filters*, for instance, have such narrow transmission bands, down to a few nanometers *bandwidth* at 50 percent transmission.

The narrower the spectral band a surface reflects, the less light it will return, and the blacker it will appear. The way in which the perceived color of a surface is determined by its reflection spectrum is not a trivial matter, and it will be dealt with later in Chapter 5.

Visual acuity

The minimum distance between two borders that can be resolved is usually given as an angle, expressed in minutes of arc (arcmin; Figure 2.28). Visual acuity is a

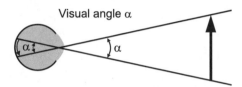

Figure 2.28   Visual angle α is the angular size of a distant object at the eye.

measure of how fine details a person can resolve at maximum contrast. Acuity can, for instance, be determined using a Snellen chart to read small letters [Figure 2.29(a)], or by a Landolt ring where the task is to indicate the orientation of the gap [Figure 2.29(b)]. In recent years, several other tests have been made available, among others a grating test for the measurement of contrast sensitivity and grating acuity. The task is to detect a pattern of thin black and white bars – a grating – from a homogeneous area. This can, for instance, be done by asking the subject to report the orientation of the grating, whether it is vertical or horizontal. Figure 2.29(b) also gives an example of Hyvärinen's (1992) optotypes (figures) that can be used to determine acuity in children.

If α, measured in arcmin, is the smallest angle that can be resolved between two adjacent black bars of the grating, then the numerical value for the visual acuity (*VA*) is given by:

$$VA = 1/\alpha(arcmin^{-1})$$

For 200–300 lx illuminance, a value of 1.0 is considered normal.

Converting a width measured in cm to an angle measured in degrees is relatively easy if one remembers that 1 cm at a distance of 57.3 cm from the eye equals 1°. An acuity

(a)

(b)

Figure 2.29 (a) A letter chart of 100% contrast used in testing visual acuity (VA). (b) Different optotypes and a grating also used to test visual resolution.

value of 1.0 means that one can resolve two adjacent lines that are 1'
(= 1 arcmin = 1/60°) apart, or about 0.17 mm apart at a distance of an arm's length
from the eye.

In subjects with normal vision, grating optotype acuity and letter optotype acuity can
be very similar. However, the two acuity forms are not identical. Gratings measure
resolution or detection, whereas letter acuity charts demand recognition of patterns.
In visually impaired subjects, large discrepancies can be found between the two
measures. This applies for instance to some children with cerebral visual impairment,
to subjects with eccentric fixation due to foveal or macular disease (Fosse *et al.*,
2001), and to amblyopic eyes (Friendly *et al.*, 1990).

The apparent reason for the differences between the two acuity measures is that one
requires detection of gratings and the other recognition of elements. The latter
demands greater cognitive resources than the former. Another explanation might be
that, as gratings are extended stimuli and as such subject to a greater spatial
integration than letters, they are more easily detected than letters (Fosse *et al.*,
2001). A third possibility is that contrast sensitivity for letters decreases faster as a
function of spatial frequency than for gratings (Herse and Bedell, 1989).

Visual acuity increases for higher light levels, as shown in Figure 2.30. Here, visual
acuity, *VA*, is plotted as a function of the luminance of the target background.
Maximum acuity for the highest luminance reaches a value of 2.0 or greater, acuity
values corresponding to the minimum angle of resolution being 0.5 arcmin or smaller.
This high acuity in younger years represents a significant acuity reserve compared
with what is regarded as normal in our society today. Visual resolution gets worse as
you move away from the fovea, as is seen in Figure 2.31.

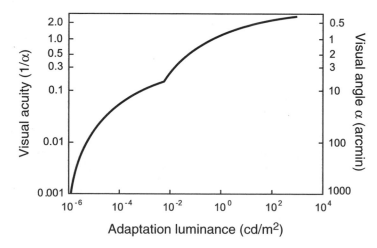

Figure 2.30 Decimal visual acuity, $VA = 1/\alpha$ (arcmin^{-1}), as a function of adaptation luminance. The curve is an average of several sets of data found in the literature.

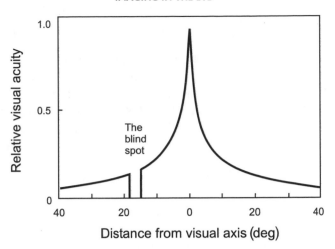

Figure 2.31 The dependence of visual resolution on eccentricity in the retina. Visual acuity is given relative to its maximum value in the fovea.

Visual acuity is best in younger years, except for infancy, and is said to decrease steadily after 40 years of age (see Figure 2.32). As a rule, upon reaching an age of 80 years, acuity is halved, but the spread of the given data indicates great variety between individuals.

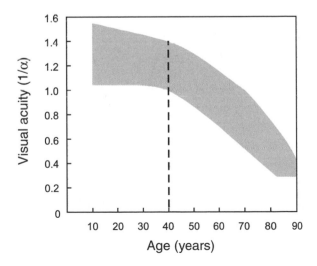

Figure 2.32 Visual acuity decreases with age, particularly after the age of 40. The gray area indicates the spread based on published data.

Contrast rendering

Owing to lens aberrations, diffraction and stray light, an optical system, such as a camera, cannot be expected to reproduce a contrast pattern with exactly the same contrast as the original. Naturally, the contrast is most severely reduced for the smaller details. The *contrast rendering factor* (CRF) can be defined as the contrast of the image divided by the contrast of the object, and this ratio gives a simple measure of the quality of an imaging system, whether it be a camera or a complex TV system.

$$\text{CRF} = 100 \times \text{image contrast/object contrast}$$

A black and white grating is a frequently used pattern to assess image quality and contrast rendering. The most commonly used grating pattern is a so-called 'sinusoidal grating'. In a sinusoidal grating the contrast in a direction perpendicular to the length of the bars varies like a sinus wave (see Figures 2.33 and 4.17). The Michelson contrast C_M of the regular grating is given in terms of luminance, L by the equation

$$C_M = (L_{max} - L_{min})/(L_{max} + L_{min})$$

Sinusoidal gratings can be considered as fairly fundamental stimuli, and they are used not only to characterize optical systems, but also for testing a person's sensitivity to contrast, the most fundamental capacity of the visual system. Measurement of

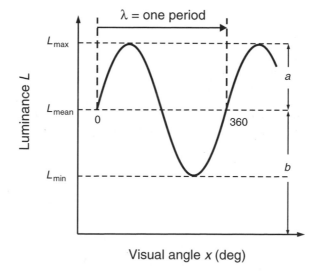

Figure 2.33 A sinusoidal curve and the definitions used to calculate spatial frequency and contrast. Spatial frequency $f = 1/\lambda$ (cycles/deg), where λ is the length of one cycle, measured in degrees. The amplitude of modulation and mean luminance are given by a and b, respectively. At a distance of 57.3 cm from the eye, 1 cm corresponds to 1°. Therefore, a grating that has a cycle length of 2 mm (0.2°) at this distance has a spatial frequency of 5 cycles/deg.

contrast sensitivity is also a useful tool to evaluate the visual ability of visually impaired people. In order to appreciate the information inherent in a contrast sensitivity function, we shall sketch some of the mathematical background that makes this test so valuable.

A square wave grating, with its sharp edges of abrupt transitions in contrast, can be described mathematically as a sum of particular sinusoidal gratings of fixed contrasts (amplitudes) and spatial frequencies (see Figure 2.34 for an explanation). The same applies to other contrast waveforms (triangular waves, saw-tooth, etc.). Sinusoidal spatial variations of contrast therefore have a particular relevance; they are more fundamental than square wave modulation or other wave variations of contrast. Thus, if we could determine an optical system's contrast rendering for all sinusoidal functions, we would be able to calculate the contrast rendering of all other distributions of contrast.

In electrical engineering, for instance in analyzing electrical signals in electronics and sound in acoustics, an analysis based on fundamental sinusoidal waves is common. For instance, the rendering of sound of a loudspeaker is characterized by looking at distortions in particular frequency bands, similar to the contrast rendering of an optical system. Most sounds, such as spoken words, are a sum of many simple, sinusoidal sound waves of different frequencies each, with their own amplitudes and phase. Sound is, of course, described in the time domain, whereas images are described in the spatial domain, but the principle of the mathematical treatment of these *modulation transfer functions* is the same.

At the bottom of Figure 2.34 the luminance distribution of a square-wave grating is shown, and above it how this waveform is obtained by adding several odd harmonic frequencies to the fundamental frequency ω:

$$L = L_o + L_{\text{mid}}[\sin \omega x + (1/3)\sin 3\omega x + (1/5)\sin 5\omega x + \cdots (1/n)\sin n\omega x]$$

Here, ω is the *angular frequency*, x is the distance in degrees along the x-axis, and n are the odd numbers 1, 3, 5, 7 etc. If the length of a period is $\lambda°$, the *spatial frequency, f,* in cycles per degree is given by

$$f = 1/\lambda(c/\deg)$$

The angular frequency is defined as $\omega = 360/\lambda = 360f$.

Figure 2.34 shows that even a sum of only the three first parts of the expression above the fundamental (the third and the fifth harmonics) approaches a square wave-modulated stimulus (more so when the test is far from the viewer). With the first eight harmonics the result is even better, and with all of them a perfect square wave is generated. The amplitude of the final square wave is $\pi/4$ of the fundamental at the top of the figure. This method of analyzing a particular waveform can be used also in two dimensions, for example to describe a pattern in a picture. The pattern can theoretically be described by a set of spatial sine functions that runs over the image in different directions.

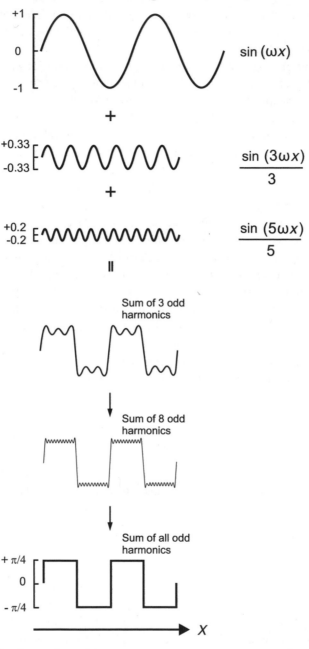

Figure 2.34 The figure shows how one can construct a square wave by summing several sinusoidal waves. The amplitude of the sinusoidal functions in the sum decreases as 1, 1/3, 1/5, 1/7,..., while the spatial frequencies increase with factors 1, 3, 5, 7,.... The figure shows the result of adding the first three and the first eight parts of the series. The square wave at the bottom has an amplitude that is 0.785 (= $\pi/4$) of the fundamental sine curve on the top.

The modulation transfer function

We have just seen how the contrasts in a picture can be broken down to a sum of sinusoidal frequencies with different amplitudes and phases. This means that the contrast rendering properties of an optical system can be characterized by measuring its reproduction of the contrasts of many sinusoidal gratings with different spatial frequencies. Figure 2.35 shows how this is represented by the modulation transfer

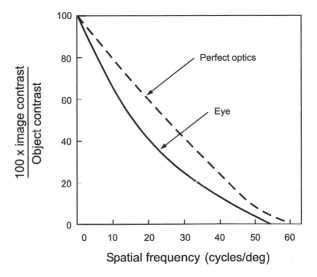

Figure 2.35 The modulation (contrast) transfer function of the human eye using a 2 mm pupil, compared with a perfect optical system with the same aperture. The dashed curve shows the image contrast of different sinusoidal gratings of 100% contrast. When spatial frequency increases, the contrast of the reproduction decreases. For perfect optics and 2 mm aperture size, diffraction leads to the loss of all contrast at 62 cycles/deg. The eye loses all contrast at a somewhat smaller spatial frequency, around 55 cycles/deg.

function (MTF). In this diagram the contrast rendering factor for an original with 100% contrast is plotted for several spatial frequencies. The examples are for perfect (only diffraction-limited) optics and for the eye.

The solid curve in Figure 2.35 applies to the human eye with a 2 mm pupil diameter. The dashed curve is for perfect optics with the same pupil size. Contrast is reduced most for the higher spatial frequencies (as the bars of the grating get narrower), and finally approaches zero at high frequencies. One can also see how even the best optical system reduces the contrast of small details, this limitation being due to diffraction in the entrance aperture. In such an optical system, 100 percent contrast in the original will be reduced to 0 percent at 62 c/deg.

The example shows that, in the eye, the contrast rendering for the same pupil size reaches zero a little earlier, around 55 c/deg, corresponding to a bar width of 0.55 arcmin (which corresponds to a grating acuity of 1.8).

MTF curves are measured empirically, and they give a good description of the quality (resolution, sharpness and contrast rendering) of optical instruments and imaging systems. Examples of such systems are combinations of camera, electronic transmission and a monitor, such as for closed circuit TV (CCTV) or TV broadcasting. Similar curves are used in elaborate tests in vision research, and particularly in low-vision research. Later we shall take a closer look at this application.

Is the image contrast always lower than the object contrast? Not necessarily. In photographs and in copies it is possible, through image manipulation, to give the reproduction a higher contrast than the original.

3 Physiology of the eye

The evolution of eyes

Animals have different ways of looking at the world. Even Charles Darwin (1809–1882) wondered how such complex organs as the vertebrate's lens eye, the facet eye of insects and the camera obscura of even lower organisms could have been invented 'by chance'. Apparently this development did not occur only once during evolution, because the complex eye of a fly is likely to have developed independently of the lens eye of higher vertebrates. It has been suggested that the eye must have been invented about 60 times in different species during evolution. The opposite hypothesis, that this diversity should have developed from a single primitive organ, is hard to believe, but in recent years evidence has accumulated that this may indeed have been the case. In 1997, the Swiss biologist Walter Gehring demonstrated that the same genes were controlling the development of the eye in the mouse and in the fly (Stöcklin, 2002). A gene from the mouse, called *Pax 6*, was shown, in some spectacular experiments, to control the growth of eyes in the fly. Since then, *Pax 6* or a very similar gene has been found in almost all species with eyes. It is believed that this gene was present in animals more than 600 million years ago, and therefore it is highly probable that this gene is some sort of control gene for the development of eyes.

The first primitive 'eye' seems to have developed from a common light-sensitive cell, with the photopigment rhodopsin as the light-responsive substance. From this simple origin, in what seems to be a natural and logical precursor for an eye, nature invented different functional systems, among them facet eyes and lens eyes, but also different receptor types. These were the prerequisites for vision under different light levels and for color vision. *Pax 6* need not be alone in controlling eye development. It is quite feasible that it has a dominant role, and that genetic networks play an important role as well.

Light Vision Color. Arne Valberg
© 2005 John Wiley & Sons Ltd

The light-absorbing pigment rhodopsin seems to be utilized everywhere in the animal kingdom, and it is likely that humans evolved from an ancestor with only one type of photoreceptor containing this pigment (Oyster, 1999). The later occurrence of three classes of cone pigments is thought to have developed from processes of gene duplication and divergence (Sharpe *et al.*, 1999). When extrapolating backwards in time from the occurrences of changes in rhodopsin in different species, one arrives at the estimate that the S- and L- (or M-) cone pigments appeared as separate molecules more than 500 million years ago, long before the first vertebrates (Neitz *et al.*, 2001). The first primitive photoreceptor structure that was the precursor of the earliest eyes probably appeared then.

The eye is not a camera

When the eye is sometimes compared with a camera (Figure 3.1), the focus of interest is on the imaging properties of the optic media and on spatial vision. Such optical

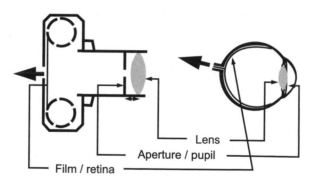

Figure 3.1   A comparison of the eye with a camera.

analogies in vision have been inspired by Johannes von Kepler, Rene Descartes and Hermann von Helmholtz. Below we shall also use this analogy, but, contrary to most popular reviews, we shall focus more on the differences than on the similarities.

The possibility of creating images without lenses, through a small circular aperture, was reported as early as the eleventh century by Ibn al-Haytham (965–1039, born in Iraq and also called Alhazen) and by Leonardo da Vinci in about 1500. Leonardo da Vinci illustrated the imaging properties of the eye by referring to what happens when:

... the images of the illuminated objects penetrate through some small round hole in a dark habitation. You will then receive these images on a sheet of white paper inside this habitation somewhat near to this small hole, and you will see all the aforesaid objects on this paper with their true shapes and colours, but they will be less, and they will be upside down because of the said intersection (Wade, 1998).

This is a description of what we call *camera obscura.* Try this yourself! It is surprising that, for example, a landscape can be imaged sharp and without distortion, in the finest shades of color, through a little hole. If you make the hole larger, the image gains on brightness, but loses its sharpness.

Johannes Kepler (1571–1630), who gave an exact geometrical explanation of the image in a camera obscura and in the eye, was a little uneasy because the image in the eye was turned upside-down. How can we see the world right when the image on the retina is on its head? Neither Kepler nor Descartes could solve this problem, and they had to hand it over to later generations.

Kepler's geometrical optics dealt with optical imaging, and he was aware that a retinal image could not justify the existence of a visual image. As mentioned earlier, the interpretation of the latter image is not determined solely by the optics; one also has to take into account neural factors, previous experience, expectations, memory and additional information from other senses. Blind people who are older when they are given the opportunity to see have difficulty in attaching meaning to optical images.

We take this as a sign that neural mechanisms, in the retina and later, play a strong role in processing the optical image. The nervous system seems able, at least in some cases, to reduce or even neutralize the effect of optical imperfections. The visual sense is more interested in picking out and amplifying elements that carry information than giving an optical replication of the world. In particular, the physiologist Ewald Hering and the physicist Ernst Mach stressed the importance of neural processes in establishing a functional contrast and form vision.

Filter properties

There are many examples showing that sense organs filter out irrelevant information. The optical properties of the eye can also be described by pointing out its filter characteristics, e.g. which physical magnitudes are transmitted and how well they are rendered in the retinal image. This image is, for instance, characterized by the spectral transmission of the eye media. Figure 3.2 shows that some non-visible, long-wavelength radiation is allowed to reach the retina, whereas most of the energy-rich short-wavelength light is absorbed. This absorption of short-wavelength light increases with age because of the yellowing of the lens.

The light adaptation of rods and cones is a regulatory mechanism that sets a low and a high limit for what is accepted as adequate illumination. The ability of the

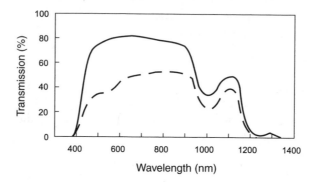

Figure 3.2 Spectral transmission of the eye media. The solid curve is for the cornea and the fluid of the anterior chamber (aqueous humor), and the dashed curve is for all the media in front of the retina (cornea, aqueous humor, lens, vitreous humor). Only about 30–50% of the light reaches the retina, depending on wavelength.

receptors to follow rapid changes in the stimulus with some delay, together with the temporal behavior of other cells, set the limits of temporal resolution. Furthermore, the spatial contrast sensitivity function defines contrast perception and poses constraints on orientation (navigation), mobility, the perception of form, and on visual resolution.

Imaging

Figure 3.3 is a schematic drawing of the eye. All translucent eye media contribute to image formation, but the cornea, as we have seen, is the most important refractive

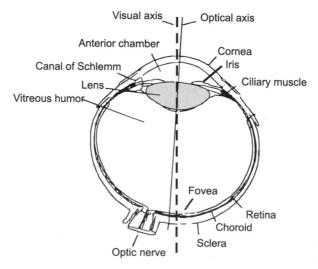

Figure 3.3 A cross section through the human eye.

surface. In a camera, all refraction occurs in the lens (or in lens combinations). A camera is empty, whereas the eye is filled with a gel (vitreous humor) that preserves the pressure and the form of the eye.

A traditional camera focuses the image on the film and modern digital cameras focus on an electronic image sensor. A sharp image is achieved by changing the distance of the lens from the film. This method is also used in fish eyes, whereas humans focus by changing the curvature of the lens, an operation that takes only a fraction of a second.

Figure 3.4 shows how the problem of image focus is solved in the animal world. In addition to all the manipulations that can be made with an elastic and movable lens (the three lower pictures, c–e), there are also different solutions for animals with rigid lenses. These animals have either a movable retina (like in the octopus) or two different retinas, one for distant viewing and another for near objects, as in scallops. In the Amazon there lives a fish (*Anablebs anablebs*) with two optic systems in each eye, making four eyes altogether. This fish lives close to the water surface, and one pair of eyes sees underwater whereas the other pair at the same time looks into the air (see Figure 3.5). The two pairs of eyes have separate pupils and retinas.

The sharpness of a photographic picture is dependent on the quality of the camera lens, its homogeneity and that it is corrected for lens aberrations. In the eye, the homogeniety of the cornea is important, whereas chromatic aberrations seem to be left to be corrected by the nervous system. The nervous system apparently also provides for some contour sharpening, a useful process because of the relatively unsharp image on the retina.

Light exposure and color

The quantity of light that falls on the film or on a digital image sensor of a camera is regulated by the size of the entrance aperture together with the duration of exposure. However, since short exposures are needed to prevent image smear, a film of less sensitivity must be used in bright daylight than at lower light levels. In the eye, the pupil size regulates the retinal intensity only by a factor of about 1/10, and the rest is taken care of by neural mechanisms. These mechanisms are more flexible and, given some time, the retina adjusts its performance easily from a moonshine level to a sunny day. At the lower end of this scale, humans cannot see colors, whereas in bright daylight we can discriminate between millions of color nuances.

Humans see images like a continuous stream, and eye movements seem to favor vision, not to hamper it as in a camera. People with normal vision do not have a series of multiple exposures as the eyes move about. Therefore, there must exist a neural mechanism that prevents smear during eye movements. It has indeed been shown that vision is strongly inhibited during fast eye movements, called *saccades*. After the

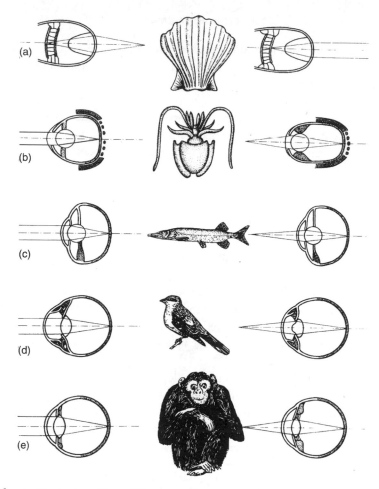

Figure 3.4 An illustration of how different species accommodate to different distances of a visual object. (a) The scallop has two retinae, one for near and the other for distant objects. (b) The octopus moves his eye lens and changes the distance to the retina. (c) Fish move a spherical, rigid lens. (d) Birds and turtles have a flexible lens, whereas primates, including humans (d), can change the radius of curvature of an elastic lens. (Reproduced from Schober and Rentschler, 1979, *Das Bild als Schein der Wirklichkeit* by permission of Moos.)

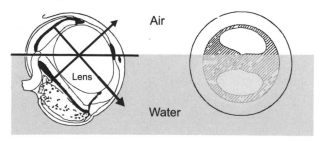

Figure 3.5 Anablebs anablebs is a 'four-eyed' fish living in the Amazon. This fish feeds at the water surface, with half the eye above water and the other half under water. It has a retina for vision in air and another for vision under water. The pupil is also divided into two parts (the rightmost picture). The arrows indicate the line of sight in the two media.

saccade, for every new fixation, the visual image seems to be refreshed. Experiments have shown that stabilization of retinal images leads to their disappearance after only a few seconds.

Vision is at its sharpest within a small, central area of the retina about twice the size of the retinal image of the moon, or about 1.5° in diameter. This part of the retina is called the *fovea*. The central 0.5° of the fovea is blind to short-wavelength blue light, and it has a narrower spectral window than the rest of the eye. Normal, three-variant color vision needs larger objects and covers the whole spectral range from 380 to 760 nm. The film and the image sensor do not have such a particular area of high resolution, allowing a colorful, sharp image over their whole area.

In contrast to cameras based on film, the eye is able to compensate for the color of the illumination, in such a way that an object appears to have a nearly constant color. It does not matter much if the illumination is bluish daylight from the northern sky or if it is yellowish incandescent light indoors; a white paper will always look white. This phenomenon is known as *chromatic adaptation* and *color constancy*. Some digital cameras have a built in compensation that is supposed to mimic chromatic adaptation in the eye, but conventional cameras based on a photochemical process to develop the film need different films indoors and outdoors. A film for daylight yields yellowish pictures when used with incandescent light.

The walls of the eye are flexible and can, together with the other eye media, produce substantial light scatter. The pigment epithelium in the back of the eye screens the photoreceptors from scattered light, and this has been regarded as an advantage of the 'inverted eye', where light passes through the transparent cell layers before it reaches the receptors. A camera has rigid, black walls and relatively little stray light.

The conical shape of the outer segment in cones [Figure 3.17(a)] makes these receptors most sensitive to direct axial light. The *Stiles–Crawford effect* characterizes the directional property of the response from light entering the eye at different parts of the pupil and being imaged on the same retinal area. Direct axial light gives the larger response in cones by a factor of about four. In rods the response may be regarded as nearly non-directional.

Visual fields and depth perception

The lens of general purpose cameras often has a focal length of 50 mm, and this allows for an image size of about 45°. A longer focal length gives a smaller angle and, if the image size is the same, the image of the object will appear magnified. This gives the impression that the camera is closer to the object–a telephoto lens. Lenses with shorter focal length than 50 mm give a wider angle and are called 'wide-angle lenses'.

The visual field of the two eyes together is about 208°. This means that, when looking straight ahead, we can see a light source up to 14° behind the eye. Yet, even with eye movements, we have a considerable blind area behind our head. Birds, on the other hand, with their eyes placed on the side of the head, have no such blind region.

The visual field of one eye is vertically divided in two parts: a temporal half and a nasal half-field, and so in the retina. Objects that are in the temporal visual field are imaged on the nasal retina, and vice versa. There is a *monocular sector* of about 30° that is viewed by one eye only because the nose is an obstacle for the other eye (Figure 3.6). If there is a damage to the upper part of the retina, for example, the

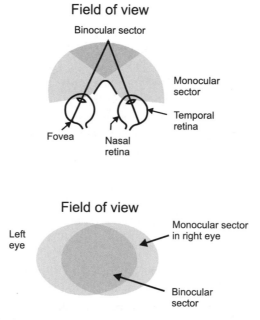

Figure 3.6 The field of view of the two eyes. In humans they together cover about 208°. The binocular sector refers to the part which is common to both eyes while looking straight ahead.

effect is seen in the lower part of the visual field, and if the damage is in the nasal retina its effect will be observed in the temporal visual field. Damage to one eye within the part of the visual field that is imaged in both eyes (the binocular field of view) can be hard to discover without examining the visual capacity of each eye separately.

When the eye is fixating a point at a given distance from the eye, objects that are situated in front of or behind this point will not be seen as out of focus before the difference in distance corresponds to about 0.15 diopters. This is called stereoacuity, and it is largest for rays close to the visual axis, e.g. for rays with a small angle relative to the axis. The eye's pupil size is as critical for depth resolution as is the aperture in a camera; the smaller the pupil size the larger the range of depth acuity.

The value of 0.15 diopters refers to a pupil diameter of 4 mm, in which case sharp depth vision reaches from infinity to 3.5 m. If the pupil size is 2 mm, sharpness is extended down to about 2.3 m and, with accomodation to 1 m, the image will be sharp between 1.8 and 0.7 m. The imaging plane in the retina can be anywhere along the length of a receptor (being about 0.08 mm = 80 μm), without having an effect on the sharpness of the visual image. This corresponds to a little more than 0.15 diopters, and a better correction than this is not necessary. Reading text is easier in high illuminance, both because the pupil then is small and because depth resolution is relatively good; the need for precise accomodation is thus less. Too small a pupil will produce image smear because of diffraction.

Horopter

When we fixate a point in space with both eyes, the eye muscles adjust so that the images of that point and all other points on a circle through the fixation point and the optic centers of the two eyes fall on *corresponding points* in the two retinas (Figure 3.7). This means that every such point has the same angular distance to the

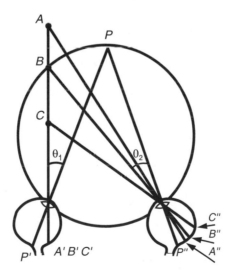

Figure 3.7 Images of objects that are located on a circle passing through the fixation point P and the optic centers of the two eyes, fall on corresponding points on the two retinae. This circle is called a horopter. The image of object B has the same angular distance to the fovea in both eyes ($\theta_1 = \theta_2$). The images of objects A and C have different angular distances to the fovea and are therefore seen at different depths.

fovea in both eyes. This circle is called a *horopter*. The images of objects that are not situated on this imaginary circle will have different distances to the fovea in the two eyes [A, B and C Figure 3.7], and this makes it possible to experience depth and to have stereo vision. The two eyes see an object outside the horopter at a slightly different angle (dependent on distance) because of the pupil distance of about 6 cm between the two eyes. Binocular neurons in the visual center of the brain are able to register this difference in angular distance on the two retinas and to use this as an information about depth.

It is possible to observe one's own horopter by using, for example, two pencils. Hold one pencil ahead of you at a distance of about 30 cm and fixate the tip. Hold the other pencil a little to the side of the first and move it to and fro. Close to the eye and far from it, you will see a double image of the latter pencil. This double image disappears at the distance of your horopter. In reality, the horopter resembles an ellipse rather than a circle.

Double images arise in binocular vision also when one squints, but the image from one eye will usually be suppressed (probably by some cortical mechanism). Failing cooperation of the two eyes and squinting at a young age is an unwanted condition, because if it persists over a long period of time vision of the suppressed eye will deteriorate. The eye will not develop normally, with the consequence that, for instance, the perception of depth is lost. It is therefore important to take the necessary medical measures to avoid squinting.

Crude judgements of distance can be made by one eye alone and without depth perception. Relative movements of far and near objects, movement parallax, perspective, object occlusion, surface structure and relative size make it possible to judge distances. However, normally functioning binocular vision relies upon a subjective visual dimension of depth and the perception of three-dimensionality that we call *stereopsis*.

Autostereograms

Stereopsis is another name for stereoscopic vision, i.e. the vision of depth that is only possible with two eyes. Each eye has a two-dimensional retinal image, each viewed from a somewhat different angle, and these two images are combined by the brain to give rise to three-dimensional depth perception. It is possible to make a *stereogram* by drawing pictures from two different viewing points, e.g. an angle apart that corresponds to the distance between the two eyes. In Figure 1.2 there is an example of an *autostereogram*, an apparent chaotic collection of points that is actually hiding two images of the same three-dimensional object. These two images represent the object in the way it would appear when viewed separately with either the right or the left eye. The three-dimensionality 'pops out' when the two pictures are in register,

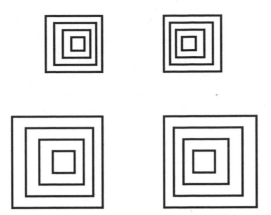

Figure 3.8 Examples of three dimensional pyramids. Try to 'look through' the figures so that the two figures fall on top of each other. After a few seconds you may see a three-dimensional image.

either by fixating in front of the paper or behind it (the two ways of fixating lead to reversal of what are the back and front of the object).

Because of the distance between the eyes, and the fact that each eye sees the same point under a somewhat different angle, an object's position relative to other objects (that are behind or in front of the first) is somewhat different in the two eyes (see Figure 3.8). In the autostereogram of Figure 1.2, the surface structure seen by one eye is overlaid by that which is seen by the other. These structures are too chaotic to be discovered by one eye fixating the stereogram, but if one fixates in front or behind the figure, the two images will fall into register and a coherent three-dimensional object can be seen.

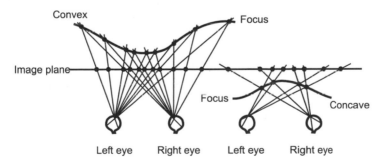

Figure 3.9 The principle of autostereograms. The figure on the left illustrates stereo vision with focus behind the two-dimensional image (parallel eyes), and the figure on the right shows stereo vision with focus in front of the page (crossed eyes). The same two points on an object will be seen under different angles by the right and left eye and therefore have different positions in the two-dimensional image (the stereogram). The pattern in the image is constructed in such a way that it allows the eyes to reconstruct the depth relief from what we perceive as two images of the same three-dimensional object, viewed from slightly different angles, in the same way that the real object would be seen by the two eyes.

We are very sensitive to the difference in position of the same substructure on the surface of an object (the difference in angle relative to the foveal direction in the two eyes). This difference angle is a measure for how far off the horopter the point is, and the brain translates this difference into a perception of depth. Figure 3.9 shows how these relations are constructed in an autostereogram. In the figure on the left, focus is behind the image, and in the figure on the right focus is in front of it. Not all people are able to see these three-dimensional images, although for some it may help to train squinting. Some people, however, completely lack the ability to see in depth. This applies, for instance, to those who squinted severely during childhood so that they developed an *amblyopic* eye, with reduced neural connections due to central suppression. However, if one eye has a severely reduced or distorted image, depth perception will also suffer.

The optic media

The cornea

The imaging media of the eye consist of several refracting surfaces (Figure 3.3), each with a slightly different index of refraction and curvature. The greatest refraction of light occurs at the corneal surface because here, at the transition between air and tissue, the difference in refractive index is the greatest. This first refracting surface of the eye is transparent for ultraviolet light (UV) down to about 290 nm. Because of its great power of more than 40 diopters, irregularities in the cornea will have major consequences for the quality of the retinal image. The cornea has fine nerve fibers and is sensitive to touch. It is kept moist by a thin tear film. Strong UV radiation can cause inflammation of the cornea, so-called 'snow-blindness'.

Behind the cornea is the *anterior chamber*. The depth of this chamber decreases with age because the lens grows. The fluid generated in the *ciliar muscles* (accommodation muscles) enters the anterior chamber and floats out of the eye through fine channels (*Schlemm's channels*). If this flux of fluid is in any way impeded, the inner pressure of the eye increases, and the higher pressure on the optic nerve and the retinal nerve cells may eventually lead to a vision deficit called *glaucoma*.

The pupil

The pupil regulates the flux of light onto the retina. Its diameter can vary between 2 mm in strong daylight and 8 mm in darkness. The illumination on the retina is proportional to the area of the pupil, and the given range of pupil diameters (from 2 to 8 mm) implies that the pupil's contribution to the regulation of retinal illumination is only by a factor of up to 1:16, even when the light level changes by more than a factor of $1:10^{12}$. To cope with the normal light levels during a day, this restriction calls for additional mechanisms of adaptation. The pupil reflex can be triggered by rods as well as by cones. Periodical light variations with a frequency higher than 2 Hz do not evoke modulation of the pupil.

When only one eye is exposed to light, both pupils will increase in size. Try to observe your own pupil reaction when you close one eye in front of a mirror. What happens?

As for the camera obscura and for lenses in general, sharpness extends over greater depths when the pupil size is small. A pupil diameter of 3 mm normally gives a good image quality on the retina because, with this size, the effects of lens aberrations and diffraction are usually minimized. For a greater pupil area, irregularities and lens aberrations will be more prominent and, for a smaller pupil, diffraction will decrease the quality of the image.

The maximum possible size of the pupil decreases with age, and this development is more pronounced in dark-adapted than in light-adapted eyes (Figure 2.10). This reduced size in the dark or at dusk leads to less light entering the retina and thus, for neural reasons, to reduced visual acuity and contrast sensitivity. However, the advantages of a small pupil should not be forgotten, i.e. the improvement of the quality of the retinal image, the reduction of stray light within the eye and greater depth of focus.

The pupil also responds to psychological factors. Widening of the pupil makes your facial expression more friendly and open. Women in the middle ages used drops of atropin–bella donna–to widen their pupils and make them look more attractive. Narrowing has the opposite effect, making you appear withdrawn and unfriendly.

The iris, the colored ring around the pupil, has the same pigment in all people; it is just the concentration that determines whether the eyes look blue or brown. Brown eyes have more pigment. Irises have distinct patterns and can be used for identification. Eye recognition technology, where passengers look into a video camera for a few seconds, is currently on trial at several international airports.

The lens

The lens is an elastic body fastened to the ciliary bodies by threads. When the ciliary muscle contracts, the tension that normally keeps the lens focused on infinity is relaxed. This reduces the size of the ring of the ciliary muscle and increases the curvature of the lens, mainly of its frontal surface, and its refracting power is thus increased. In this way, accommodation is changed from infinity at rest (in practice more than 6 m) to near. To accommodate from far to near takes about 0.35 s. In darkness the ability to accommodate is not as good as at high illuminance.

The lens becomes more rigid with age, and its loss of elasticity makes accommodation difficult. At a young age one is able to focus on a near point shorter than 10 cm from the eyes, whereas, at the age of 45, the near point has moved to between 25 and 90 cm, far enough away that you may need glasses to read the newspaper. After 70 years of age, the near point is more than 1 m away [see Figure 2.21(a)]. The lens absorbs UV light that is transmitted through the cornea in the wavelength range between 290 and 400 nm. The lens becomes yellowish with age and thus transmits less blue light. Compared with a 20-year-old, at 80 years of age the retina receives only about one-tenth of the short-wavelength light around 400 nm.

Cataract refers to an ailment where the lens has become less transparent and more diffusing, resulting in a blurred retinal image. Cataract occurs more frequently in older people. It is possible that exposure to infrared radiation (IR) over a long period can lead to cataract, as in the case of glass blowers. Congenital cataract is linked to a gene defect. Usually, the cataract can be removed by replacing the defect lens with an artificial one (see also section on Vision Loss, p. 202).

The vitreous

The vitreous is a gel-like substance that fills the eye and keeps its form by means of an inner pressure. In the vitreous there are fine threads that, when disrupted, can form curled shadows on the retina. These threads are easily seen when you suddenly raise your eyes to view something above. You will observe diffuse, curled threads that slowly sink downward in the field of view.

The sclera

The sclera is seen as 'the white of the eye'. It is an opaque outer membrane that holds the eye together. At the front of the eye it is transformed to a transparent tissue, the cornea, that allows light to enter the eye.

The retina

When we look at the retina through an *ophthalmoscope* (Figure 3.10) we see an orange background, called the fundus, and a network of arteries. This is the imaging plane of the eye and corresponds to the film in a camera.

> The ophthalmoscope was first used by Helmholtz in 1850, and it is still the most frequently used instrument to diagnose diseases of the retina. The principle is shown in Figure 3.10. In 1886, Jackman and Webster (1886) took the first image of the living human retina, and subsequently the fundus camera has been improved several times. The development of the scanning laser ophthalmoscope (SLO) in 1979 was an important step in achieving better images of the fundus. A laser beam scans the retina, and the reflected light is measured with a sensitive detector. The SLO makes it possible to measure important parameters in real time, such as for instance the thickness of the nerve layer, edemas and dynamic changes in blood flow. It is also possible to make

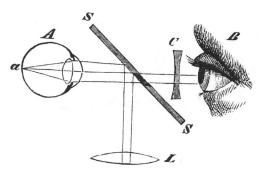

Figure 3.10 An ophthalmoscope is used to look at the back of the eye. Light from a source passes the lens, L, is reflected from a mirror, S, and illuminates an area, a, of the patient's retina. Light scattered and reflected back from the retina enters a hole in the mirror and produces an image on the retina of the examiner, B. The lens, C, is necessary only if the patient or the examiner has a refraction error. The hole in the mirror separates the light paths for illumination and imaging and prevents corneal reflections from reducing the contrasts of the retinal image seen by the examiner.

video recordings of fixation and eye movements during reading (Trauzettel-Klosinski *et al.*, 2002). This is of help in rehabilitation of reading functions when the fovea (macula) does not function because of illness, as in age-related macular degeneration (AMD).

The retina contains a dense mosaic of light-sensitive cells, the *rod* and the *cone* photoreceptors. The photoreceptors are in the deepest layers of the retina, and light must first pass through rows of other, transparent cell layers (*ganglion-*, *amacrine-*, *bipolar-* and *horizontal* cells) before reaching the receptors. Functionally, there are these five main cell types in the retina, and in the retinal transmission of information the receptors are the first units and the ganglion cells the last. The ganglion cells contact higher brain centers via their axons that form between 1 and 2 million nerve fibers. The optic nerve leaves the eye at about 17° to the side of the fovea. These are the nerve bundles that form the connections between the ganglion cells in the retina and the cells in the *lateral geniculate nucleus* (LGN), the next station in the pathway to higher brain centers. The projection of this retinal area is called the 'blind spot' because images that fall on it are not seen (Figure 3.11)

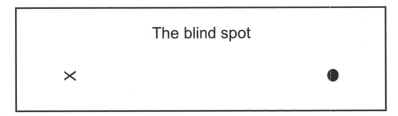

Figure 3.11 The blind spot of the eye. Close your left eye and fixate the cross. Move the book closer or farther away until the black spot disappears. The image of the spot is now projected onto a place on the retina where the nerve fibers leave the eye and where there are no photoreceptors.

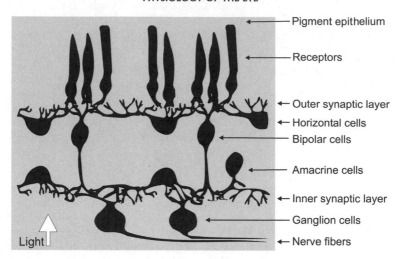

Pigment epithelium

Receptors

Outer synaptic layer

Horizontal cells

Bipolar cells

Amacrine cells

Inner synaptic layer

Ganglion cells

Nerve fibers

Light

Figure 3.12 A schematic cross section of the retina showing the functionally most important cell types. Light enters from below and passes through several layers of cells before it is absorbed in the receptors at the back of the eye. Bipolar cells belong to the direct pathway between receptors and ganglion cells, whereas horizontal and amacrine cells provide lateral connections to receptors and ganglion cells, respectively.

Figure 3.12 shows a simplified picture of the retinal cells and their interconnections. Only the ganglion cells are nerve cells in a strict sense since they are the only cells with action potentials. The different parts of a nerve cell are the *cell body* (*soma*), the *dendrites* and the *nerve fiber* (*axon*), and we shall later learn more about their function. For now it is enough to say that the dendrites are fine branch-like continuations of the cell body. Their many *synapses* (not shown in the figure) are the points of contact with the axon terminals of other cells, from which they receive incoming signals. The signals travel from the dendrites to the cell body where they are summated, and the cell sends a new signal down its axon to the next cell in the pathway. The nerve fiber, for instance the axon of a ganglion cell, leads signals from the cell body to the synaptic terminal, where information is transmitted to other cells' dendrites via synapses. In ganglion cells, these signals are conducted along the nerve fiber by means of a number of action potentials (nerve impulses). These nerve fibers are surrounded by an isolating sheet of myelin that consists of fatty molecules with gaps in the wrapping at regular distances, called the nodes of Ranvier. This arrangement maintains the amplitude of the traveling nerve impulses. The action potential generated at one node elicits current that flows passively within the myelin segment until the next node is reached. This local current flow then generates a nerve pulse in the unmyelined segment, and the process is repeated along the length of the axon.

Photoreceptors: rods and cones

The retina is less than half a millimeter thick, and it is fixed to the eye at two locations: where the optic nerve leaves the eye and with a saw-tooth-formed seam behind the lens. The retina is separated in layers, and the light-absorbing receptors are farthest back towards the pigment epithelium. The light-sensitive substance of a receptor is its photopigment, situated in its outer segment. In front of the receptors in the light path we find a network of other cell types.

> 'Receptor' is the general name of an element that communicate stimulation to the organism. Receptors transform one type of energy to another, and they are usually adapted to a specific stimulation. In the skin we have mechano-receptors that respond to touch and pressure, but also *cold receptors* and *warm receptors*. The mechano-receptors adapt very fast to the stimulus, and the perception of a constant pressure diminishes over time. Pain receptors, on the other hand, adapt slowly. Other organs have other receptors that respond to other physical or chemical changes. In the retina, a photoreceptor's absorption of light quanta leads to a series of chemical processes that in the end give rise to a change in electric potential. This change in potential is transferred to the next cells in the retina.

Schultze (1866) found that the owl and other nocturnal animals only had rods in their retinas, whereas birds which are active in the daytime, like the hen, had a large majority of cones. Based on these findings, the *duplicity theory* was developed by the German physiologist J. von Kries. This theory states that the cones are the receptors for daylight vision, whereas the rods are designed for vision in the dark, even though in the human visual system, which is specialized for daylight vision, 95 percent of our photoreceptors are rods. Loss of rods leads to night blindness.

At night we can see faint light that is not visible during the day. In the dark-adapted eye, sensitivity has a maximum at 507 nm, whereas the light-adapted sensitivity, represented by the *photopic luminous efficiency curve*, V_λ, has a maximum at 555 nm. The difference in maximum sensitivity of the rods and cones at these two wavelengths is about a factor of 100 in favor of the rods. However, when dark-adapted, the absolute sensitivity of the rods for wavelengths longer than 650 nm is somewhat lower than for the L-cones. In red light, one can thus use the sharper cone vision without losing the high rod sensitivity.

In the human retina there are about 130 million receptors (120×10^6 rods and 6×10^6 cones; see Pirenne 1967). They all respond to light absorption by generating an electric potential that is conveyed to other nerve cells. The density of rods and cones is shown in Figure 3.13(a) and (b). In the central fovea the cones have a density of about 150 000 cones/mm^2, corresponding to an inter-receptor distance of about 0.0025 mm.

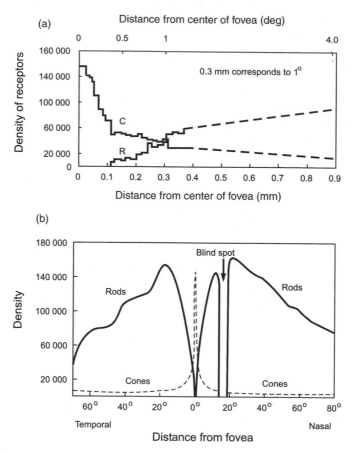

Figure 3.13 The density of cones (C) and rods (R) in the primate retina (number per mm²) as a function of the distance from the fovea in mm (a), and as a function of distance in degrees (b).

The fovea

The *fovea* is a small pit in the retina at the back of the eye of about 1.5 mm (5°) in diameter, centered on the larger *macula lutea* of about 10° diameter, also called the yellow spot because it normally has a weak yellowish color. The fovea is the location at the retina where vision is most acute. The rod-free central part – the *fovea centralis* – is about 0.2–0.3 mm in diameter (somewhat smaller than 1° visual angle) and contains only cones (Figure 3.13). A still smaller part, only about 0.3–0.5°, contains no short-wavelength S-cones, causing a central 'blue blindness'. The post-receptoral cells are moved to the side in the fovea, and the cones are thus more directly exposed to light (Figure 3.14).

The representation of the fovea in the cortex is magnified; this means that the cortical area allocated to an image falling on the fovea is enlarged relative to the area

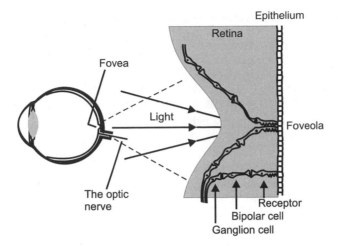

Figure 3.14 In the central foveal pit (the foveola) bipolar and ganglion cells are shifted to the side so that light has more direct access to the photoreceptors (horizontal and amacrine cells are removed from the picture to make it clearer).

it occupies in the retina. Whereas the central fovea comprises only about 0.01 percent of the total area of the retina, it covers about 8 percent of the primary visual cortex, or about a 1000 times larger area (Azzopardi and Cowey, 1996).

In the fovea, the distance between the cones is about 2–2.5 μm, or a visual angle of less than 0.4 arcmin (5 μm corresponds about 0.7 min, depending somewhat on the length of the eye). The density of rods (in numbers per mm^2) increases as one moves away from the fovea, whereas at the same time the density of cones decreases. At a distance of 1° to the side of the fovea there are about as many rods as cones per unit area [Figure 3.13(a)].

Humans normally have three cone types, each type with a different pigment and spectral sensitivity. A cone photoreceptor is named after the spectral region it covers: L-, M- and S-cones for long-, middle- and short-wavelength sensitive receptors. Earlier, these receptor types were called red-, green- and blue-cones (R-, G-, B-cones), but since these notations are prone to the misunderstanding that it is the excitation of cones that determines the color of a stimulus, they are better not used. The spectral sensitivity of the light-sensitive pigment in a cone type is relatively well known, and all of them will be described in more detail later. The wavelength for which the sensitivity is maximum is 560 nm for L-cones (corresponding to greenish-yellow), 530 nm for M-cones (corresponding to yellowish-green), and 425 nm for S-cones (corresponding to violet). In recent years, individual differences of several nanometers in these maxima have been found (Bowmaker, 1991; Stockman and Sharpe, 1999).

Based on psychophysical evidence, it was assumed for a long time that the relative number of cones in the retina is in the proportion of 32:16:1 between L-, M- and

S-cones. However, recent studies based on cone isolation techniques have demonstrated large individual variations. One method used is to determine the flicker detection ratios of isolated L- and M-cones, as well as their relative amplitudes using the *electroretinogram* (ERG). In addition, more direct methods like the imaging of the retina through adaptive optics after selective pigment bleaching have revealed surprising variations, not only in the relative numbers of cone types, but also in their spatial distributions in the retina. Such pictures often show clusters, or 'islands', of one cone type surrounded by areas of another type (Roorda and Williams, 1999). Figure 3.15 is a reproduction of such a distribution 1° away from the central fovea.

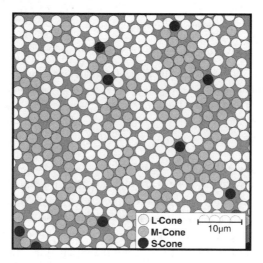

Figure 3.15 An example of the distribution of different cone types, L, M and S, in the human retina. The drawing closely resembles distributions found by Roorda and Williams (1999), using adaptive optics. See also Color Plate Section.

Although the relative excitation of cones cannot be directly associated with the color perceived, the three different cone types make *trichromatic color vision*, i.e. color matches with three variables, possible. If one type is lacking, for instance the *L*- or the *M*-cones, the result is color vision with only two variables, called *dichromacy*. This leads to color confusions, most commonly in a failure to distinguish between red and green. Color vision deficiencies will be described in more detail later.

The rods respond more slowly than the cones to light stimulation. In rods the effect of absorption of light quanta is summed within about 100 ms. This enables the rods to detect faint light, but it has the side effect that rod signals cannot be modulated in time faster than about 10–15 Hz. The cones respond much faster, with a summation time of between 10 and 20 ms at higher light levels. The rise time is short and they also return much faster to the resting potential as soon as the stimulation is over. This allows for better temporal resolution, and measurements of flicker sensitivity have shown that cones can follow up to 80–90 Hz (Lee *et al.*, 1990).

Night vision is without colors, and we cannot discriminate fine detail. An object is in fact seen better when we focus a little to the side of it because the central fovea is free of rods. The sensitivity in the dark is higher a little to the side of the fovea, where the density of rods is highest [see Figure 3.13(a, b)]. A weak light, being just visible in the periphery of the visual field, will disappear when we try to fixate it directly on the fovea. People that have lost their central vision, like the many elderly with AMD, must learn to fixate to the side of the fovea, and this requires extensive training.

Daylight (cone) vision is called 'photopic vision' and night (rod) vision 'scotopic vision', and between there is a range of 'mesopic vision' where both cones and rods are active (see Figure 3.16). A luminance of white light above about 3 cd/m^2, is

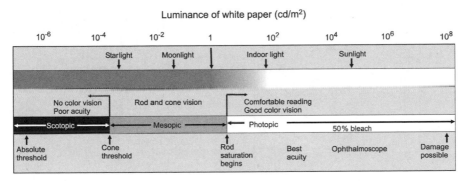

Figure 3.16 The eye can operate over about 14 logarithmic units of intensity. Rods are active for the first six decades of luminance above absolute threshold and the cones for eight to 10 decades in the higher range, with about four decades of overlap in between.

regarded as photopic, and a luminance below 0.001 cd/m^2 is scoptopic, while the mesopic range is from 0.001 to 3 cd/m^2. Normally, rods will not differentiate between light above 10–20 cd/m^2, provided the pupil is not very small. The cone threshold is about 0.1 troland, which corresponds to about 0.001 cd/m^2 with a 7 mm pupil diameter. Figure 3.16 shows the intensity range over which the visual system operates. Rods are active for the first six decades of luminance above absolute threshold and the cones for eight to 10 decades in the higher range, with about four decades of overlap in between.

Light absorption

A photoreceptor responds to light by absorbing light quanta in its photopigment (also called excitation) and by converting this event to an electrical potential difference. The pigment molecule of a rod is rhodopsin (also called visual purple), consisting of opsin (a protein) and retinal, whereas in the L-, M- and S-cone types there is another photopigment consisting of a different opsin bound to retinal (a lipid vitamin A derivative). It is the opsin's structure that determines where in the spectrum the attached retinal absorbs light, and this gives rise to the different cone absorption spectra. The absorbing pigment is located in disks that are stacked one after the other in the receptor's outer segment, perpendicular to the light rays [see Figure 3.17(a)]. This is an effective arrangement for capturing the light. It is analogous to placing pieces of the same color filter one after the other in a light path, thus increasing the chance that light quanta that escape absorption in one filter will be caught by the next.

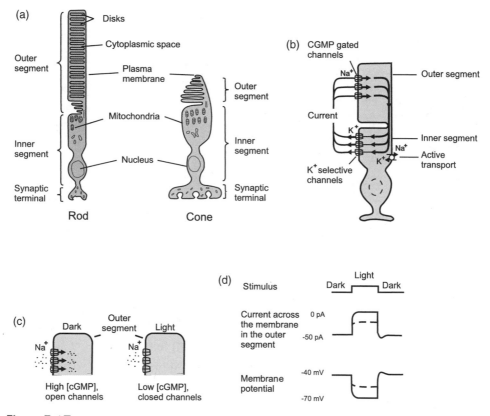

Figure 3.17 (a) Rods and cones differ in their size and shape, and in the arrangement of the membrane disks in the outer segment. (b) In the dark there is a current of Na^+ ions into the receptor's outer segment through cGMP-controlled sodium channels. The current of K^+ ions out of the cell is not driven by cGMP. An active pump in the inner segment maintains the concentration of the two ions at a stable level. (c) When the receptors are exposed to light, cGMP levels are reduced and the cGMP-controlled channels close. (d) A dark current of 50 pA is turned off when the receptor is illuminated with strong light and the membrane is hyperpolarized from -40 to -70 mV. For light intensities in between, the membrane is polarized to somewhere between -40 and -70 mV. (Modified from Kandel *et al.*, 2000.)

There are several hundred such thin free floating disks or membrane plates in a rod's outer segment, whereas in a cone each 'filter' is formed by folding the membrane. The pigment molecules are embedded in this membrane. When retinal absorbs light, the pigment molecule changes its form from being bent to becoming a long straight chain, a process called photoisomerization, and the photopigment molecule eventually breaks into its two parts. One says that the photopigment has been 'bleached'. The higher the intensity, the more pigment is bleached. This process is accompanied by a change in color of the rod pigment, from rose to pale yellow.

The receptor disks are short-lived and have a lifetime of about 12 days. The old disks at the tip of the receptor are shed into the pigment epithelium, where they are normally dissolved. However, in some cases this destruction process does not function normally, and pathological conditions may occur, often leading to severe degenerative processes in the receptors.

Rods contain more photopigment than the cones, and they thus trap more light. A medium's ability to absorb light is characterized by its optical density. As we have seen previously, density is the logarithm of inverse transmission, and, as for color filters, the optical density of photoreceptors changes with wavelength. The optical density of a cone is related to the axial length of its outer segment and the concentration of photopigment. Typical values for maximum optical density for foveal cones is about 0.5 for L- and M-cones, and about 0.4 for S-cones (Vienot, 2001).

Rods amplify their signals more than cones. Whereas as little as one absorbed light quantum is sufficient to obtain a signal from a rod, about 100 quanta are required to obtain a comparable response from a cone. Even so, natural light levels from dusk to dawn do not provide enough light for one photon to be captured per rod in all rods over their integration time. Fewer quanta are needed to bring a rod to maximum response or to response saturation, whereafter it is not possible to increase its response by further light increments. Therefore, in daylight rods are not affected by normal contrasts, whereas the cones are easily adapted to daylight and are modulated by small contrasts.

Main differences between rods and cones

Rods

- High sensitivity.
- More photopigment than the cones.
- High amplification; response to a single light quantum.
- Response saturation at normal daylight light levels.
- Slow response, long integration time; low temporal resolution.
- More sensitive to interocular stray light than the cones.

In addition, the rods provide low spatial resolution; converging signal pathways, and they are absent in the central fovea. They have all the same pigment and are thus not able to provide for hue differentiation as in chromatic vision.

Cones

- Lower sensitivity than rods.

- Less photopigment than rods.

- Lower amplification.

- The response reaches saturation only at high intensities.

- Fast response, short integration time; high temporal resolution.

- More sensitive than rods for light in the axial direction.

The cones enable high spatial resolution; they have their highest density in the fovea and show less convergence than rods. There are three types of cones, each with a different pigment. Cones contribute to both achromatic and chromatic vision.

Genetics of photopigments

People with normal trichromatic color vision have L-, M- and S-cone receptors with different pigments. The difference in spectral sensitivity of the pigments can be traced back to three genes that determine the sequence of the amino acids of the cone opsins. The production of photopigments for the different types of photoreceptors seems to be independent of the making of the photoreceptor structure, at least for L- and M-cones A color-deficient person who is unable to distinguish red from green (a dichromat) is likely to lack one of the normal L- or M-cone pigments. Yet his or her ability to see fine details, as characterized by his or her visual resolution, is completely normal. Therefore, in the absence of evidence of other differences between L- and M-cones, one is led to conclude that such a color-vision-defective person must have a normal number and spacing of functioning cones. Thus, these two cone types merge into one group containing one type of photopigment (either the L- or the M-type). The question of what determines the functional properties of a cell, such as the different types of photoreceptors, and how such cells are organized into functional units, like in the receptive fields of ganglion cells, still remains a scientific challenge. Environmental factors also influence the development of the nervous system, particularly at its higher levels, and this aspect will be discussed later. Here we shall focus upon inherited properties as they apply to the photoreceptors. In order to facilitate the appreciation of the recent achievements with regard to visual pigments, let us review some the essential principles. Over the past decades, intensive research has disclosed much of the genetic basis of the production of visual pigments (Nathans *et al.*, 1986; Neitz *et al.*, 1993; Sharpe *et al.*, 1999). It has led to the surprising finding that also people with 'normal' color vision possess variants of the fundamental pigment types.

Genes are the main factors in determining cell structure and function, and they also determine the amino acid sequences of photopigments. Opsin interacts with the chromophore retinal to tune the spectral sensitivity of the resultant visual pigment. The production of a photopigment protein is a complicated process, in which DNA (deoxyribonucleic acid) plays an important role. DNA is found on the chromosomes in the nuclei of all cells in the body. A DNA molecule is built up of two long strands composed of bases, or nucleotides, of which there are four different kinds: thymine (T), adenine (A), cytosine (C) and guanine (G). Long strands are wound together into a double helix with pairs of bases forming bridges between the twisted strands. The code for protein synthesis is a triplet of successive nucleotides along a chain, for instance ACG (a sequence of adenine, cytosine and guanine). Each set of three adjacent nucleotides constitutes a *codon* which specifies a particular amino acid.

In this way the sequence of nucleotides determines a sequence of amino acids. Only some parts of the DNA chain contain the genetic information necessary to produce pigment. This information is found on a limited number of portions of the gene, called exons. The synthesis of a pigment protein is determined by the tandem codon sequence within the exons and the number and positions of the exons on DNA. The S-cone pigment gene comprises five exons, and the M- and L-cone pigment genes each comprise six exons. The regions between the exons, called introns, are not involved in coding the protein.

The exon code regions are translated into a proper sequence of amino acids, the 'building blocks' of proteins (the L- and M-photopigments are made up of 364 amino acids in sequence). A visual pigment gene, as a part of a DNA molecule, consists of several hundred codons, or triplets of nucleotides linked together. The gene encoding the rod photopigment, rhodopsin, has been located on chromosome 3, and the gene specifying the S-cone pigment is located on chromosome 7. The L- and M-cone pigments have almost identical amino acid sequences and their genes are located on the X chromosome. Of the 364 codons in the L- and the M- cone pigment genes, only 15 are different. Of these, only three codons (codon 180, 277 and 285) are believed to contribute to the spectral differences between the L- and M-cone pigments (Sharpe *et al.*, 1999). The S-cone sequence seems not to vary much in the human population, but the M- and L-cone opsin genes come in several hybrid forms.

A substitution of the amino acid serine for an alanine residue at codon 180 on exon 3 produces a slight 3 nm shift towards longer wavelengths in the spectral sensitivity curve of both L- and M-cone pigments. Among human L-cone pigment genes, approximately 56 percent have serine and 44 percent have alanine at position 180. Thus, there are two L-cone pigments about equally distributed among the normal population. Among M-pigment genes, approximately 6 percent have serine and

94 percent have alanine (Sharpe *et al.*, 1999). These and other forms of polymorphism in the normal L- and M-pigment genes are sources of variation in normal color vision, as are also individual differences in optical density of the lens and of the macular screening pigment.

Signal generation

Ionic currents

We need not go into the finer details of the chemical processes that lead to receptor potentials. These are still the object of scientific inquiry, and only a brief summary of the main features will be given (Figure 3.17). The absorption of light by the pigment of a photoreceptor triggers a cascade of reactions that finally lead to alterations in the current of electrically charged particles – the ion currents – across the receptor's cell membrane, and thus to a change in its potential relative to the magnitude in the dark. The polarization over the membrane increases with increasing light intensity, and this potential change is conveyed from the receptor to the next level in signal processing. A substance, called cyclic guanosine monophosphate, or cGMP, has the task of transferring information about the light absorption in the pigment disks to the cell membrane.

cGMP controls the ionic current across the cell membrane by opening some special ionic channels, the so-called 'cGMP-controlled channels', in the outer segment of the receptors. In the dark, the concentration of cGMP is high, and many such ionic channels (for example for sodium, Na^+) are open. This leads to a current of Na^+ ions into the receptor's outer segment. At the same time there is a current of K^+ ions (potassium ions) out of the cell, such that the potential inside the receptor is about -40 mV relative to the outside. The magnitude of the current is about 50 pA ($= 50 \times 10^{-12}$A; see Figure 3.17(d)).

When a receptor is illuminated and the pigment absorbs light, this eventually leads to a reduction in the concentration of cGMP. The cGMP-controlled Na^+ channels are closed, whereas the K^+ channels, which are not affected by cGMP, continue to stay open. There is a net outflux of positive charges, creating a deficit of positive charges inside the cell. As a result, the membrane potential becomes more negative. This is called *hyperpolarization*. In strong illumination, all cGMP-controlled channels may be closed, and the potential approaches -70 mV. For lower intensities of the light, the receptor is hyperpolarized to a level between -40 and -70 mV.

In the dark, the receptor sets free relative large quanta of a signal substance – a transmitter – in the synaptic cleft between the receptor and the next cells in the signaling pathway – a *bipolar cell* and the *horizontal cells*. This release of transmitter substance keeps the next cells on a resting potential in the dark. Whenever the receptor is hyperpolarized by absorbing light quanta, the amount of transmitter decreases, and the potentials of the next cells subsequently change. These changes in potential are traveling from cell to cell, and the result is the transmission of an electric signal, the magnitude and sign of which is dependent on the cell type and the local illumination of the retina. As seen in Figure 3.18, post-receptoral cells respond with either hyperpolarization or depolarization.

Hyperpolarization and depolarization of retinal cells

Figure 3.18 presents a simple scheme of how the different cell types in the retina react to light absorbed by the receptors. These examples are highly schematic, and the details can deviate from one species to another, but the logical principles are similar.

The left column of Figure 3.18 shows that receptors, horizontal cells and bipolar cells respond to a small light spot by slow, negative or positive potential changes, whereas amacrine and ganglion cells in addition have a train of short nerve pulses, the action potentials or spikes. The right column exemplifies how a ring of light (an annulus), added around the central spot, reduces the former responses of receptors and horizontal cells. In the bipolar cells, the addition of an annulus can even lead to a response of the opposite polarity; if the center spot alone leads to depolarization (left column), simultaneously adding a ring of light might lead to hyperpolarization. Whereas the potential of the receptor is merely reduced by adding the ring, a bipolar cell displays an inverted potential relative to the excitation elicited by a small spot, most likely brought about by the horizontal cells. Typically, the resulting *receptive field* of a retinal cell (defined as the retinal area for which light stimulation causes a response in the cell; see section on Ganglion Cells, p. 116) consists of a relatively small center and an antagonistic surround structure. Thus, the task of the immediate surround is to reduce the signal caused by the center stimulus alone.

As demonstrated in the figure, in amacrine and ganglion cells one observes short nerve impulses in addition to a small depolarization. In excitatory cells this is a pure increase of the response in impulses/s when the local light is turned on. With simultaneous illumination of the center and annulus, the result for the cells of Figure 3.18 is a reduced response (but see Figure 3.29). Other cells will show increased firing to a decremental stimulus in its receptive field center. These different responses are usually called ON- and OFF-responses, although the important thing here is not that the light is switched on or off, but the particular direction of the change of light intensity (the sign of the contrast) leading to a particular response. We therefore prefer to call these reactions *Increment* and *Decrement* responses. The alterations in the response of a cell that occur after adding a steady annulus are due to cross-connections within the retina and lateral influences. These processes are important for establishing contrast and color vision, and we shall return to these functions later.

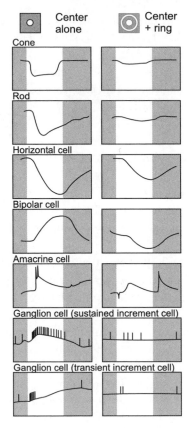

Figure 3.18 A schematic overview over the responses of some retinal cell types when they are stimulated with a small light spot (left column), and the same spot with a ring around it (right column). Rods, cones and horizontal and bipolar cells respond with slow potential changes (hyperpolarization) while the reaction of ganglion cells is a firing of action potentials. In photoreceptors and in horizontal cells the center + ring stimulus leads to smaller responses (less hyperpolarization) than the center stimulus alone. The ring thus inhibits the center response. In bipolar cells, however, the center + ring stimulus leads to a polarization opposite to that of center stimulation alone, and the ganglion cells shown here respond with a reduced number of nerve impulses.

Horizontal cells

Figures 3.12 and 3.19 show schematic images of the functional elements of the retina and their main connections. On the top of the figures, on the outer retina, we find rods and cones. They convey their signals first to bipolar cells and then to the ganglion cells. This we may call the direct signal pathway. However, horizontal cells and amacrine cells (the latter are removed in Figure 3.19 to gain clarity) can modify these signals. These are cells that integrate stimulation of extended areas of the retina and mediate this laterally to the direct pathway.

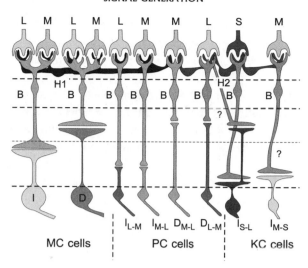

Figure 3.19 The main cell types in the retina of primates and their interconnections. The figure shows L-, M- and S-cones, H1- and H2-type horizontal cells, various bipolar cells (B), and the ganglion cells. MC cells that project to the corresponding layers of the LGN are large with large-area dendritic fields. The smaller PC cells connect to L- and M-cones and project to PC layers in the LGN. There are two types of MC cells, increment (I) and decrement (D) cells, while PC cells come in two I-cell types and two D-cell types. Cells that are connected to S-cones (KC) are less well studied, and a possible model is shown to the right in the figure. The dendritic tree of I- or D-cells branches in different layers of the retina, I-cells in the *inner* and D-cells in the *outer plexiform layer*. The bistratified KC cells with S-cone excitation seem to branch in both of these layers, while the KC cell with S-cone inhibition is likely to be monostratified. The colors used in the figure serve to separate the cell types from each other and have little to do with color coding (modified from Lee and Dacey, 1997). Rods and amacrine cells have been left out. (See also color plate section.)

Horizontal cells make lateral connections between receptors, and they have electric contacts with each other. In monkeys, one finds two types of horizontal cells, H1 and H2. Both sum the signals from many receptors over an extended retinal area (*spatial summation*), and it was suggested that they contact every cone in their vicinity. However, more recent data indicate a certain cone selectivity (Dacey *et al.*, 1996). Horizontal cells modify the cone signals at the *cone pedicle* (Figures 3.12 and 3.19), before the signals reach the bipolar cells, and they may well contribute to the antagonistic surround of the receptive fields of bipolar and ganglion cells.

An H1 cell has a large network of contacts with the receptors, and it has the largest dendritic tree of the two horizontal cells. A foveal H1 cell contacts between 11 and 14 cones, but only a few rods. A foveal H2 makes contact with only about seven cones and with 350–500 rods. It is not quite clear from which cone types the inputs of the two horizontal cells come from. Dacey *et al.* (1996) found that the H1 cells are hyperpolarized for light increments, and that they barely have any response to pure chromatic changes. They received activating inputs form the L- and M-cones, but none responded to a selective activation of S-cones. H2 were also hyperpolarized by light increments, but in contrast to H1 they received excitatory inputs from all three cone types. Neither H1 nor H2 responded to chromatic stimuli of equal luminance.

From these studies it seems that the horizontal cells in primates are hyperpolarized for light of all wavelengths. Hence, they are not *spectrally opponent*. A spectrally opponent cell would hyperpolarize for some wavelengths and depolarize for others. Such opponent horizontal cells were first found in fish by the Swede Gunnar Svaetichin (1956).

If horizontal cells contribute to the antagonistic surround of bipolar and ganglion cells, one would suggest that H1 modifies the signals in the pathways that involve only L- and M-cones, whereas H2 also includes S-cones.

Bipolar cells

Bipolar cells are the next step in the direct route of information transmission from the receptors to the ganglion cells. As we have seen, their inputs are modified by the horizontal cells which sum the activity of several receptors surrounding the central ones, and since the horizontal cells are connected to each other, a concentric antagonistic surround results. Thus, the bipolar cells may be excited by light in the center and inhibited by light in the surround of its receptive field, or the other way around. When center and surround are illuminated simultaneously, activation and inhibition may to a large extent cancel each other, and the cell's activity will be less than if only the center was illuminated (Figure 3.18).

In the primate retina, there are several subgroups of bipolar cells. The *ON-center bipolar cells* contact *ON-center ganglion cells*, and *OFF-center bipolars* are connected to *OFF-center ganglion cells*. This suggests an important role of these systems in coding for the contrast of an object. Electrophysiological recordings have shown that ON- and OFF-center cells have segregated and parallel signal pathways up to the visual cortex. Later we shall argue that it is perhaps time to rename these cell types and call the ON-cells for *Increment cells* (*I*-cells) and the OFF-cells *Decrement cells* (*D*-cells).

As shown in Figure 3.20, light that hyperpolarizes the receptor will activate (depolarize) an *ON*-center bipolar and inhibit (hyperpolarize) an OFF-center bipolar. Light leads to less transmitter substance being released by the receptor, and this neurotransmitter (being glutamate) has a different effect on OFF- and ON-bipolar cells. Stimulating the surround by light leads to hyperpolarization of the horizontal cells (not shown in the figure) and to an inhibitory effect from them on the signal from the receptor to the bipolar cell. This lateral effect of horizontal cells is one of depolarizing the receptor potential, and it is thus opposite to the effect of direct excitation by light. The net receptor response is then conveyed further to the bipolar cell.

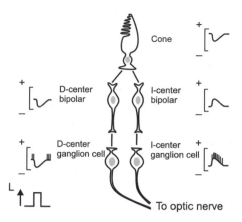

Figure 3.20 One type of bipolar cells is activated by a luminance increment in the center of its receptive field (I-center cells) and another type is activated by a decrement (D-center cells). In the literature these cells are known as ON- and OFF-center bipolar cells. At this early stage, visual information is already divided into two parallel pathways activating either I-center or D-center ganglion cells. When a cone is hyperpolarized by light, the I-center bipolar is activated by depolarization while the D-center bipolar is inhibited (hyperpolarized). The activation of I-center bipolars leads to an increased number of action potentials in I-center ganglion cells (modified from Kandel *et al.*, 2000).

It is also possible to distinguish anatomically between ON- and OFF-bipolar cells from the depth of the layer of contacts with the dendrites of the ganglion cell. An ON-bipolar cell contacts an ON-center ganglion cell (I-center cell) closer to the ganglion cell body than is the case for an OFF-cell. Figure 3.19 shows these two layers of synaptic contacts, the layer closer to the receptors being that of OFF-center bipolar and D-center ganglion cells.

The size of the contacting area of the bipolar's axon seems to depend on its type and to which cell type it is connected – if it is a *tonic* or a *phasic* ganglion cell. Those bipolars that have the largest field, the *diffuse bipolars*, probably propagate their signals to big, *phasic ganglion cells*, also called *parasol ganglion cells*. The bipolars with the smaller fields, *the midget bipolars*, make contact with the smaller, *tonic ganglion cells*, also called *midget ganglion cells*. In primates, the foveal midget bipolars have connection to only one cone, and their dendritic tree is often smaller than a cone pedicle. A diffuse bipolar on the other hand, receives inputs from many cones. One does not know if these latter contacts are made at random, in such a way that all the cone types within the area of the receptive field (and the dendritic tree) affect the activity of the bipolar cell, or if connections are made selectively with a single cone type.

Rods are connected to a special type of bipolar. It is assumed that there exists only one type *rod bipolar cells*, the depolarizing *ON*-center bipolar.

Amacrine cells

Not much is known about the function of amacrine cells in primates. They constitute another layer of a network of lateral connections, often through electrical contacts. They seem to be particularly sensitive to temporal changes of the stimulus. ON and OFF amacrines modulate the message presented to the ganglion cells by the direct pathways, via bipolars. Apparently there are several tenths of amacrine cell types that can be distinguished by properties like the diameters of their dendritic fields and the layers of their stratification.

Ganglion cells and other spiking neurons

The ganglion cells are the last functional cell types in the retina; they send their signals to the higher brain centers via the optic nerve. Whereas at the early stages of retinal information processing signals are conveyed by means of slow, graded potentials, this is changed in the ganglion cells. In a ganglion cell the magnitude of a graded potential, received from the bipolars, is transformed to a train of short nerve impulses that travel up the optic nerve to the lateral geniculate nucleus (LGN). It is the number of impulses per second (a frequency code) that informs the brain about the degree of retinal activation. Before going into detail as to the function of ganglion cells, let us introduce some general features of spiking neurons. The mechanism for generating action potentials will be treated more thoroughly later in this chapter.

Spiking nerve cells (neurons) are the essential functional units of the brain (of which the retina is considered to be a part). Neurons select, modify and forward incoming signals; they are small computational units that can be modeled by linear summation and subtraction of incoming signals. The human brain has between 10^{10} and 10^{12} nerve cells. In the visual centers alone, there may be approximately 500 million such cells. The nerve cells of the retina and the brain are different from other cells in the body in morphology, because of the ability of their membrane to generate electrical signals and the existence of *synapses*. In the synapses the electrical signals are transmitted from one cell to the next, normally by means of a chemical substance – a *neurotransmitter*. Commonly a nerve cell receives signals from between 1000 and 10 000 synapses.

The nerve impulses can be registered and counted by placing a microelectrode outside the cell, as illustrated in Figure 3.21. The electrode is often a thin tungsten needle only a few thousand millimeters in diameter surrounded by a glass pipette (as an isolator). The thin needle is a conductor enclosed in glass, with only the sharp end exposed and making contact with the medium surrounding the cell. The tungsten electrode is connected to an amplifier, an oscilloscope and a data-collecting unit (a computer). This computer synchronizes the stimulus and the recordings. With the electrode close to the cell, it is possible to register the changes in electrical potentials and thus count the impulses that stand out against the background noise. When the

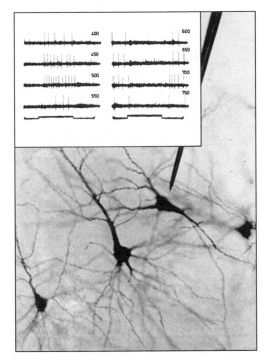

Figure 3.21 A microelectrode (a thin metal needle enclosed in an isolating glass pipette) that is positioned close to a ganglion cell can pick up every nerve pulse that is generated by the cell. The inset shows the reaction of a retinal ganglion cell when its receptive field is illuminated by light of different wavelengths. The lower trace shows the response of a photocell to the light stimulus being switched on and off. Light of wavelengths around 500 nm activates the cell, while red light of wavelengths above 600 nm inhibits it completely. This could therefore be an 'M–L' ganglion cell.

electrode is outside the cell, the cell is not so easily damaged, and one can register its activity over a relatively long time, up to several hours if one is lucky.

A ganglion cell receives activating signals from a bipolar cell of the same type. When its receptive field center is illuminated, an ON-center bipolar responds by depolarization, and it in turn depolarizes the ON-ganglion cell that elicits a train of nerve impulses. Both horizontal and amacrine cells have modified the receptor signal before it reaches a ganglion cell. A nerve fiber leads from the ganglion cell to the LGN in the *thalamus* (see Figure 2.23). It appears that the numbers of ganglion cells in the retina and cells in the visual cortex decrease with age, with about a 30 percent reduction from 20 to 80 years of age.

Receptive fields

A nerve cell within the visual system, for instance a bipolar cell or a retinal ganglion cell, responds only if light is exciting a small number of receptors within a well-defined area of the retina. This area may be called the cell's territory, or its *local* or *classical receptive field*. Receptive fields are defined as the small area on the retina where light stimulation of the photoreceptors causes a cell's response (for example a change in the firing rate of a ganglion cell). In the retina, the shape of receptive fields is simple and circular, whereas for cells later in the visual pathway they become more complex. As we shall see, receptive fields are well suited for the coding of stimulus contrasts. The simple retinal receptive fields are the 'building blocks' of more complicated receptive fields at later levels, making them able to register and discriminate contrast, angular orientation, direction of movement, of form, color, etc.

Light falling on the retina outside the local receptive field does not evoke a response in the cell, even though it has been repeatedly demonstrated that peripheral lights may indirectly influence (modulate) the cell's sensitivity and responsiveness to stimuli within the classical receptive field. Such lateral processes extend to much larger areas of the retina than the classical receptive field. In recent years, it has therefore become customary to distinguish between a cell's 'classical receptive field', consisting of a small center and a relatively narrow surrounding annulus of the opposite effect, and a more 'global surround' (Valberg *et al.*, 1985a, 1991a; Allman *et al.*, 1985).

A ganglion cell responds to light falling on its local receptive field by generating action potentials. These pulses can be translated into sound using a loudspeaker or displayed on an oscilloscope, as in Figure 3.21, after they have been picked up by a tiny *microelectrode* placed either inside or near the cell. When probing a cell's sensitivity and response by means of a small spot of light traversing its receptive field, one finds that the spatial extension of receptive fields is smallest for cells in the fovea and increases for more eccentric cells. In the fovea, a receptive field center comprises only one or a few receptors, and thus has an extension of only a few minutes of arc. In the peripheral retina, receptive fields (including their classical surround) can become several degrees in diameter (1° is about 0.250–0.3 mm on the retina). For cells in the peripheral retina the extension of the receptive field corresponds roughly to the cell's dendritic field.

The discovery of a neuron's receptive field was made in 1938 by H.K. Hartline (Nobel Prize winner, 1967), who recorded the activity of ganglion cells in the frog. Hartline found that these cells were sensitive to light that fell within a limited retinal area (Hartline, 1940). The next important discovery came when Kuffler (1953) and Barlow (1953), independently of each other, found that in cats and frogs the receptive fields were actually divided into two fields with opposite effects. For instance, if in one cell type a stimulated central field elicited an excitatory response when the light was turned on (an increment response), this area would be surrounded by an antagonistic ring that responded with increased firing whenever the light was turned off (decrement response). This concentric arrangement into two antagonistic parts was found only for relatively high levels of background illumination. When the adaptation illumination decreased, the decrement response of the surround diminished, and for really low adaptation levels it totally disappeared. The increment response of the center could, however, be recorded for all levels of illumination. In addition to this cell type, there was also the opposite type that responded with a decrement response in the center and an increment response in the surround.

These response characteristics gave rise to the notations of ON- and OFF-center cells, respectively. This has since become a common notation, but it is important to note that the smallest change of contrast, a small increment or decrement, will also give rise to a response. ON-cells respond with increased firing rate to an incremental contrast, and OFF-cells are inhibited by the same luminance increment, but increase their firing rate for a small decrement of the local luminance. *Therefore, it is the local contrast of a stimulus that is of importance for these cells, not the overall light being switched on or off. We will therefore call these cells I-cells and D-cells, in order to indicate their function more precisely.*

Figure 3.22 shows examples of sensitivity profiles of the center and the near surround of a classical receptive field when traversing the receptive field with a small

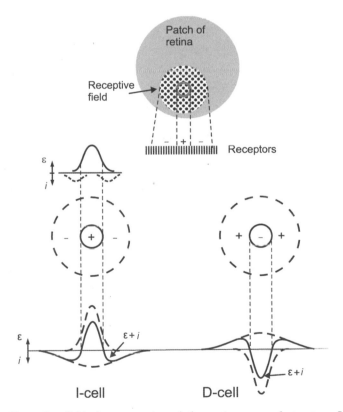

Figure 3.22 Receptive fields have an antagonistic center-surround structure. I-cells have an excitatory center and an inhibitory surround, while D-cells have an inhibitory center that is antagonized by the surround. The sensitivity profile can be found by testing the responsivity (or the sensitivity) of a cell to a small spot of light at different positions along the diameter of the receptive field. The lower part of the figure shows a model of the spatial sensitivity profiles of I- and D-cells. ε stands for excitation and i inhibition.

spot of light. In the left half of the figure we find the response of an I-cell, and to the right that of a D-cell. In the fovea, where the center field contains only one cone, the spatial sensitivity profile on the top left of Figure 3.22, where the annulus does not extend into the center, represents the more likely profile.

The size of the receptive field center increases with distance from the fovea. With increasing center size (and eccentricity), the antagonistic process from the one cone type(s) in the surround is likely to also be present in the receptive field center; only in the fovea may it form an annulus around the center as in the top-left part of Figure 3.22. For some cells, particularly those with S-cone inputs, one often finds that the excitatory and inhibitory receptors cover about the same area of the retina. This gives a receptive field with coextensive excitatory and inhibitory fields, without spatial separation of center and surround mechanisms. Before we continue to explore the highly important consequences of the structure of receptive fields, we shall take a look at a much celebrated mathematical model of the receptive field.

A linear model of receptive fields

The spatial sensitivity of a classical receptive field, with a restricted spatial extent, can be modeled by the sum of two exponential functions, one activating the cell and the other inhibiting it. Let us assume that the activating sensitivity and response of a cell can be described by a Gaussian function $\varepsilon(x)$ of the distance x from its receptive field center:

$$\varepsilon(x) = Ae^{-\alpha x^2}$$

Let us further suppose that the inhibitory surround mechanism diminishes the sensitivity of the center by an amount $i(r)$

$$i(x) = Be^{-\beta x^2}$$

This results in overlapping excitatory and inhibitor mechanisms as shown in Figure 3.23. The constant B of maximum inhibition is smaller than the magnitude A

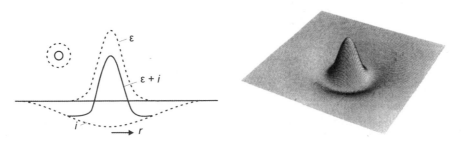

Figure 3.23   The spatial sensitivity of a receptive field modeled in two and three dimensions as the difference between two Gaussian functions.

of maximum activation ($B < A$). The result of the sum $q(x)$ of activation and inhibition is:

$$q(x) = \varepsilon(x) - i(x)$$

This function is shown by the dashed curve in Figure 3.23, with activation plotted along the positive y-axis and inhibition along the negative.

To describe the larger global surround, the above Gaussian function converges too fast to zero, and therefore another, more slowly converging, inhibitory function has been used:

$$I(x) = B/(ax^p + 1).$$

B gives the maximum inhibition, a determines how rapid the function falls off with distance, and the exponent p determines the 'weight' of the flanks. Smaller p means a slower convergence towards zero. For a value of p between 2 and 3, this function has given a good description of some experimental data where an outer surround affected the responses of LGN cells to stimulation of the classical receptive field (Valberg *et al.,* 1985a; 1991a; Wold, 1992).

Parvocellular, koniocellular and magnocellular cells

Since 1982 we know that there are two main types of ganglion cells, the *tonic parvocellular cells*, or PC cells, and the *phasic magnocellular cells*, or MC cells, in the primate retina (Kaplan and Shapley, 1982; Hicks *et al.*, 1983). In old world primates, the PC cells project to four parvocellular layers in the lateral geniculate nucleus (LGN, part of the thalamus; see Figures 3.24–3.26), and the MC cells send

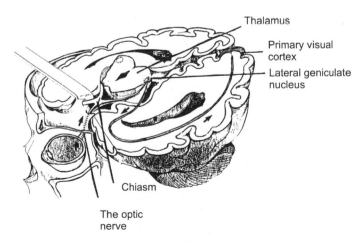

Figure 3.24 The pathway of information from one eye, via the optic chiasm, to the LGN in the thalamus and further to the primary visual cortex.

Figure 3.25   The lateral geniculate nucleus (LGN) of the macaque monkey (one for each brain half) contains several layers. Each of the upper four parvocellular layers contains small PC cells, and the two lower layers contain the larger MC cells. The koniocellular layers with KC cells are found between and below the PC and MC layers. Each layer is supplied by only one eye.

their signals to two magnocellular layers. The names refer to the magnocellular cell bodies being larger than those of parvocellular cells. These to cell classes form independent neural networks in the retina. As for I- and D-center cells (which are subgroups of PC and MC cells), the pathways from PC and MC cells are separated all the way through the geniculate to area V1 (former area 17) in the visual cortex (Figure 3.26). These are two of the many parallel and functionally independent pathways from the retina to the cortex. Signals from MC cells travel along nerve fibers with a speed of about 15 m/s, and in the somewhat thinner fibers of the PC cells the speed of signal transmission is about 6 m/s.

Tonic ganglion cells with excitatory S-cone inputs were previously integrated into the PC cell group, but newer data (Hendry and Yoshioka, 1994; Martin *et al.*, 1997) indicate that they belong to a separate *koniocellular pathway*. The retinal koniocellular cells, or *KC cells*, project to cells in interlaminar layers between and below the PC and MC cells in the geniculate (Figure 3.26).

Figure 3.27 displays examples of silhouettes of PC and MC cells in the macaque monkey (*Macaca fascicularis*). One sees that not only are MC cell bodies much

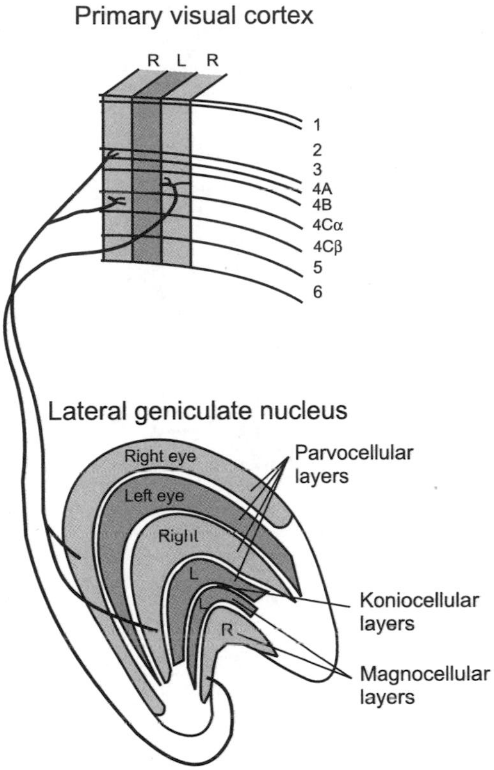

Primary visual cortex

Figure 3.26 Cells in the left half of the retina (activated by stimuli located in the right visual field) send their impulses to the LGN in the left half of the brain. The right half of the LGN represents the left visual field. MC and PC cells in the LGN project further to the layers $4C\alpha$ and $4C\beta$ in the primary visual cortex. KC cells project to layers 2/3 and to 4A. The layers of the LGN that are specific for each eye project to corresponding eye dominance columns in the cortex.

larger than for PC cells, but their dendritic tree is also much larger. The diameter of the dendritic tree of an MC cell is about three times that of a PC cell at the same eccentricity. Only about 10 percent of the ganglion cells in the retina are MC cells, whereas about 80 percent are PC and KC cells. The remaining 10 percent form no homogeneous group, and it is assumed that some may govern eye movements and others the pupilary reflex.

Cell counts performed by Perry *et al.* (1984) and Grünert *et al.* (1993) on macaques indicated that the ratio of the number of MC and PC cells is about the same over the whole retina (about 1:8). However, data from the human eye reported by Dacey (1993) estimate the percentage of PC cells in the peripheral retina to be about 45 percent, and that their number increases to make out about 95 percent in the central retina.

There are about 100 times more photoreceptors in the retina than there are ganglion cells and nerve fibers. On average, therefore, signals from many rods and cones

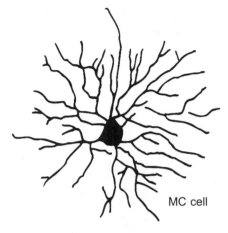

Figure 3.27   A retinal MC cell (also called a parasol cell) and a PC cell (also called a midget cell) with their dentritic trees. These cells were located at about the same distance from the fovea. The area of the dendritic tree of an MC cell is about 10 times that of a PC cell. The examples are from the macaque monkey.

converge onto one ganglion cell. However, in the central fovea – the fovea centralis – a bipolar cell and a ganglion cell have contact with one cone in their receptive field center (the surround receiving inputs from several photoreceptors). Anatomical studies (Kolb and Lipetz, 1991) have indicated that in the fovea there is a slight divergence in that there are two bipolars and two ganglion cells connected to each cone. This probably means that an I-center cell and a D-center cell share the same cone (see also below).

Figure 3.28 shows the results from measurements of the size of the dendritic trees of human MC and PC cells as a function of their distance from the fovea. The two cell types clearly form two distinct groups, with no overlap between them. In this respect, there seems to be a clearer distinction between the two cell types in humans than in monkeys. Similar measurements in the monkey retina have shown some degree of overlap. Since PC cells outnumber MC cells by about the same factor as the ratio of MC to PC dendritic field area, retinal coverage (the area covered by all their receptive fields) is about the same for both networks.

A ganglion cell, like a bipolar, is very sensitive to contrasts and less so for the absolute light level. An MC cell responds best to contrast, i.e. to a difference in luminance of center and surround of its receptive field, for instance when a contrast edge is moving across its receptive field (see Figure 3.32). MC cells have transient

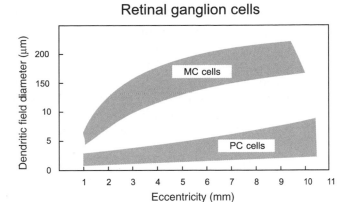

Figure 3.28 A representation of the sizes of the denritic fields of MC and PC cells in humans, as a function of their eccentricity in the retina. The dendritic fields of MC cells are distinctly larger than those of PC cells at all distances from the fovea.

responses to such a stimulus; they respond rapidly and only for a short period of time, with a particular high sensitivity to luminance contrast. The I- and D-types of MC cells have a spectral sensitivity that is very similar to the human luminous efficiency curve, V_λ (see Figure 4.18). The reason for this is that they respond to the summed inputs from M- and L-cones (Lee *et al.*, 1988).

The axons of MC cells are somewhat thicker than for PC cells, and as we have already mentioned, the signal velocity is higher. With simultaneous stimulation of MC and PC cells in the retina, the signals of MC cells will arrive about 7–10 ms before the others in visual cortex (V1).

Cone-opponent cells

MC cells are non-opponent in that they add the inputs of L- and M-cones. PC and KC cells are cone opponent, meaning that they receive excitatory as well as inhibitory cone inputs, but from different cone types. If one cone type, say the L-cones, excites the cell and causes a higher firing rate, the excitation of another cone type, say the M-cones, inhibits this activity. Such a cell is denoted a 'L–M' cell, and in the simplest case of an I-cell, L-receptors feed the excitatory receptive field center and the M-receptors the inhibitory receptive field surround. If it is the other way around, with M-cones having an inhibitory input to the center and L-cones exciting the surround, we have a D-center cell. Thus, one encounters both I- and D-center cells with an 'L-M' cone opponency. A cell activated by M-cones and inhibited by L-cones is called an 'M–L' cell. The abundance of 'L–M' and 'M–L' cells is about the same (in macaques). Since the L- and M-receptors are excited about equally by white light, the response of such opponent PC cells to achromatic light covering the whole receptive field is naturally weaker than for the preferred chromatic contrast, with the receptor type feeding the center being dominant.

The inputs of the receptor types that form the excitatory and inhibitory parts of an antagonistic receptive field can have unequal weights, usually with the center input being the stronger. An opponent I-center cell responding to light increments is less inhibited than a D-center cell with the same cone opponency. We shall consider some consequences of this fact later on when dealing with neural correlates of color vision. There are thus four types of PC cells with opponent L- and M-cone inputs; both 'L–M' and 'M–L' cells have I- and D-variants.

The S-cone pathways to the ganglion cells seem to be different from the PC pathways of midget cells. Valberg *et al.* (1986b) recorded from a substantial number of cells that could be modeled as opponent 'M–S' cells (yellow ON cells). Such monostratified cells have been identified only recently (Dacey and Packer, 2003), while bistratified blue ON/OFF-cells with inhibitory L- and M-cone bipolar inputs to the receptive field surround had been identified earlier (Dacey and Lee, 1996). Together, the four PC-types and the two classes of KC cells with excitatory and inhibitory S-cone inputs seem to constitute the majority of primate opponent retinal ganglion cells.

In recordings from the primate retina and LGN one sometimes encounters I- and D-cells with properties intermediate between those mentioned here. For instance, the opponency of a tonic PC cell can be so weak that it is virtually absent under normal adaptation, giving rise to what appears to be a non-opponent cell. It can either be an I-cell when the excitatory input dominates, or a D-cell when excitation is absent. Usually, selective chromatic adaptation reveals a hidden opponency in these cells. This makes one wonder if what we have here regarded as distinct classes are in reality the extreme ends of a continuous distribution. However, we shall see that the dendritic fields of cells of the same type do not overlap, and this together with the fact that they branch in different depth layers of the retina may be taken as evidence for the existence of distinct types.

Even though a model of opponent cells assuming a pure cone center and a pure cone surround fits the experimental data reasonably well (Reid and Shapley, 1992), mixed cone inputs to the receptive field surround of opponent cell have been reported. Such mixed surrounds could originate in horizontal cells that indiscriminately sum the activity of all receptor types in its vicinity. This may appear as a simple way of constructing the receptive field surround, and up until now findings of horizontal cells that contact only one cone type have not been reported. The H1 horizontal cells sum L- and M-cone inputs, whereas H2 make contact with all three cone types. When considering the recent discovery of a relative small receptive field size of H1 horizontal cells (Packer and Dacey, 2002), together with the uneven distribution and occasional clustering of cones shown in Figure 3.15, it seems that both pure and mixed surrounds are possible. Pure surrounds, consisting mainly of the same cone type that feeds the center, also seems theoretically possible, resulting in a non-opponent cell. It is conceivable that a gradation of opponency from non-opponent to opponent may exist, depending on the proportion of L- to M-cone inputs to the surround. It is, however, still not quite clear how and where the surround structure and the opponency originate. We will come back to these problems later.

From a functional point of view it would be of interest to know if rods converge to contact MC cells more than on PC cells, but we know more about how ganglion cells are connected to cones than to rods. Rods seem to project both to MC and PC cells, and electrophysiology seems to suggest that they have greater input to the MC cells than to opponent cells. For instance, Purpura *et al.* (1988) found that at low luminance MC cells had a higher contrast sensitivity than PC cells, and they accepted this as evidence that rod vision is mediated via MC cells. Lee *et al.* (1997) also found that MC cells are more strongly affected by rods than PC cells at low luminance.

Other functional aspects of tonic opponent PC and KC cells and of phasic non-opponent MC cells will be discussed in several places in this book.

Differences between PC, KC, and MC cells

PC cells

- 70–80 percent of all cells; small dendritic fields.

- Sustained responses.

- Relatively thin axons; conduction velocity about 6 m/s.

- Project to the four parvocellular layers in LGN.

- L- and M-cone-opponency ('L–M'; 'M–L'), excited by a narrow spectral band and inhibited by the rest of the spectrum.

- Increment- and decrement-center cells (I- and D-center cells).

- Strong and weak opponent inhibition.

- Respond well to the proper chromatic contrast.

- Respond well to achromatic contrasts for small objects.

- Inferior temporal resolution compared to MC cells.

KC cells

- Strong S-cone input.

- Large cell bodies.

- Response behaviour much as for PC cells.

- Project to the intralaminar koniocellular layers of the LGN.

- Most likely: increment, 'M–S' cells and 'S–L(+M)' increment/decrement cells.

- Relative large receptive field centers.

MC cells

- About 10% of all cells; large dendritic fields.

- Transient responses.

- Relatively thick axons; conduction velocity of about 15 m/s.

- Project to the two magnocellular layers of the LGN.

- Add inputs from L- and M-cones; have a V_λ-like spectral sensitivity.

- I- and D-cell subgroups.

- High sensitivity to achromatic contrasts.

- High temporal resolution.

- Support stereoscopic vision.

- Sensitive to stimulus movement.

- Respond to a contour between red and green for equiluminant stimuli.

Coding of contour and contrast

Activation and inhibition in I- and D-center cells

The structures of receptive fields are well suited for the detection of contrast, and let us therefore take a closer look at how this is done. Imagine the common situation where the luminance of an object is slightly different from its surroundings; the object may be slightly lighter than the surround (positive contrast), or slightly darker (negative contrast). If the object is so small that it is imaged on the center of the cell's receptive field, an I-center cell will be excited by a bright spot while a D-center cell will be inhibited by the same stimulus [see Figure 3.29(a)]. This inhibition manifests itself as reduction related to the maintained activity (which is the activity without stimulation). With a large patch of light covering both center and surround of the receptive field, as in Figure 3.29(b), or with diffuse illumination, the excitation of an I-center cell is less than in Figure 3.29(a). A D-cell is excited by a dark gray or black object projected on its receptive field center, but less so for a diffuse light decrement. Also, the inhibition of a D-center cell to a diffuse light increment

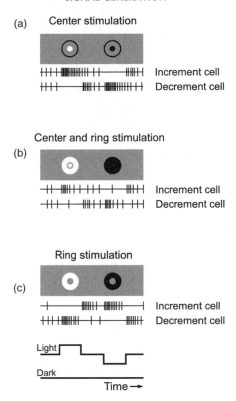

Figure 3.29 Schematic drawing of the responses of I- and D-types of PC cells for three different light stimuli. The stimulation is either an increasing or a decreasing intensity of a small spot of light projected on the center of the receptive field (a), a large field projected on both center and surround (b), or an annulus projected on the surround alone (c). I-cells are activated by a light increment in the center, while D-cells are activated by a light decrement in the center. This response is strongly diminished when the stimulus covers both center and surround. Stimulation of the surround alone gives a weak response with the opposite sign in I- and D-cells (after Schiller, 1992).

[Figure 3.29(b)] is not so strong as when stimulating the center alone [Figure 3.29(a)]. This is so because simultaneous stimulation of the center and the surround leads to the surround subtracting from the response of the center, but it seldom leads to complete annihilation of the center activity. A light increment located on the annulus inhibits I-cells and activates D-cells, whereas an annular decrement activates I-cells and inhibits the D-cells [Figure 3.29(c)].

I-ganglion cells are thus activated by positive contrasts (when the local luminance increases) and the nerve fibers transmit excitatory signals to the visual cortex. D-cells signal excitation when the luminance is decreasing. The strength of these signals depends on the size of the stimulus area and the magnitude of the contrast. The absolute luminance level is of lesser importance, provided that the luminance is in the photopic range (over a few cd/m^2).

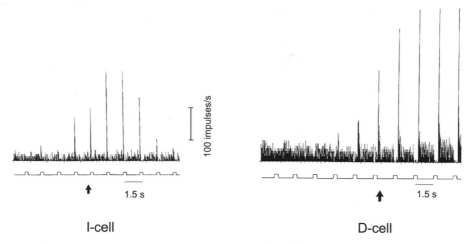

I-cell   D-cell

Figure 3.30 Examples of response histograms for two MC cells of the I-center type (a) and the D-center type (b). The stimulus intensity increases in equal logarithmic steps from left to right in the diagram. The length of each line in the histogram corresponds to the sum of nerve impulses within a time interval of 15 ms. Stimulus duration was 300 ms, with a period of 1.5 s. MC cells respond for only a short time interval of about 50 ms (from Valberg and Lee, 1989).

Figure 3.30 shows response histograms from a phasic magnocellular I-center and a D-center cell as a function of contrast. Stimulus is on for 300 ms and off for 1.2 s, as indicated by the traces below the histograms. Contrast increases in 0.3 log unit steps towards the right as in Figure 4.14(a). The length of each vertical thin bar represents the number of spikes within a time interval of 15 ms. The I-cell to the left responds for a short time to the onset of light, and it increases its response as a function of contrast up to a certain level and decreases thereafter. The D-cell is inhibited when the light increases and responds with a short burst of spikes for a short time after the offset of light.

Receptive fields of different types of ganglion cells overlap on the retina, and therefore it is possible that the same cones that excite an I-cell will inhibit a D-cell. This is possible, even if the receptor uses the same transmitter substance, because I- and D-center bipolar cells have different postsynaptic receptors. The same signal substance therefore has different effects on the two cell types.

There seems to be no overlap of the receptive fields and of the dendritic trees for PC cells of the same type (Dacey, 1993; and Figure 3.31). The dendrites of a PC cell end at the border of its neighbor's territory, and the result is a mosaic of receptive fields of cells of the same type. This applies for instance to 'L–M', I-cells. The other type of 'L–M' cells, the D-cells, have their own mosaic of receptive fields. The latter mosaic can overlap with that of I-cells, but not within itself.

Every cell can, like a receptor, forward its signal to many other cells (divergence), and theoretically it can be a member of many different functional units or networks. Such networks may, for example, contribute to the perception of color, to the polarity

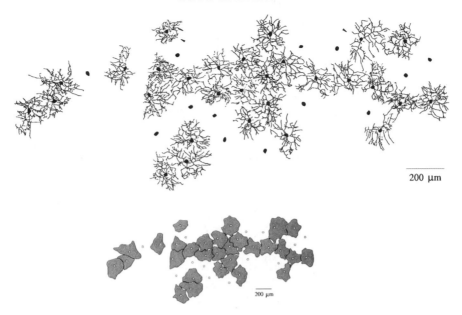

200 µm

200 µm

Figure 3.31   (a) A regular mosaic of retinal PC cells of the same type and their dendritic network, 12 mm from the fovea. The data are for humans. (b) Shading of the dendritic fields shows that neighboring fields do not overlap and that the mosaic has a covering factor close to 1. (Reproduced from Dacey, 1993, *J. Neurosci.* **13**, 5334–5355 by permission of Society for Neuroscience.)

of contrast, to the perception of the direction of movement, to depth perception, angular orientation, and so on. However, to maintain information derived by early functional units, their outputs need to be kept separated and forwarded in parallel to the higher brain centers.

Mach bands

In the cat, ON-center cells are activated when the center of the classical receptive field is illuminated and inhibited by light in the surround. OFF-center units respond in the opposite fashion by being inhibited by light in the center and activated by light in the surround. Such response patterns are common for the phasic Y-cells and the tonic X-cells of the cat.

Based on these findings, Günther Baumgartner and Richard Jung from the University of Freiburg in Germany developed a hypothesis that the excitation of the cat's ON-cells is correlated with the perception of brightness (B-system), whereas the response of OFF-center cells is correlated with the impression of darkness and blackness (D-system; Jung *et al.*, 1952).

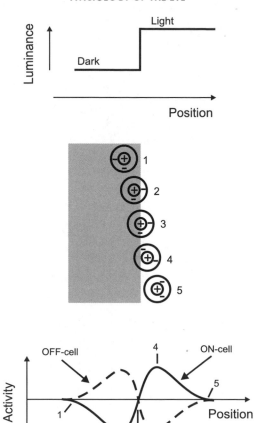

Figure 3.32 An example of neural enhancement of border contrast due to a luminance step. The contrast accentuation performed by a cell is a result of the organization of its receptive field into a center field and an antagonistic surround. The example shown here is from the cat where an ON-cell that lies in the shadow (1) has a certain firing rate. In situation (2) where the eye has moved a little, this cell is inhibited more than in (1) because some of its surround is now on the bright side of the luminance step. In position (3), the center is illuminated and the activity has increased with respect to (2). In (4), where some of the inhibitory surround still is on the dark side (inhibition is therefore not maximal), the activity has reached a maximum. In (5) the inhibitory surround is stimulated more than in (4) and activity is reduced. The same reasoning applies to the OFF-cell, only with the opposite sign. The description above applies equally well for a series of five neighboring cells in a stationary eye. If ON-cells provide information on lightness and OFF-cells on darkness, these two systems together will have twice the effect of one when encountering a border. This is the background for the theory of a 'B'-system for signaling 'brightness' and a 'D'-system for 'darkness'.

Figure 3.32 reproduces the responses of ON- and OFF-units when their receptive fields are in different positions relative to a contrast border between a white and a dark area. One can imagine that when an eye is moving to looks across the border, the neighboring receptive fields are moved over the border. However, it is also possible to regard Figure 3.32 as an image of what happens when the eye is stationary and we are looking at the responses of many cells with adjacent receptive fields in different positions relative to the border. Their activity will depend on the position, as Figure 3.32 demonstrates. The two types of cells respond in opposite ways; while the responses of the ON-cells increase, those of the OFF-cells decrease. An ON-cell has a maximum firing rate when the entire center is on the bright side of the border, with a small portion of the inhibitory surround in the dark (position 4). In this case the inhibition by the surround is less and the response will be larger compared with when the whole surround is illuminated (position 5). The OFF-cell has a maximum firing when a portion of the activating surround is positioned over the brighter side of the contour (position 2).

In this way Figure 3.32 illustrates a possible contour-enhancing mechanism, giving some of Jung and Baumgartner's arguments for the existence of B- and D-systems. When the B-system signals 'brighter', the D-system will signal 'less dark', and when the D-system signals 'darker', the B-system will signal 'less bright', thus both pulling in the same direction. However, without additional processing, this hypothesis cannot account for the perception of homogeneous gray, white and black in extended fields. A physiological enhancement of contours (contrast gain) like that of Figure 3.32 needs not be related to lightness and contrast inside an extended area, unless supplemented by some 'filling in' mechanism that accounts for area contrast. Particularly for transient MC cells, which are the most sensitive to the luminance contrast of stimuli of low spatial frequency, it is necessary to look for such auxiliary correlates of the lightness (gray) of a homogeneous field, whereas sustained opponent cells already respond to colors within extended areas and less to borders (Figure 7.3).

The response patterns of the ON- and OFF-cells in Figure 3.32 amplify the response difference due to a luminance contrast, a behavior that correlates well with our experience of Mach bands [Figures 1.8 and 3.33(a)]. The German physiologist Ewald Hering was already postulating in the nineteenth century that the nervous system needed to have such contrast-enhancing mechanisms in order to compensate for the relative unsharp borders imaged on the retina. It is not clear which cell system is responsible for Mach bands in humans, nor is it known where in the visual pathway they arise, but the MC cell system may be as good a candidate as any other.

Figure 3.33 describes another contour-enhancing mechanism in the faceted eye of the horseshoe crab, *Limulus.*

Recurrent inhibition of the Limulus eye

A relative simple explanation of contour enhancement and contrast gain goes back to classical investigations of H.K. Hartline (1940) on the faceted eye of the horseshoe crab, *Limulus.* The process in primates is likely to be more complex than in the *Limulus,* but *Limulus* has long served as model of a simple biological solution to the problem of contour enhancement.

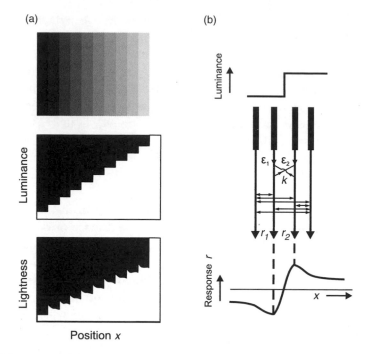

Position *x*

Figure 3.33 (a) Examples of Mach bands (contour accentuation) when luminance changes in steps like a staircase. Below is the spatial luminance distribution and how it appears to the viewer. (b) A neural mechanism in the eye of the horseshoe crab *Limulus* provides contour enhancement. Neighboring ommatidia (light sensors) interact through recurrent inhibition, with the strength *k* that depends on the distance between the ommatidia. The spatial response profile when each ommatidium is illuminated in turn is similar to the luminance profile on top of the figure. When all ommatidia are illuminated simultaneously with the same luminance step, the response of each one of them is distributed like that shown on the bottom of the figure. This is a result of all receptor units interacting mutually with each other and is thus a property of the neural network.

When one receptor (one ommatidium) is illuminanted in the *Limulus* eye, action potentials are elicited in the corresponding nerve fiber. There is no convergence of receptor signals, as in the human eye, and inhibitory influence is transmitted via lateral connections between the fibers. In Figure 3.33, ε_1 denotes the activity in impulses/s of fiber 1 when the corresponding ommatidium is illuminanted alone. ε_2 is the corresponding activity in fiber 2. r_1 and r_2 are the activities when both receptors are illuminated simultaneously. r_1 depends on ε_1 and r_2:

$$r_1 = \varepsilon_1 - i_{21}$$

where the inhibition, i_{21}, of receptor 1 from 2 is:

$$i_{21} = k(r_2 - r_{20})$$

This means that lateral inhibition is initiated only when the response r_2 has exceeded a threshold value, r_{20}. *k* is a constant of proportionality that gives the strength of inhibition. The expression is analogous for r_2. We simplify the equation above by setting the thresholds $r_{20} = r_{10} = 0$,

$$r_1 = \varepsilon_1 - k\, r_2$$
$$r_2 = \varepsilon_2 - k\, r_1$$

and find the following expression for r_1:

$$r_1 = \frac{\varepsilon_1 - k\,\varepsilon_2}{1 - k^2}$$

r_2 is obtained by interchanging the indices 1 and 2. The lateral inhibition, expressed by the magnitude of k in these equations, is largest for receptor units close to each other and decreases with the distance x between them; $k = k(x)$. An ommatidium is inhibited more the closer it is situated to an illuminated neighboring ommatidium, and the more the latter is illuminated. In this way, receptors in a dark, or in a weakly illuminated area, will be strongly inhibited by a bright area. The inhibition is stronger the closer they are to the border of an illuminated area (Figure 3.33).

The interaction has also a reverse effect in that the receptors at the bright side of the border give the highest firing rate near the contrast border. Figure 3.33 shows how this firing rate, r, in neighboring fibers depends on the geometry of illumination, $L(x)$, in the *Limulus* eye. This spatial distribution of activity gives rise to a contour sharpening that, to some degree, compensates for scattered light and unsharp imaging.

In Figure 3.33(b) the upper curve reproduces the response of each ommatidium illuminated in isolation with the luminance L_1 and L_2. The response, ε, is then proportional to the luminance. The bottom curve in Figure 3.33(b) shows the response of neighboring receptor units when they all are illuminated by the spatial luminance profile shown in the top curve.

Example 1

At a distance far from the luminance step, the response to a constant illuminance is:

$$\varepsilon_1 = \varepsilon_2 = \varepsilon$$
$$r_1 = r_2 = r = \varepsilon/(1 + k)$$

With $k = 0.5$, we get a response that is reduced by the same factor everywhere

$$r = 0.67\,\varepsilon$$

Example 2

In this case we introduce a contrast border (a luminance step) between receptor 1 and receptor 2:

$$\varepsilon_2 = \varepsilon_1 + \Delta\varepsilon$$

According to the response equations for receptors 1 and 2 above, the responses for receptors 1 and 2 then become:

$$r_1 = \frac{\varepsilon_1}{1+k} - \frac{k\Delta\varepsilon}{1-k^2}$$

$$r_2 = \frac{\varepsilon_1}{1+k} + \frac{\Delta\varepsilon}{1-k^2}$$

The difference in response across the contrast border then becomes:

$$\Delta r = r_2 - r_1 = \frac{\Delta\varepsilon}{1-k}$$

For $k = 0.5$ at the border, we now obtain:

$$\Delta r = 2\Delta\varepsilon$$

This gives a contour enhancement for receptors close to each other on either side of the border (they have the largest magnitude of the inhibitory constant k). The constant decreases with distance, and the response differences therefore also decrease the farther away the receptors are from the border (Figure 3.33).

Communication between nerve cells

We have already mentioned the different types of functional nerve cells in the retina, situated between the receptors and the ganglion cells, and in the following section we shall take a closer look at how they communicate. A nerve cell's reaction to a certain stimulation is influenced by other cells with which it exchanges information. This interaction occurs through synaptic transmission and subsequent potential changes in

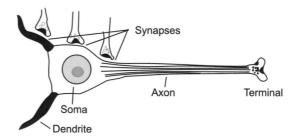

Figure 3.34 The figure shows the most important functional units of a nerve cell: dendrites, axon, terminal, nerve endings and synapses.

dendrites that are conducted toward the cell body (Figure 3.34). If these potentials are strong enough, in some cells they give rise to action potentials that propagate along the membrane of the nerve fiber. In addition, one also finds transport of material from the cell body to the terminal of the nerve fiber and back again. The arrival of electric nerve pulses in the presynaptic membrane of the terminal of a neuron leads to the release of molecules of the transmitter substance. These molecules are stored in a number of small vesicles in the terminal. The transmitter, for instance the chemical substance glutamate, is transmitted over *the synaptic cleft* to receptors on the dendrite or on the cell body of the next neuron. The synaptic cleft separates the *presynaptic membrane* from the *postsynaptic membrane* (see Figure 3.35). The effect of the transmitter substance is to either open or close ion channels in the postsynaptic membrane, thus eliciting an electric signal in the receiver cell that either amplifies (excites) or inhibits ongoing activity. The more vesicles that are emptied in the synaptic cleft, the stronger is this effect. Often several nerve fibers end on the same dendrite, and the cell can thus receive and add the electrical potentials that arise as a consequence of excitation and inhibition from many sources.

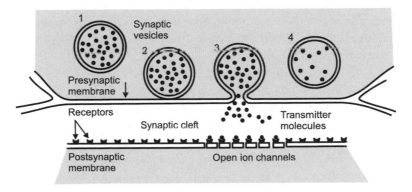

Figure 3.35 A chemical synapse where vesicles filled with transmitter molecules (e.g. glutamate) are emptied into the synaptic cleft. These molecules activate receptors in the postsynaptic membrane, which in turn open sodium channels for Na^+. Na^+ moving into the postsynaptic cell causes depolarization.

Sometimes it is possible from the shape of the vesicles to tell if they contain an excitatory or inhibitory transmitter. Round and oval vesicles usually contain an excitatory transmitter, and elongated, cigar-like vesicles contain transmitter molecules that hyperpolarize the membrane and inhibit activity. A vesicle can contain about 5000 molecules of a transmitter. After it is emptied, it is soon refilled again. Figure 3.35 displays schematically the transmission of transmitter in a synapse: (1) a vesicle adjacent to the presynaptic membrane establishes contact with it (2) fuses with it and empties its content (3). The molecules are taken up by the receptors on the postsynaptic membrane. The first vesicle separates from the presynaptic membrane (4) and is refilled with transmitter.

Signal transduction

Graded potentials and action potentials

We have already mentioned a few ways in which nerve cells transmit information. They can use two types of electrical signals, either *graded potentials* or *action potentials* ('spikes'). The graded postsynaptic potentials, or local field potentials as they are also called, are conducted passively and are used over short distances (1–2 mm). They arise in the synapses between cells that communicate exclusively by means of such potentials, including photoreceptors, horizontal cells and bipolar cells. As we shall see below, local dendritic field potentials are also important for spiking neurons. Graded potentials make it possible for the nerve cells to easily summate the influence (a) over time in one synapse (provided the signals follow each other within relatively short intervals), and (b) spatially from many synapses. In this way one obtains temporal and spatial integration in the dendrite. In a spiking cell, the resulting potential converges towards the axon hillock located at the beginning of the axon, which is the cell's trigger zone for action potentials. This is the integrating part of the spiking neuron, and whenever the potential is large enough to reach a certain threshold it triggers a nerve pulse.

The membrane in the dendrites and in the soma behaves like a leaky capacitor, and therefore the amplitude of the graded potentials elicited at the synapse decreases with distance from it. The electric resistance of the membrane and its capacitance determines the temporal course of the potential that reaches the cell's axon, and amplitude and form determine how many nerve pulses will be generated and transmitted away down the axon. The nerve pulses that travel down the axon towards its terminal do not decrease in amplitude with distance due to an intrinsic amplifying mechanism mentioned earlier (p. 100).

In many neurons, the signal receiving parts (dendrites, soma) are spatially separated from the place where nerve pulses are generated. Dendrites and soma have relatively low electric sensitivity, and this part of the membrane thus has a relatively large range of passive potential changes.

Chemical synapses and ion channels

A spiking neuron adjusts its firing rate according to the sum of inputs it gets from excitatory and inhibitory synapses. Both electric and chemical synapses exist. In the chemical synapses a signal substance diffuses across the synaptic cleft (as in Figure 3.35), affects the ionic channles of the post-synaptic membrane, and causes a transient change in the permeability for ions, mainly Na^+ and K^+. This is achieved through opening and closing of the ionic channels. Acetylcholine, glutamate and aspartate are activating transmitters, whereas dopamine, GABA (γ-aminobutyric acid) and glycine exhibit an inhibitory effect. The equilibrium potential of the spiking cell (the inside being held at about -65 to -70 mV) is altered by this process.

A steady equilibrium potential of $-70\,mV$ requires that a steady current of positive Na^+ ions (sodium) into the cell (from a high to a low concentration) is balanced by a steady current of positive charges, the K^+ ions (potassium), out of the cell, also from high to low concentration. An active ion pump controls the process and provides for ions inside and outside the cell. The pump maintains the ion concentration on both sides of the membrane in that Na^+ is pumped back out of the cell and K^+ is pumped into the cell.

In excitatory synapses, the transmitter selectively increases the permeability for Na^+, for instance, so that more Na^+ ions can flow into the cell and give rise to depolarization, for example a decrease in the negative membrane potential. The probability of an ion channel opening depends on the concentration of transmitter at the receptor site and not on the membrane potential. The channels are opened by the transmitter and not by the potential difference. The depolarization of the postsynaptic cell is proportional to the amount of transmitter that is carried over without eliciting an action potential. The amount of transmitter is, however, not exactly proportional to the number of incoming nerve pulses to the presynaptic membrane. Therefore there is not an exact proportionality between incoming firing rate in impulses/s from the presynaptic cell and the degree of depolarization of the postsynaptic potential (PSP). The amount of transmitter released by each nerve pulse is different for high and low firing rates.

As mentioned earlier, the passive electrical cable properties of a neuron is the physical basis for its spatial and temporal integration of local PSPs. When the sum of excitatory PSPs, called EPSPs, has changed the membrane potential at the beginning of the axon from being $-70\,mV$ in equilibrium, to about $-40\,mV$, the threshold is reached for triggering a nerve pulse in the nerve fiber (Figure 3.36). Inhibitory PSPs,

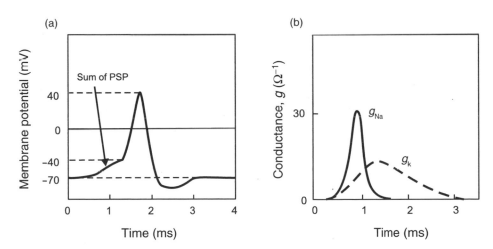

Figure 3.36 An action potential results when a sufficient number of EPSP builds up in a cell. The action potential is initiated when the cell is depolarized to the threshold value of about $-40\,mV$, after which the current of Na^+ ions into the cell accelerates and the potential increases rapidly. After some time this rapid depolarization is retarded by K^+ ions that move out of the cell. When the Na^+ current has started to decrease, the K^+ current is still rising, eventually bringing the action potential back to its resting level of $-70\,mV$ or a little less. The whole process takes less than 2 ms.

IPSPs, are also summated passively and lead to closure of sodium channels and thus to hyperpolarization of the cell membrane (K^+ ions move out of the cell and make the inside more negative). The resulting potential that triggers a nerve impulse is the result of many summed EPSPs and IPSPs from the whole cell, from its dendrites and its soma.

The model of Hodgkin and Huxley

In several classical electrophysiological experiments between 1945 and 1952, Alan Hodgkin and Andrew Huxley (Nobel Prize winners in 1963) demonstrated that the occurrence and the conduction of a nerve impulse – an action potential – is related to a time-dependent variation in the permeability of the membrane to Na^+ and K^+. When several EPSPs converge and are summated in the trigger zone, the permeability to Na^+ is increased so that there is a current of many Na^+ ions into the fiber, which becomes depolarized. Upon reaching a threshold of about -40 mV, the depolarization immediately causes other, potential-dependent Na^+ channels to open. The result is an accelerating and self-amplifying process, with a stronger current of Na^+ ions into the fiber resulting in an even stronger depolarization and the opening of more Na^+ channels. The opening of K^+ channels (the increasing of K^+ permeability) occurs much slower but will after a while counteract the depolarization (since K^+ diffuses out of the cell).

Figure 3.36(a) illustrates the temporal change of an action potential, and Figure 3.36(b) shows how the conductivity (the permeability), g_{Na} and g_K, for Na^+ and K^+ changes as a function of time. The action potential moves along the nerve fiber with a velocity of between 5 and 25 m/s, depending on axon thickness and cell type. The duration of a single action potential – a spike – is about 1 ms, and it takes about 1 ms before the next spike is generated (*refractory period*). This means that a cell can maximally generate about 500 impulses per second.

The processes described above lead to an action potential that travels like a wave along the axon. In myelinated nerves, where the nerve fiber is covered with electrically isolating myelin with regular gaps in between ('the nodes of Ranvier'), the impulse will 'jump' from one gap to the next and maintain its amplitude. The transmission of information from one cell to the next does not happen by means of changes in the amplitude or the form of the action potentials, but through a modulation of the cell's firing rate, i.e. in the number of impulses/s that travels towards the axon's terminal. A particular cell's firing rate in impulses/s is the result of a sum of excitatory and inhibitory synaptic processes, and this is again dependent on the type of stimulus and its strength. In the next chapter we shall return to the functional aspect and see what consequences such processes and the properties of receptive fields might have for the coding of contrast and contour.

Patch clamp

With the patch clamp technique it is possible to measure the current through a single ion channel. This technique was developed by Erwin Neher and Bert Sackmann at the Max-Planck Institute for Biophysical Chemistry in Göttingen. Neher and Sackmann received the Nobel Prize in Medicine and Physiology in 1991 for inventing the method and for their use of it.

A glass pipette with a small opening and filled with physiological salt water and a small concentration of a transmitter substance is pressed gently against the membrane of a cell. With a somewhat smaller pressure applied to the inside of the pipette, it will stick to the membrane and thus significantly reduce the electrical noise. A metal electrode in contact with the salt solution will conduct the current caused by the opening of ion channels under the tip of the pipette. The currents, I, that are measured from a single channel, lies in the range of 2–3 pA ($1\ pA = 10^{-12}\ A$), depending on the membrane potential. With a membrane potential, V, of -70 mV, this means that the resistance, R, is about $3 \times 10^{10}\ \Omega$ ($R = V/I$). The opening of one channel results in a depolarization of about 0.3 μV, and a depolarization of 30 mV (needed to reach the threshold for an action potential) therefore requires that about 100 000 channels are opened at the same time ($30 \times 10^{-3}/0.3 \times 10^{-6}$).

4 Sensitivity and response

Psychophysical sensitivity

The dark-adapted retina is very sensitive to light, and a visual sensation can be provoked by intensities that correspond to only a few rods absorbing one light quantum each. The sensitivity, s, of a measuring device (e.g. an ammeter) to external influence is the ratio between the instrument's response, R, and the magnitude, I, of the physical stimulus that provokes this response:

$$s = R/I$$

In this example from physics, the readings of the instrument depend on the voltage applied and the resistance of the circuit. For a certain voltage, a small resistance gives a larger reading than a large resistance. In an extreme case the reading of a particular ammeter may be so low that it is difficult to measure any current at all, in which case the sensitivity of the meter is low. On the other hand, if the meter's sensitivity is very high, it will have difficulty differentiating between large currents.

In acoustics, the response, R, can represent a subjective impression of the loudness of a sound, and the stimulus magnitude, I, would typically be the physically measured sound pressure in air. In photometry, which is the art of light measurement, the sensitivity, s, may be the ratio between a subjective impression of brightness and the physical radiance of the light measured in W/m^2. However, subjective impressions are always qualitative and difficult to quantify. It is easier to measure the magnitude of the physical stimulus that gives rise to the smallest subjective impression, i.e. to measure the physical energy or power of a stimulus that is barely detected by the individual. This stimulus magnitude is called a *threshold*. At threshold, R can be given the constant value 1.0. In a comparison of stimuli of different wavelengths, for instance the physical stimulus magnitude, I, is varied for each wavelength until the

Light Vision Color. Arne Valberg
© 2005 John Wiley & Sons Ltd

subjective detection threshold magnitude is reached. At this threshold, all intensities are regarded as being equally strong for the subject.

This psychophysical sensitivity is influenced by many conditions, relating specifically to the test subject, such as whether the person is awake and alert, whether adapted to the situation, interested in the task and concentrated, or whether easily distracted or tired. Even after taking all such factors into account, psychophysical measurements of thresholds display a variable degree of uncertainty for each individual as well as differences between individuals. Both types of data tend to have a Gaussian distribution.

Absolute visual sensitivity is given as the inverse energy, or the smallest number of photons that gives a light impression. This requires that the person be completely dark-adapted (see below) and that stimulation is applied to that part of the retina which has the highest absolute sensitivity.

It is also valuable to compare stimuli directly with each other with respect to their effectiveness in evoking a visual impression, with or without a reference stimulus. In Figure 4.1 the left half, A, is a reference light with a certain wavelength λ_0, and an

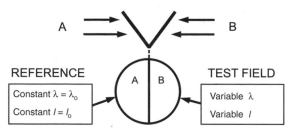

Figure 4.1   A bipartite photometric field can be used to compare a stimulus B with a reference stimulus A. I denotes intensity and λ wavelength.

intensity I_0. In the right half field, B the wavelength is changed from λ_0 to λ, and the intensity, I, is varied until field B looks as bright as field A. Making such comparisons for wavelengths in the entire spectrum, relative to the same reference (λ_0, I_0), leads to the derivation of the eye's *relative sensitivity* to spectral lights. The method can, for instance, be used to determine the effectiveness of every wavelength in the spectrum in evoking the same impression of brightness in the dark-adapted eye. This procedure is used to determine a spectral luminous efficiency function, V'_λ, of humans in the dark, which has a maximum sensitivity at 507 nm (Figure 4.2). When the eye is dark-adapted we see no colors. In this condition it is therefore possible to achieve complete equality between fields A and B. The sensitivity of the light-adapted eye, however, needs to be determined in a different manner because color differences make it impossible to achieve equality between two wavelengths (see section on Photometry, p. 165).

The sensitivity to differences between two similar stimuli, for instance to increments or decrements in reflectance on a gray scale, is also an example of a visual

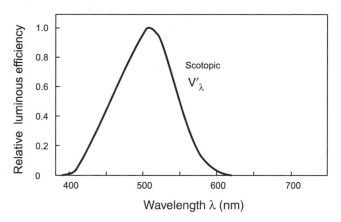

Figure 4.2 The scotopic luminous efficiency curve of the dark adapted human eye. Maximum sensitivity is at 507 nm.

threshold. Yet another example is the smallest wavelength difference, $\Delta\lambda$, that can be detected as a color difference at a specified spectral location. The experimental setup of Figure 4.1 can also be used for these tests. In the latter case, one starts with the wavelength λ_o in both fields A and B and after a complete color match has been achieved, the wavelength of field B is changed to $\lambda_o \pm \Delta\lambda$ until a color difference is barely detected. To avoid wavelength discrimination being contaminated by lightness differences, this experiment should be carried out with equiluminant stimuli, where isoluminance is determined in a separate experiment.

The results of such experiments tell us how good we are in discriminating between physical stimuli, i.e. it gives us charactersitics about ourselves as measuring devices. In the case of wavelength differences, Figure 4.3 shows that we discriminate the smallest difference in the yellow part of the spectrum, around 570 nm. Under optimal conditions, a person can distinguish different yellow nuances for wavelengths differing by a single nanometer.

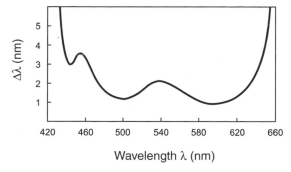

Figure 4.3 The ability to discriminate between colors in the spectrum. $\Delta\lambda$ is the smallest wavelength difference that can be discerned as a chromatic difference when luminance is constant.

Vision in daylight and in the dark

The retina can adjust its operating range to very different levels of illumination and lighting conditions. The notation *adaptation* is used both for the process of adjustment ('to adapt to a new condition') as well as for the end state ('to be adapted to a condition'). During the adaptation process, the visual system changes its sensitivity to light over time. Depending on the light level, we might talk about adaptation to light and *photopic vision*, or about adaptation to the dark and *scotopic vision*. After being adapted to a bright light, it may take about an hour or so to become completely dark-adapted, whereas the adaptation from darkness to daylight takes only a few seconds. The size of the pupil plays only a minor role in this process, which operates over more than 10 log units of light intensity.

An important feature of photopic vision is *chromatic adaptation*, a process of adjustment to the prevailing color. The visual system can compensate and neutralize chromatic changes in illumination such that the perceived color of a reflecting surface is fairly constant. White surfaces, for instance, look white even if the illumination changes from bluish daylight to yellowish incandescent light indoors (but try taking pictures under both conditions with the same roll of film). Owing to chromatic adaptation, the color of illuminated objects changes less than one would expect from the spectral distribution of their reflected light. This is called *color constancy*, and it serves a practical purpose in that it aids the recognition of objects by means of color under a wide variety of illuminations. The mechanisms behind the different adaptation processes are still not fully known (see the Chapter 5).

Adaptation to darkness

The time course of dark adaptation depends on the intensity and duration of the previous light adaptation. If you are starting from low light levels, complete dark adaptation is relatively fast, whereas a light-adapted eye needs more time to 'get used to' the dark. The rod-free fovea has only a very limited capacity for dark adaptation, and the foveal dark adaptation process as a function of time is characterized by a single monophasic curve. For stimuli that are large enough to also stimulate rods during central fixation, dark adaptation goes through an initial fast phase attributed to the cones. When the cone sensitivity has reached a maximum, or a little before that, threshold will be determined by the slower adapting rods (Figure 4.4).

In Figure 4.4, relative thresholds are plotted instead of relative sensitivity (the inverse of thresholds). Whether the physical parameter varied is adaptation luminance, retinal illuminance or quantum flux does not matter for the shape of these curves because for the same stimulus, these units are proportional to each other. In determining such adaptation curves, the experimenter changes the stimulus magnitude *I*, and the only task of the subject is to say *yes* or *no*, meaning *seen* or *not seen*.

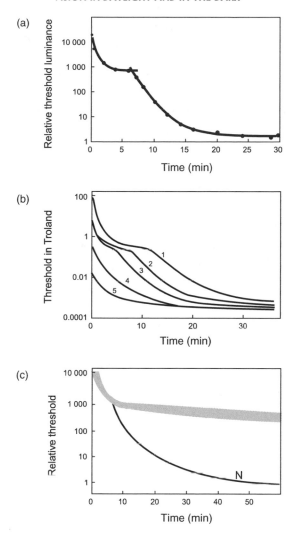

Figure 4.4 (a) The threshold luminance for the detection of light as a function of time in the dark. The first rapid phase, lasting between 5 and 10 min, is due to the dark adaptation of cones and is followed by dark adaptation of rods. The exact form of the curve depends on the level of light adaptation before the experiment started. (b) Dark adaptation curves after different levels of initial light adaptation. Curve 1 is for light adaptation to 400 000 td, curve 2 for adaptation to 40 000 td, 3 for 2000 td, 4 for 4000 td, and 5 for 260 td. (c) The shaded curve indicates dark adaptation curves for persons with no rods and who therefore suffer from night blindness. N is the normal curve.

The threshold criterion can for instance be defined as that intensity of the stimulus that results in *yes* more frequently than *no*, say in 55 or maybe 75 percent of all the presentations. The curve in Figure 4.4 (a), showing how thresholds depend on time in the dark, was obtained using short-wavelength, violet light. The light intensity at detection threshold was measured in terms of relative luminance. As time passes, the subject is able to detect spots of increasingly lower light intensity, and threshold decreases to reach a minimum value. Dark adaptation is rapid during the first few minutes, and in the first minutes [the branch to the very left of Figure 4.4(a)], the subject can see a violet hue at threshold. After a few minutes in total darkness the curve flattens out, and after another minute or so, the threshold again starts to fall rapidly with time. Now the chromatic impression has disapeared, and the light has no hue. The first phase can last for 5–10 min, depending on the foregoing light adaptation; this branch of the curve is ascribed to the activity of cones. After this initial phase, the rods take over as the most sensitive, threshold-determining elements. Achieving maximum sensitivity in the dark takes between 30 min and 1 h, depending on the level of light exposure before starting dark adaptation.

After exposure to strong light, a significant amount of the rod rhodopsin photo-pigment will be broken down (bleached). With a lower concentration of pigment, fewer light quanta will be absorbed, and the sensitivity will be reduced. This process of changing the concentration of rhodopsin in the rods is called *photochemical adaptation*. In normal daylight the rods are non-functioning, with sensitivity being low owing to significant bleaching; their output signals are not modulated by normal variations in retinal illuminance, and they cannot transmit information about contrasts. If most of the rhodopsin has been broken down by light exposure, it takes about 45 min to regenerate it fully and to obtain maximum sensitivity. When rods are shielded from light, retinal returns to the outer receptor segment from the pigment epithelium and contributes to the regeneration of rhodopsin.

Figure 4.4(b) shows the course of dark adaptation after preadaptation to different retinal illuminances, measured in photopic troland (td).

$$\text{troland} = A \times L \, (\text{td})$$

where troland is the unit for retinal illuminance, A is the area of the pupil in mm^2 and L is the luminance of the object in cd/m^2. One sees from the curves of Figure 4.4 that the cone branch disappears at about 0.1 photopic troland. This is also regarded as the absolute threshold value for the cones. Below this value cones are non-functioning. In addition to the slow adaptation of the photoreceptors, illustrated here over a span of about 4 log units of light intensity, there is also a fast neural adaptation that is important under normal lighting conditions. It will be described later.

The dark adaptation curve of a subject suffering from night blindness is shown as a hatched curve in Figure 4.4(c). When, as in this case, night blindness is caused by the absence of rods in the retina, sensitivity will depend on cones only. Anomalies in the dark adaptation curve may also result from vitamin A deficiency.

The time course of dark adaptation is different for young and old as illustrated in Figure 4.5. While older subjects are less sensitive than younger ones at all light levels, the difference is greater for dark-adapted eyes. In this particular case, using short-wavelength light, the difference between old and young can be between 1:100 and 1:1000.

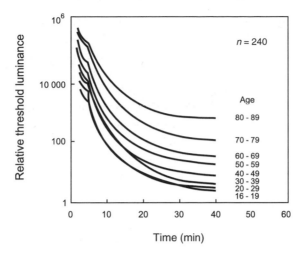

Figure 4.5 Dark adaptation curves for different age groups using short-wavelength light. The curves show relative threshold luminance as a function of time in the dark. Older persons are less sensitive to light than younger ones, and the difference is largest for the dark-adapted eye. A dark-adapted 20-year-old is about 200 times more sensitive than an 80-year-old (the latter needs about 200 times more light at threshold).

In curves of dark adaptation measured with long-wavelength red light there is no evidence of a rod-branch (see Figure 4.12). Consequently, using long-wavelength red light of moderate intensity in the dark-adapted eye allows for L-cone vision, without disrupting the state of dark adaptation for the rods. White light has about the same dark-adaptation curve as yellow light. The difference in sensitivity between the rods and cones is dependent on the state of dark adaptation and has been denoted the photochromatic interval (Lie, 1963).

Maximum sensitivity

Maximum absolute sensitivity for detection of light is found between 10 and 20° to the side of the fovea, where the density of rods is the highest. The minimum energy that gives rise to light detection in the dark for a young eye is close to the physical limit, namely the simultaneous absorption of a single quantum of light by each of only a few rods. The lowest energy that can be recognized as light in a dark-adapted eye (at 507 nm, at which the rods are the most sensitive) was found to be between 3.3 and 6.6×10^{-17} J measured at the cornea (Pirenne, 1967). At 507 nm, the energy, $E = h\nu$, of a light quantum is 3.92×10^{-19} J. This means that between about 50 and 150 quanta reach the cornea at absolute threshold. Let us assume that (i) about 50 percent of the light is absorbed in the eye media before it reaches the retina (see Figure 3.2), and that (ii) so many of the incoming quanta are lost between the rods that only 20 percent of the quanta reaching the receptor level are absorbed. This means that only between 5 and 15 quanta are actually absorbed at absolute threshold. Early data and statistical methods, based on the comparison of experimentally determined 'frequency of seeing' curves and summated Poisson distributions, gave five to seven quanta as the best estimate (Hecht *et al.*, 1942). With so few quanta absorbed, the likelihood of the same rod being hit by two quanta simultaneously is vanishingly small. Therefore, we must assume that one quantum absorbed in each of between 5 and 15 rods is a sufficient condition for light detection. This is said to compare to perceiving the light of a candle in absolute darkness at a distance of 20–25 km.

Light adaptation and contrast

In practical life it is more interesting to know a subject's sensitivity to contrast at daylight levels than the absolute threshold intensity. Contrast sensitivity can be determined with stimuli arrangements like those in Figure 4.6 for measuring

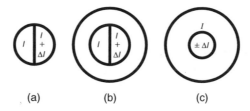

    (a)   (b)   (c)

Figure 4.6 Different configurations of stimuli used to measure a difference threshold, ΔI. Thresholds can be measured relative to a comparison field with or without a surround (a, b), or relative to a background of intensity I (c). The ratio $\Delta I/I$ is called the Weber ratio or the Weber fraction.

difference thresholds. One can, for instance, measure the smallest detectable intensity change, ΔI, in a field of intensity I, or one can measure the smallest detectable intensity difference between a reference field of intensity I and a test field of intensity $I \pm \Delta I$. Thresholds determined relative to a light background (Figure 4.6(c)), give relative sensitivities resembling those of Figure 4.7. In this case one measures the *increment threshold*, $\Delta L = L - L_B$, relative to a steady background with the luminance L_B. The ratio $\Delta L/L_B$ is called the '*Weber ratio*' (after the German physiologist E. H. Weber, 1795–1878). The index 'B' for background is often omitted, and the ratio is simply written $\Delta L/L$. The inverse ratio, $L/\Delta L$, is taken as a measure of the relative sensitivity, i.e. the increment sensitivity for a particular background of luminance L.

    Figure 4.7 demonstrates that the sensitivity for discriminating small light increments on a background increases as the background luminance, L, increases up to

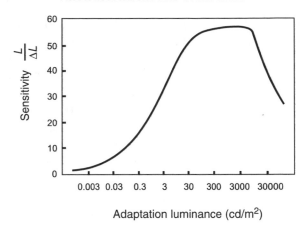

Adaptation luminance (cd/m²)

Figure 4.7 Threshold sensitivity (the inverse Weber ratio) plotted as a function of background luminance L. Increasing adaptation luminance leads to a smaller Weber fraction and a higher sensitivity, up to about 50 cd/m². For a higher luminance, up to about 10 000 cd/m², the Weber ratio is constant.

about 50 cd/m². As luminance increases up to this level, we are able to distinguish weaker and weaker contrasts. For higher luminance levels, the Weber ratio at threshold stays constant for a relatively large range of luminance. It is in this range, between 50 and 10 000 cd/m², that we are able to distinguish the smallest contrasts. Here Weber's law holds that:

$$\Delta L / L_B = \text{constant}.$$

Weber's law says that the Weber ratio is constant at threshold, and we see from Figure 4.7 that the value of this constant is about 0.02. This means that ΔL is proportional to L over a large range of daylight luminance levels. The value of 2 percent contrast refers to a particular experiment and a medium test field size. We shall later take a closer look at how contrast sensitivity depends on stimulus size.

Weber's law behavior is assumed in many sensory domains. It was originally a psychophysical finding, and only in recent years has it been possible to relate it to electrophysiological measurements of receptor responses. In a later section, 'Adaptation of Cones' (p. 160), we shall see that the responses of cones give a physiological foundation for Weber law behavior in photopic vision.

Purkinje's phenomenon

Purkinje's phenomenon illustrates the gradual transition from cone to rod vision as the eye becomes dark adapted. Let the fields A and B in Figure 4.1 be two isoluminant monochromatic lights viewed in a dark room, with λ_A being about

450 nm and λ_B being 580 nm. To begin with, when the intensities of fields A and B are fairly high, A will look bluish and B yellowish. We then decrease the intensities in A and B equally until both fields are so dark that they have lost their color. During this process A will gradually look brighter than B. At really low luminance levels, field B will disappear while A is still grayish. To restore the subjective impression of equal brightness, the intensity in B must be increased about 100-fold.

In nature this phenomenon can be observed at dusk. As darkness comes at the end of the day, the eye becomes increasingly sensitive to short-wavelength light and less sensitive to long-wavelength red and orange light. If initially, in daylight, a blue-green flower has the same brightness as a red flower, the blue-green flower will become brighter than the red flower as darkness sets in. This phenomenon was first described by Purkinje (1823), whose name has since been associated with this perceptual shift. The phenomenon can easily be explained by the shift from cone vision in daylight to rod vision at night. The two different spectral sensitivity curves are displayed in Figure 4.8. The scotopic spectral sensitivity curve V_λ' has the same form as the spectral sensitivity curve of the rods, with a maximum sensitivity at 507 nm. The photopic luminous efficiency curve V_λ is a weighted sum of the M_λ and L_λ sensitivities of the M- and L-cones.

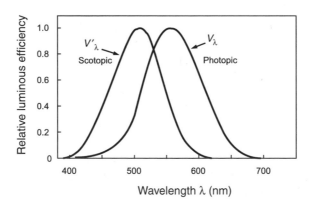

Figure 4.8   Photopic and scotopic spectral luminous efficiency curves for daylight and night vision. The photopic curve, with maximum at 555 nm, is the basis of photometric measurements of light.

Linear and nonlinear response

In a linear detector there is proportionality between stimulus magnitude, I, and response magnitude, $R = s \times I$. Irrespective of proportionality factor, s, if the stimulus strength doubles, the response magnitude will also double.

While there is no such proportionality between light intensity and response of a photoreceptor (its hyperpolarization), the first step in this process, the absorption of light in the receptor's pigment, is typically a linear function of the intensity. In this first step, light absorption transfers energy from light quanta to the receptor's pigment, resulting in receptor excitation. After a cascade of intermediate steps, this eventually leads to hyperpolarization of the photoreceptor. When the retina is illuminated by light of a wavelength λ, the receptor's excitation, S, is the product of the receptor's sensitivity, s_λ, for light of this wavelength and the light intensity, I_λ:

$$S = s_\lambda \times I_\lambda$$

For broad-band light this multiplication must be performed for every wavelength (or for small wavelength intervals $\Delta\lambda$) and summed over the spectrum:

$$S = \Sigma s_\lambda I_\lambda \Delta\lambda$$

For a linear detector, the excitation, S, can be directly replaced by the response $R = s \times I$. As above, R means output or response, and I is a general symbol that denotes the input (the stimulus magnitude). For the photocell used in a lux-meter to measure illuminance, R is actually the current in milliamperes, but the lux-meter is wavelength-calibrated so that the reading is illuminance, with the unit lux. I is the radiant flux per unit area (irradiance in $\mu W/m^2$), and s is the constant of proportionality between output and input. The photocell is linear within certain limits, and for the same stimulus (same spectral distribution) the readout is therefore proportional to the power of the incoming radiant flux.

The receptor potential, V, that is eventually generated by the excitation (light absorption), S, is not proportional to S (or I). An increase in light intensity, and the resulting proportional increase in excitation, leads to a change in the receptor potential that depends on the degree of polarization prior to the new excitation. The relationship between light intensity and receptor potential is approximately logarithmic over a large intensity range. This nonlinearity between stimulus intensity and the polarization response is carried over to the next cells in the visual pathway. However, the relationship between the magnitude of the receptor potential and the firing rate of a ganglion cell may be approximately linear.

In a limited sense the eye can be viewed as a transducer of radiant energy, or a detector, like that in Figure 4.9. In cases where the output is nonlinearly related to the input, it is useful to apply an expression of *differential sensitivity*, or *gain*, that is valid within a small intensity interval, ΔI, and a response interval, ΔR:

$$s = \Delta R / \Delta I$$

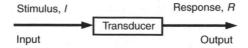

Figure 4.9 A transducer is a light-sensitive element, such as a photoreceptor in the eye and the detector in a light-measuring instrument that transforms the light input, I, to a different output. For instance, electromagnetic radiation of a certain power in Watts may be transformed to an output response, R, measured in Volts.

This expression for sensitivity (or gain) is more general than that for the sensitivity of a linear system, defined earlier. In a linear system the two definitions give the same result. The difference between the two definitions is made clear in Figure 4.10. For responses that are nonlinear, it is necessary to choose a fixed response criterion, i.e. a constant value of ΔR, when calculating gain. One sees from Figure 4.10 that

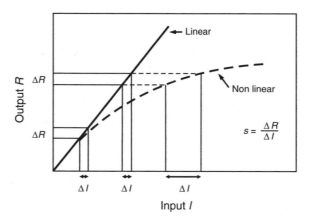

Figure 4.10 The relationship between an input, I, and an output, R, in a linear and a nonlinear system. In both cases, sensitivity (or gain) can be defined as $s = \Delta R / \Delta I$.

there is little difference between a linear and a nonlinear *intensity–response curve (I–R curve)* for small intensities to the left of the figure. Therefore, receptors and other cells can be treated as quasi-linear systems for low intensities, close to threshold.

Figure 4.11 shows schematically a receptor's polarization as a function of light intensity. In Figure 4.11(a) the dashed *I–R* curve of Figure 4.10 is drawn for stimuli of different wavelengths (but here on a logarithmic *x*-axis). This leads to *I–R* curves for different wavelengths that are displaced parallel to each other on the abscissa. Because they are from the same receptor, all the curves look the same. The parallel shift reflects the receptor's *spectral sensitivity*, or the effectiveness of the different wavelengths in exciting it. The fact that all wavelengths can give the same response in

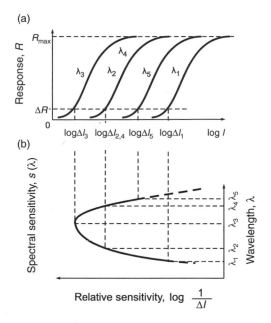

Figure 4.11 Determining the relative spectral sensitivity (b) from intensity–response curves (*I–R* curves) of an arbitrary nonlinear detector (a). Relative sensitivity is defined as $s = \Delta R / \Delta I$, where a constant response magnitude, ΔR, is set to 1.0 for all wavelengths.

a receptor only by adjusting the intensity of the light is a direct consequence of the *univariance principle* (see p. 160).

If we want to determine the receptor's *relative* spectral sensitivity [as in Figure 4.11(b)] from the *I–R* curves of Figure 4.11(a), the magnitude of ΔR does not matter. Using the general definition of sensitivity above, we can determine the relative spectral sensitivity, s_λ, as demonstrated in Figure 4.11(b): The relative sensitivity is inversely proportional to the threshold intensity, ΔI, that causes a certain constant change, ΔR (the threshold criterion) in the response.

Spectral sensitivity

Energy-based sensitivity

Until now we have treated the concept of sensitivity in a general way, and the symbol I has been used for the unspecified intensity of an arbitrary physical stimulus. More specifically, for the eye, the spectral sensitivity can be expressed as a function of wavelength as:

$$s_e(\lambda) = \Delta R / \Delta \Phi$$

where $\Delta\Phi = \Phi_\lambda\Delta\lambda$ (W). Φ is here the radiant flux in Watts (W), and the index 'e' indicates that energy units are used rather than quantum units.

In a psychophysical experiment, the threshold response, ΔR, in the expression above is held constant, and the radiant flux, $\Delta\Phi$, is changed for each wavelength until the threshold response is reached. The criterion for ΔR may be a 55 percent detection rate for a stimulus presentation, or it might be a judgement of 'equally bright' to a comparison field, as in Figure 4.1. The spectral sensitivity is then inversely proportional to the threshold power $\Delta\Phi = \Phi_\lambda\Delta\lambda$ that gives the constant criterion response ΔR. The scotopic luminous spectral sensitivity, V'_λ, of the human dark-adapted eye was determined from psychophysical experiments such as this.

Quantum-based sensitivity

The spectral absorption curve for rhodopsin, the pigment of the rods, tells us how large a fraction of the incident light of each wavelength is absorbed. Rod excitation corresponds to the number of rhodopsin molecules that are affected, which is proportional to the number of quanta that are absorbed, irrespective of wavelength. A similar reasoning applies to the cone pigments. It has been demonstrated that it is *the quantum-based spectral sensitivity derived in psychophysical experiments that corresponds to the spectral pigment absorption in rhodopsin*. This confirms that the sensitivity of the eye depends on the quantum absorption in the receptors, which can be estimated from the quantum flux at the cornea after correction for the absorption in the eye media (cornea, lens, vitreous, etc.).

The relationship between spectral sensitivity expressed in inverse energy units, $s_e(\lambda)$, and the sensitivity expressed in inverse quantum units, $s_q(\lambda)$, is the following:

$$s_q(\lambda) = hc\, s_e(\lambda)/\lambda$$

Here $s_q(\lambda) = \Delta R/\Delta N$, and $\Delta N = N_\lambda\Delta\lambda(s^{-1})$ is the number of quanta per second in the light stimulus. The derivation is made by the expression $E = h\nu = hc/\lambda$.

Since hc is constant, the relative quantum sensitivity $s_q(\lambda)$ will be proportional to $s_e(\lambda)/\lambda$. Compared with the energy sensitivity $s_e(\lambda)$, this means that the quantum-based spectral sensitivity is twice as large at 400 nm as at 800 nm. The maximum for the quantum-based spectral sensitivity will therefore be shifted towards shorter wavelengths relative to the energy-based sensitivity.

Action spectra of the cones

A spectral sensitivity curve that is derived by varying the intensity of light for each wavelength until the same physiological reaction (ΔR) occurs in the receptor is called an *'action spectrum'*. Action spectra are of great importance in vision science

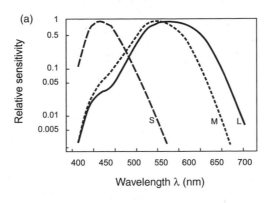

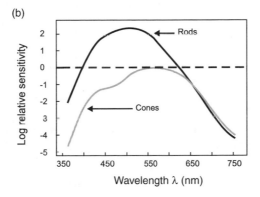

Figure 4.12   (a) Relative spectral energy sensitivity for the three cone types of the human retina. L denotes the long-wavelength sensitive cone, M the middle-wavelength sensitive cone, and S the short-wavelength sensitive cone (courtesy of J. H. Wold). (b) Energy-based spectral sensitivity, $1/\Delta E$, for the dark-adapted human eye. The heavy black curve represents the sensitivity of rods $8°$ above the fovea, and the thinner gray curve shows the sensitivity of the cones in the fovea. The curves are averages of the results for 22 subjects. The sensitivities are expressed relative to the maximum sensitivity in the fovea. The cone curve has a higher sensitivity than the rods for wavelengths longer than about 650 nm (modified from Pirenne, 1967).

because action spectra that are measured psychophysically can be compared with those measured electrophysiologically.

The three energy-based spectral sensitivities $L(\lambda)$, $M(\lambda)$ and $S(\lambda)$ of the three types of human cones are shown in Figure 4.12(a). These sensitivities can now be measured by different methods, psychophysically and electrophysiologically, and the correspondence between them is relatively good. The wavelength of maximum sensitivity for L-cones is about 560 nm, 530 nm for the M-cones, and 425 nm for the S-cones, although minor deviations from these values have been found and explored in recent years (Sharp and *et al.*, 1999; see the section 'Genetics of photopigments', p. 108).

In Figure 4.12(b), the spectral sensitivity of the rods is compared with that of L-cones in darkness. Surprisingly, L-cones have higher sensitivity than the rods above about 650 nm, although the maximum sensitivity for rods is about 100 times higher than for the L-cones.

As previously mentioned, the excitation of rods and cones is proportional to light absorption. Thus we find that the following expression holds for a receptor's excitation M:

$$M = c \int M(\lambda)\Phi(\lambda)\,\mathrm{d}\lambda$$

Here, $M(\lambda)$ is the relative energy-based spectral sensitivity of an M-receptor, $\Phi(\lambda)$ is the power of the radiation in W/nm, and c is a constant of proportionality. This is a general expression that is valid for all receptors. In other words, each receptor is excited like a simple photocell, or a light meter, with each receptor having its own spectral sensitivity.

Response

We have already mentioned some nonlinear aspects of intensity-response curves (I–R curves) of receptors (Figures 4.10 and 4.11), and now we shall see how such curves can be modeled. The receptor potential can be measured in millivolts as a function of the intensity of the incident light. The intensity can be given in a variety of units, in radiometric units of radiance and irradiance, in photomteric units of luminance or retinal illuminance of the stimulus imaged on the retina, or as the number of quanta. When plotted as a function of intensity as in Figure 4.10, the I–R curve has an initial linear portion, after which it starts saturating. On a logarithmic x-axis, I–R curves may look like those in Figure 4.13. These curves have a middle range that can be approximated by a straight line. Within the region described by this line the relationship between intensity and response is logarithmic, $R = a \log I/I_0 + b$, where I_0 is threshold intensity, a is a constant of proportionality (the slope of the line) and b is another constant giving the response for $I = I_0$.

In Figure 4.13 examples of modeled cone responses are shown for some wavelengths as a function of the logarithm of relative luminance. The family of curves has been derived from the relative spectral sensitivity and excitation of each receptor type, using the Naka–Rushton response equation below. The magnitude of the parallel shifts relates to the spectral sensitivity when luminance is used as intensity variable. One sees that the parallel shift from one wavelength to the next is much smaller for L- and M-cones than for an S-cone. The reason for this is that spectral luminance sensitivity follows the sum of the sensitivities of L- and M-cones, whereas S-cones do not contribute to luminance defined in this way. Taking the I–R curve for the M-cone as an example we have:

$$V/V_{\mathrm{max}} = M^n/(M^n + \sigma_{\mathrm{M}}^n)$$

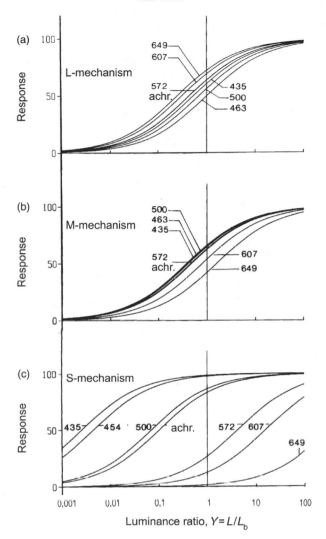

Figure 4.13 Theoretical luminance responses at different wavelengths for three human cone mechanisms, using the Naka–Rushton intensity–response relationship and the three spectral sensitivities of Figure 4.12(a). The stimuli are provided by narrow-band interference filters. All curves have been normalized to the same maximum value. Using relative luminance as the intensity variable only leads to a small separation of the curves along the *x*-axis for L- and M-cones, while the response curves of the S-cone mechanism are spread out over more than 5 log units.

M is here the excitation of an M-cone, calculated by the integral above. Corresponding expressions can be used for the other receptor types.

Experiments have shown that an exponent $n = 0.7$ is a good choice for primate cones. The response is normalized to the maximum value $V_{\max}$, and σ_M is a constant that gives the intensity value at half the maximum response. This constant will later be shown to play an important role in the description of adaptation.

Receptors and the univariance principle

The excitation of a photoreceptor is proportional to the number of photons, each with an energy, $E = h\nu$, that are absorbed. For each receptor, the same response can be obtained by all wavelengths (or frequencies) in the spectrum by merely changing the intensity of the light. For each of the pigment systems, every absorbed quantum of light will contribute equally to vision, and response magnitude is determined by how many quanta it absorbs with no regard to their wavelengths (Naka and Rushton, 1966). This is called the *principle of univariance*.

A consequence of the univariance principle is that light of one wavelength and intensity can be interchanged with light of another wavelength and intensity without this resulting in any change (modulation) of the receptor's response. The requirement is that the quantum catch is the same for the two wavelengths. This fact can be used to inactivate or neutralize the response of one receptor type in an experiment, provided that the spectral sensitivity of the receptor type is known. In recent years, this method, referred to as 'silent substitution', has found frequent application in vision science.

The accuracy of such experiments depends on our knowledge of the spectral distribution of the stimulus and the spectral sensitivity of the receptor type for the particular stimulus parameters used (size, eccentricity, etc.). Moreover, in recent years it has become evident that some receptor pigments, particularly of the L- and M-cones, are available in more than one form, each with a somewhat different spectral sensitivity. In addition, the spectral absorption and transmission of the eye media may be slightly different from person to person, thus modifying the spectral sensitivity of the receptors when the incident light is measured at the cornea. This may cause some problems for the silent substitution technique.

Adaptation of cones

The eye is able to adjust to light levels that span a range of about 12 log units (12 decades), shared more or less equally between rods and cones. Out of this vast range, the pupil regulates light intensity on the retina over only about one decade. For each wavelength, the *I–R* curve of a cone covers only about two to three decades (see Figure 4.13), corresponding to an intensity ratio higher than 100, but less than 1000. At most, ganglion cells can generate between 500 and 1000 impulses/s. Consequently, without some additional adaptation mechanism, none of these cells, can cope with and differentiate between stimuli over the whole adaptation range of the eye. The effect of this additional, and to a large extent unknown, mechanism is to shift the operating range of receptors and cells so that they become capable of responding optimally to stimuli of luminances that bracket the adaptation luminance. Later we shall take a closer look at the consequences of this intriguing adaptation shift.

Adaptation processes are very important in everyday vision, and although they have been the subject of quantitative psychophysical studies for a long time, it has been hard to identify the physiological mechanisms. Not so many years ago, it was assumed that bleaching of the pigments in the receptors played an important role in adaptation. However, under normal daylight conditions, photochemical bleaching of cone pigments is very low – less than 10 percent of the pigment in the cones is bleached – and it has become clear that the neural contribution to adaptation is by far the greatest.

Let us look at some of the factors that are important for the adaptation of cones and which regulate their excitation and response. The adaptation luminance is that which corresponds roughly to the midpoint of an *I–R* curve. For higher intensities, beyond the range where the logarithmic relationship between intensity and response applies [and where the *I–R* curve approximates a straight line in Figure 4.11(a)], an *I–R* curve displays a compressed response and flattens out. In this range, where the response to a stimulus luminance approaches saturation, a given small light increment will not give rise to a significant response increment. However, this response compression should not be regarded as adaptation in the same sense as the receptor's adjustment to a particular light level by shifting its operation range on the intensity axis.

Typically, an *I–R* curve illustrates the response (the receptor potential) of a receptor as a function of the intensity of a relatively small stimulus embedded in a surround which sets the adaptation condition. When the adaptation level changes, for example by changing the illumination of the visual scene, this leads to a displacement of the *I–R* curve towards the new adaptation level. In this way, the same contrasts will be reproduced about equally in the receptor's response, regardless of the illuminance level, and lightness constancy applies, i.e. the surface of an object will have the same lightness, independent of the level of illumination. One might well imagine that such a mechanism is also at play for chromatic adaptation and color constancy.

The main factors that determine a receptor's sensitivity and response are neural adaptation and pigment bleaching. Both of these processes lead to the receptor adjusting to the prevailing light level. In the model of receptor response given by the hyperbolic response equation, these processes are characterized by a change in the half-saturation constant σ. Such changes lead to a parallel shift of the *I–R* curve along the *x*-axis. In this way the *I–R* curve always spans about the same range of a little more than 1 log unit of negative and of positive contrasts on either side of the intensity of the adapting light.

Figure 4.7 illustrates that contrast sensitivity (threshold discrimination) improves as the adaptation luminance increases up to 50 cd/m^2, after which discrimination is fairly constant. This may be understood by examining the cone responses in Figure 4.15(b). For low adaptation values (leftmost curve) the response to a stimulus of a luminance corresponding to the adaptation luminance (given by the intensity at which the short horizontal bar crosses the curve) falls below the mid-point of the *I–R* curve. The mid-point is where the discrimination is best (it is the steepest part of the curve). This implies that the threshold for detecting contrast will be the smallest

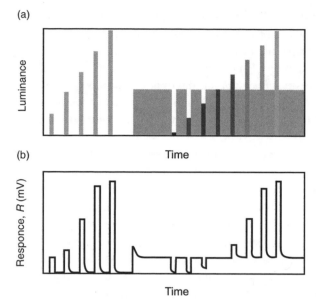

Figure 4.14   An adaptation experiment. (a) The stimulus intensities and temporal sequence for a stimulus series where a test-light of increasing luminance and short duration (marked in red) alternates with a coextensive fixed-luminance, adaptation field that is either dark (left) or white (right). (b) The responses of a cone to the stimulus series in (a) (modified from Valeton and Van Norren, 1983). See also Color Plate Section.

(and contrast sensitivity the highest) for a stimulus of the same luminance as the adapting light.

In an early, and now classical study of the adaptation of cones by Boynton and Whitten (1970), and in later investigations by Valeton and Van Norren (1983) in the Netherlands, the interplay between different kinds of adaptation mechanisms was illucidated. Below we shall summarize some of the salient points of Valeton and Van Norren's experiments. In this elegant experiment, the receptor potentials of macaque monkeys were measured by means of an extracellular microelectrode. The time course of the stimuli and the resulting potentials are shown schematically in Figure 4.14. In one experiment, the receptor potentials were measured as a function of the intensity of a disk 2.3° in size isolated in a dark surround (left-hand side of Figure 4.14). In another experiment, an adaptation background of the same size as the stimulus was introduced and the stimulus was modulated in luminance above and below that of this background (right-hand side of Figure 4.14). Examples of the recordings are shown in Figure 4.15. Figure 4.15(a) shows the polarization of a cone relative to the polarization caused by the adapting background. In Figure 4.15(b) the receptor potentials are plotted as a function of the intensity of the test field for several adapting background intensities numbered 2–6 in the figure (corresponding to the log intensity of the adapting backgrounds marked by short horizontal bars).

(a)

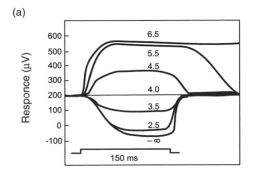

(b)

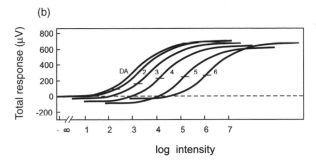

log intensity

Figure 4.15 (a) The relative response polarization of a cone for different relative stimulus intensities above and below the adaptation level of 10 000 td. Hyperpolarization has been given a positive sign. (b) The response of a cone as a function of stimulus intensity for different adaptation backgrounds. The adaptation to backgrounds of higher luminance level (the adaptation level being marked by a short horizontal bar on each curve) shifts the I–R curve towards higher intensities (towards the right in the diagram). The curve DA to the far left is for the dark-adapted eye, while curve 4 was obtained from the responses in (a) (redrawn from Valeton and Van Norren, 1983). Such adaptation shift of the intensity–response curve can provide the basis for lightness constancy.

Curve 4 shows the result of Figure 4.15(a), with a background of 10 000 td (4 log units). Assuming a 3 mm pupil diameter, 10 000 td gives about 1 400 cd/m². One interesting result of this experiment was that I–R curves had the same shape for every background adaptation luminance, provided that the response was plotted as a function of total intensity, i.e. as function of the sum of the background intensity, I_B, and the test field's intensity, I_T:

$$I = I_B + I_T$$

The position of the curves, however, depends on the luminance, I_B, of the background.

The experiment showed that all I–R curves had the same shape, independent of background luminance, and that, after normalization, they could be described by the same response equation:

$$V/V_m = I^n/(I^n + \sigma^n); \quad n = 0.74$$

where V_m is the maximum potential and

$$I = I_B + I_T.$$

The magnitude, σ, is an adaptation constant that reflects how much a curve is shifted along the log intensity axis for each background luminance. This adaptation factor is a product of several multiplicative adaptation processes:

$$\sigma = \sigma_{DA}\sigma_{\alpha}\sigma_{\beta}$$

σ_{DA} reflects the position of the I–R curves on a log-intensity axis in the dark-adapted state. σ_{β} is a factor that can be related to pigment bleaching. The more pigment is bleached, the greater the adaptation and the larger the value of σ_{β}. σ_{α} is a neural adaptation factor, the only unknown magnitude in the expression above. This factor can be quantified in the experiment by determining how much the experimental curves are shifted along the x-axis when I_B is changed. From such experiments one has drawn the conclusion that, for a retinal illuminance below 10 000 td, the neural adaptation, expressed by σ_{α}, is larger than the adaptation caused by pigment bleaching in the cones. Only with adapting background illuminance higher than 100 000 td does bleaching achieve the same importance as neural adaptation. In Figure 4.15(b) the response curves for different adaptation states are shown as a function of $\log I$. The actual value of the background intensity is indicated by the short horizontal line on each curve. For intensities greater than 4 log td in Figure 4.15(b), the I–R curves have about the same slope K in the middle intensity range for $\log I = \log I_B$, which means that the Weber ratio, dI/I, is constant in this region:

$$\text{slope } K = dV/d(\log I) = I\, dV/dI$$

The ratio, dV/dI, we recognize as the expression of differential sensitivity or gain. This means that the slope of a curve is proportional to differential sensitivity and inversely proportional to the Weber ratio. The steeper the curve is, the smaller the Weber ratio, and the higher the sensitivity for small changes in stimulus intensity.

The receptor potential elicited by the adapting background light (marked with short horizontal bars in Figure 4.15(b)) follows an I_B–R curve that is much shallower than those for the smaller, 2.3° stimulus. With an adapting surround added to the stimulus and background configuration of these experiments, one would expect a more effective adaptation and somewhat different positions of the curves on the x-axis. However, the relative shift with changing adaptation luminance would not be expected to differ much from Figure 4.15(b).

Photometry

The 'art of light measurement' is an old discipline. In 1729, Pierre Bouger described important principles of photometry for the first time (Bastie, 1999). Later Johann Henrick Lambert (1760) laid its theoretical foundation, and Joseph Fraunhofer (1814) was apparently the first to compare the perceived brightness of spectral lights in the solar spectrum. However, as von Helmholtz (1911, p. 162) emphasized, such brightness comparisons do not provide information about the objective intensity of lights. Therefore, he said, this procedure does not belong to physical photometry, but rather to physiological optics. One could say that objectivity was regained through the adoption in 1924 of the spectral luminous efficiency function $V(\lambda)$ of the International Comission on Illumination (CIE; Commission Internationale de l'Eclairage). $V(\lambda)$ represents the normalized spectral sensitivity of the human eye, and is the spectral weighting function for radiation aimed at in light measuring instruments.

The perception of lightness and brightness

At the end of the nineteenth century, an understanding of the term *'brightness'* was considered essential to photometry. However, in order to arrive at an appropriate photometric measure of light, which was the more important task for illumination engineering, this aim had to be abandoned. Studies of brightness were taken up by the CIE much later, and they resulted in $V_b(\lambda)$ curves describing the direct heterochromatic brightness matching for monochromatic stimuli (CIE, 1988).

In 1987, CIE defined the term 'brightness' as 'the attribute of visual sensation according to which an area appears to emit more or less light' (CIE, 1987) That brightness is a perceptual dimension of light is totally unproblematic at scotopic light levels where a stimulus is devoid of chromatic color. At photopic levels, direct brightness comparisons are relatively easy for light sources of nearly the same color, but the greater the color difference, the greater the problem (Ives, 1912). Some researchers have experienced such overwhelming difficulty in comparing the brightness of different color stimuli that they have, in despair, suggested that the task is tantamount to comparing sound with odor. Also von Helmholtz doubted his ability to equate different colors in terms of brightness, and wrote:

> Personally, I must repeatedly declare that I have hardly any confidence in a judgement of the equality of heterochrome brightnesses, only on what is greater and lower in extreme cases ... On my part, I have the definite sensory impression that in heterochromatic brightness comparisons, the concern is not comparison of a magnitude, but rather of a combination of two (magnitudes), brightness and color glow (Farbengluth), of which I do not know how to make a simple sum, and which I also cannot yet define scientifically. (von Helmholtz, 1896, pp. 439–441.)

A patch of much higher luminance than its surround looks like it is emitting light, and its appearance is associated with the so-called 'unrelated colors' of Willhelm Ostwald ('unbezogene Farben'; see below). A direct brightness comparison of such colored lights and of light sources is difficult, but it is even more so in the case of reflecting surfaces. The appearance of reflecting surfaces seen under daylight conditions is quite different from that of self-emitting lights. A reflecting field of lower luminance than its surround is normally perceived as the solid surface of an object, and its color belongs to the class of *related colors* ('bezogene Farben').

> The same color stimulus is perceived quite differently when presented in the object or in the aperture mode, i.e. as a reflecting surface (embedded in a brighter surrounding) or as a self-emitting light source. In the first mode it is said to be a related color (related to the lighter surroundings), and in the second it is an unrelated color, also called a void color. An example is the moon viewed against the dark sky at night, which has an unrelated color, or viewed against the bright blue sky in the day, having a related color. While unrelated colors appear self-luminous, the related colors appear to reflect more or less light. The term *fluorence* was coined by Evans and Swenholt (1967) to characterize the peculiar perceptual phenomenon of color glow that occurs for fluorescent materials and other colored surfaces in the transition zone between apparent reflection and self-luminosity.

The intensity dimension of reflecting surfaces has been given many different names. It has been characterized in terms of value (Munsell), lightness, Dunkelstufe (Deutsche Industrienormen, DIN) or white–black content (Hering, Natural Color System NCS), all entities that denote some relevant perceptual aspect. Reflectance factor and relative luminance are adequate physical specifications of a surface's luminance relative to a white surround, but the relationship of these physical magnitudes to the perceptual attribute of colors is often complex and not well understood.

One of the clearest and most provocative statements on this issue was made by Hering (1964, p. 31):

> The prevailing view according to which the 'intensity' of the corresponding 'light sensation' that is called the brightness or whiteness of the correlative area of the visual field must necessarily increase along with the increasing light intensity of an external object and its retinal image, belongs to those prejudices which are especially obstructive to an understanding of the facts of vision.

And he continues (p. 39):

> . . . a color (simply) as a quality has no magnitude or intensity, and only its difference from another color can be quantitatively treated.

Following Hering's recommendation, the NCS, unlike the Munsell system, does not use a direct correlate to 'intensity'. Instead, in the NCS object colors are ordered according to their relative amounts of chromatic, white and black contents (as is also the case in Wilhelm Ostwald's color system, although for a different reason).

For our purpose, however, '*lightness*' can be defined as 'the attribute of color perception by which a non-self-luminous body is judged to reflect more or less light' (ASTM, 1991). Lightness thus signifies a *relative* intensity quality of reflecting surfaces (related colors). In Hering's system, whiteness is the dimension that replaces lightness (Hering, 1964; Evans, 1964; Heggelund, 1992). Blackness induced by spectral surrounds into a darker, achromatic central area depends on the surround luminance and not on its brightness (Werner *et al.*, 1984; Cicerone *et al.*, 1986). One is therefore tempted to call blackness an appropriate inverse 'intensity' dimension, at least for related, achromatic colors. These and other aspects related to the problem of the dimensionality of color perception still occupy vision researchers (Volbrecht and Kliegel, 1998; Heggelund, 1991).

The physical measurement of light

A strict definition of light might be: 'electromagnetic radiation that evokes a visual impression'. This is radiation in the wavelength range between about 380 and 780 nm. When we speak of radiation outside these wavelengths as light, like *ultraviolet light* (UV) and *infrared light* (IR), this is just a manner of speech. Neither UV nor IR gives any visual impression of light.

The physical, radiometric description of electromagnetic radiation that is of importance for vision uses concepts and magnitudes such as wavelength, λ, measured in nanometers, and spectral distributions of

- radiant flux, Φ_e (W)

- irradiance, E_e (W/m^2)

- radiance, L_e [W/(sr $\cdot$ m^2)]

The radiometric units are given in the brackets, and the index 'e' after the symbol means that these symbols are based on energy units. The same symbols are used for the corresponding photometric magnitudes, but then with the index 'v', for visual.

Leaving the difficult questions of brightness perception aside, the most relevant photometric units have been defined as a weighted sum of the radiometric spectral energy distribution. The spectral weighting factors are proportional to the spectral luminous efficiency function, V_λ, of the human eye. V_λ is defined between 380 and 780 nm, and only where V_λ has a value different from 0 may we speak of light in the proper meaning of the word. The weighted photometric magnitudes and units that correspond to the radiometric ones are:

- light flux, Φ_v (lumen, lm)

- illuminance, $E_v = \Phi_v/A$ (lm/m^2 or lx)

- luminance, $L_v = I_v/A$ (cd/m^2)

In addition we have the fundamental magnitude

$$\text{light intensity}, I(\text{candela}, \text{cd}) = \Phi_v / \Omega (\text{lm/sr})$$

for a point source.

In these expressions A stands for area (in m^2) and sr for a solid angle measured in *steradians* (a steradian is part of the surface of a sphere when the whole sphere has a solid angle Ω of 4π steradians viewed from its center). Luminance was in earlier days called light density (*Leuchtdichte* in German). Figure 4.16 illustrates the relation between some of the photometric concepts.

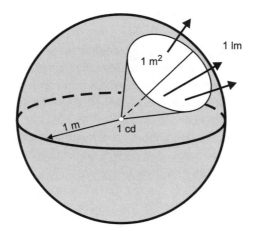

Figure 4.16 The interdependence of some photometric magnitudes and units. When a homogenous point source with intensity of 1 cd is positioned at the center of a sphere, a luminous flux of 1 lumen (lm) passes through 1 m^2 of the sphere's surface at a distance of 1 m from the source. The illuminance 1 m away from the center is 1 lm/m^2 = 1 lx.

An important magnitude in visual science and visual physiology is

$$\text{retinal illuminance}, T = L_v \cdot A (td)$$

Here A represents the area of the pupil in mm^2 and td stands for *troland*.

Wyszecki and Stiles (1982) give data on the pupil diameters for 12 subjects for a range of retinal illuminance. Average pupil diameter for 10 000 td is 3 mm, and 2.5 mm for larger values. For 1 000 td the value is about 5 mm, and for 1 td and lower the diameter is between 6 and 7 mm. Data differ from subject to subject, and below 10 000 td all the values given above have a distribution of about ±1.5 mm.

Luminance and illuminance are the two photometric magnitudes that are the most important for visual performance. Luminance of an area is measured with a luminance meter. For a constant pupil size, the luminance of a surface tells us how much light reaches the retina from that surface when the object is imaged through the eye media. The luminance ratio of that field relative to its surroundings correlates with the perception of its lightness – if it is black, gray or white. The luminance distribution within the field of view contributes to the eye's momentary light adaptation.

Illuminance tells us how much light flux enters a unit area, and this magnitude is measured by a *luxmeter*. The illuminance, E, from an artificial light source depends on the distance, r, from the source to the illuminated area. E is proportional to $1/r^2$ (if the distance is doubled, the illuminance will be less by a factor $\frac{1}{4}$).

Luminance-meters are much more expensive than luxmeters, and they are therefore rarely used, except by professionals. However, there are some rules of thumb for how to estimate luminance of a given surface from the illuminance. For a diffuse surface that reflects the light equally in all directions, the relationship between illuminance (measured in lux) and luminance (measured in cd/m^2) is relatively easy since the luminance is the same in all directions. Such a surface is called a *Lambertian surface*, and its luminance L is

$$L = \beta \cdot E/\pi$$

where β is the reflection factor. For polished or specular reflecting surfaces the relationship between illuminance and luminance depends on the direction of the incident and the reflected lights, and on the angle of viewing. The formula above does not apply to such surfaces.

Recommendations for the illuminance in work spaces depend on the visual task that is required. In a normal office, about 500 lx is recommended in the working area, whereas very precise visual tasks may need up to 2000 lx. In living rooms 50 lx is regarded sufficient as general illumination, but reading and other tasks will need additional light. All these values apply for the average young observer, with normal vision. Elderly people and people with a visual impairment generally will require better illumination.

In comparison, illuminance outdoors on a sunny day may be around 100 000 lx, and the luminance of a white sheet of paper with a reflection factor $\beta = 0.8$ will be about 25 000 cd/m^2 (according to the formula above). To convince yourself that physical luminance and perceived lightness are different entities, take a piece of coal (or some other black object, like black velvet) with you out into the bright sunlight and compare its blackness with that inside, in the darkest corner of the room. No doubt, the piece of coal appears blacker outside, despite the fact that it has a much higher luminance outdoors than indoors.

Spectral weighting functions

We have already mentioned that the photometric units used in light measurements depend on the sensitivity of the human eye. Photometric magnitudes are derived from the radiometric magnitudes by weighting the spectral radiation differently for each wavelength. Under photopic conditions, e.g. for luminances above a few cd/m^2, these spectral weights are given by V_λ, the CIE relative spectral luminous efficiency function. Since 1924 V_λ has been fundamental to all technical light measurement. Luminous efficacy,

$$K = \Phi_v/\Phi_e (\text{lm/W})$$

is the ratio between light flux and the corresponding radiant flux, and it gives a weight to be used when one wants to convert from radiant flux, Φ_e, in Watts to light flux, Φ_v, in lumen for a certain spectral distribution (or to convert from radiance, L_e, to luminance, L_v). Figure 4.8 shows the relative spectral weights, V_λ, which are proportional to K_λ for each wavelength. V_λ has the greatest weight at 555 nm, where it is normalized to 1.0, and where a radiant flux of 1 W corresponds to 683 lm, the light efficiency being 683 lm/W. This factor is called maximum spectral light efficacy,

$$K_m = 683 (\text{lm/W})$$

The weighting factors for the other wavelengths, $V_\lambda = K_\lambda/K_m$, decrease rapidly towards the ends of the spectrum, and are practically zero at 380 and 780 nm. In transforming radiant flux to light flux, one must consider the spectral distribution of the light and calculate an integral over the whole visible spectrum:

$$\Phi_v = 683 \int \Phi_e(\lambda) V(\lambda) \, d\lambda$$

This integral (or sum) treats the eye as a linear detector. All instruments that measure light in photometric units, like for example luxmeters, have a sensor that, in combination with a proper spectral filter, is excited proportional to this integration. The linearity impied by the formula simplifies the technical measurement, but creates problems for those who expect photometric magnitudes to correlate in a simple way with our subjective and qualitative impressions of the intensity (brightness or lightness) of light and color.

The CIE 2° luminous efficiency functions

Hermann von Helmholtz's opinion that the equalisation of brightness of two colors was based simply upon a judgement of the minimum difference between them was later taken up by Erwin Schrödinger (1920b). However, experiments showed that for large color differences this method was not additive and therefore not a suitable basis for physical measurements (Abney and Festing, 1886). Fortunately, in dynamic flicker experiments researchers found a way of avoiding such obstacles as non-additivity. According to Le Grand (1968), Rood (1899) was the first to suggest that photometry should be based on the flicker method. The important works of H. E. Ives promoted this approach (Ives, 1912; König, 1929). Figure 4.17 illustrates the main features of the flicker photometric method.

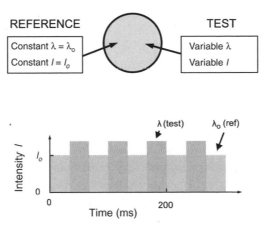

Figure 4.17 In heterochromatic flicker photometry a reference light (green in the figure) is alternated rapidly with a test light. The alternation frequency is usually between 10 and 20 Hz. The intensity of the test light is adjusted until the impression of flicker is minimized or disappears altogether. When this happens, the two lights have, per definition, the same luminance for the person in question. See also Color Plate Section.

The ideal method for determining the photopic luminous efficiency curve, V_λ, is to use a psychophysical method called 'heterochromatic flicker photometry'. In such measurements two lights of different wavelengths, λ_0 and λ_1, are alternated rapidly in a $2°$ field. The alternation is so rapid that the color difference between the two wavelengths disappears (this will be the case at about 15 Hz; see Figure 4.17). The task for the human subject is to adjust the relative intensity of the two lights until the impression of flicker disappears or reaches a minimum. At the intensity ratio where this happens, the two radiances are said to have equal luminance (in this case equal subjective luminance). This same procedure is then performed for all wavelengths; λ_0 is compared with λ_2, λ_3, etc. until the whole spectrum has been covered.

One may think that this is a strange way of determining which radiances have the same photometric intensity, especially when applied to the intensity of static lights and surfaces. One reason why a flicker method is best suited for physical photometry is because it is practical: it makes life easier for lighting engineers – albeit at the expense of the understanding of the consumer. The great advantage of the flicker method is that it leads to linearity, a property that was an absolute requirement in photometry. In a linear system the law of transitivity holds, together with the laws of additivity and proportionality.

> Transitivity (in a flicker experiment) means that if a light A is equal to light B, and B equal to C, then A must also be equal to C with respect to the intensity that gives minimum flicker. Transitivity does not hold when the subjective brightnesses of two static fields of different colors are compared in a setup like that of Figure 4.1. In that case A = B, and B = C, but A ≠ C. (see the section 'Additive Color Mixtures' in Chapter 5, p. 216).

As we have mentioned earlier, a direct, subjective comparison of the lightness or brightness of two different stationary color stimuli is very difficult, and the task is almost impossible if the color difference is large. When one is forced to make such a comparison, the result deviates significantly from that obtained with flicker photometry. Monochromatic blue (about 470 nm), green (about 500 nm) and red lights (about 700 nm) of the same luminance as a white light will always appear much brighter than white in such a stationary comparison. It appears that the higher saturation of narrow-band spectral lights makes them appear brighter. The difference is particularly great for the highly saturated short- and long-wavelength lights. When compared with white light of the same luminance, the subjective difference of brightness is much less for mid-spectral yellow lights around 570 nm.

The history of $V(\lambda)$

Historically, we may distinguish between static and dynamic methods for establishing photometry: direct heterochromatic brightness matching and the minimization of flicker. Despite the problems with heterochromatic brightness matching, in deriving the 1924 CIE 2° luminous efficiency curve, both methods were used, including a refinement of the first method based on a comparison of monochromatic stimuli differing only slightly in wavelength (the step-by-step method). The resulting CIE $V(\lambda)$ of 1924 is thus a hybrid curve where the results of different research groups and different methods have been combined. It follows a proposal by Gibson and Tyndall (1923), who based their smooth and symmetric curve on a suggestion by the Illuminating Engineering Society (IES) of the USA and on later experiments. The 1924 $V(\lambda)$ is thus based on the following methods and data (n = number of subjects):

- flicker experiments (427–746 nm, $n = 125$; Coblenz and Emerson, 1918);

- direct heterochromatic matching (620–770 nm, $n = 9$; Hyde and Forsythe, 1915);

- step-by-step heterochromatic matches (500–660 nm, $n = 29$; Hyde *et al.*, 1918; 430–740 nm, $n = 52$; Gibson and Tyndall, 1923).

The final curve of Gibson and Tyndall (1923, their Table 3) is a compromise of all methods and does not correspond to an overall average of these data sets. From 400 to 500 nm the curve follows the IES data, from 510 to 540 nm those of Coblenz and Emerson, from 550 to 690 nm those of Gibson and Tyndall, and above 690 nm again those of IES and Coblenz and Emerson. This patching of data and wavelength regions, and particularly the lack of good data for the shortest wavelengths, has led to a function that is not easily related to a particular physiological mechanism.

Technical photometry is based on this 1924 $V(\lambda)$ and the following definition of luminance:

$$L_\mathrm{v} = K_\mathrm{m} \int_\lambda L_\mathrm{e}(\lambda) V(\lambda)\, \mathrm{d}\lambda$$

Here, $L_\mathrm{e}(\lambda)$ is the spectral radiance distribution, and K_m is the maximum photopic spectral luminous efficacy. The integral is taken as being between 360 and 830 nm. $V(\lambda)$ is identical to the spectral tristimulus value $\bar{y}(\lambda)$ of the CIE 1931 standard colorimetric observer, thus making a colorimetric measurement also a photometric one.

Unfortunately, the spectral sensitivity of the 1924 $V(\lambda)$ is too low at short wavelengths, below 460 nm. Gibson and Tyndall were aware of this problem, but decided to keep the IES values unchanged and instead regarded the proposal as preliminary in this region. In an attempt to remedy this, Judd (1951a) proposed a modification of the 1924 $V(\lambda)$. The result was close to the original flicker photometric data of Coblentz and Emerson (1918). Ultimately, after a revision at and below 410 nm by Vos (1978), a new function, $V_\mathrm{M}(\lambda)$, was adopted by the CIE in 1988 as a supplement to the 1924 curve (CIE, 1990). Figure 4.18 compares 1924 $V(\lambda)$ with 1988 $V_\mathrm{M}(\lambda)$, and the lower diagram shows the difference in the logarithmic values. This latter curve has been recommended for a long time for use in visual psychophysics, but this corrected function has not yet been implemented in any lux meter or luminance meter. Industry has been reluctant to develop photometric instruments of such a precision that the difference between the 1924 $V(\lambda)$ and 1988 $V_\mathrm{M}(\lambda)$ sensitivities really matters (Schanda, 1998).

Therefore, caution is advised when using such instruments to measure lights with a major contribution from the short wavelength region. Such instruments are not suitable at all for short wavelength narrow band lights. This difference in spectral sensitivity is not of major importance for measurements of broad-band daylight or light from incandescent lamps, where the flux in the blue–violet range at short wavelengths is small compared with the rest of the spectrum. Although the difference amounts to about 1 log unit around 400 nm, and is highly significant for monochromatic lights, it adds up to only 0.5 percent in the luminance of the equal energy spectrum. In practice, with fluorescent light sources, this difference may be larger (CIE, 1998), and with blue LEDs (light emitting diodes) and the blue phosphor of

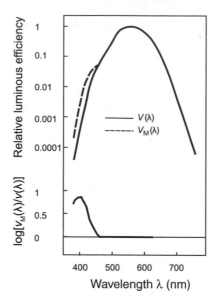

Figure 4.18 The 2° spectral luminous efficiency function for photopic vision, $V(\lambda)$, defined by the CIE in 1924, is here compared with the revised function, $V_M(\lambda)$ defined by CIE in 1988. The lower plot shows the logarithm of the ratio of the two curves. The difference is significant only for short wavelengths, below 460 nm. $V_M(\lambda)$ corresponds better to coherent psychophysical data than $V(\lambda)$ does, and is therefore preferred when high precision is required (courtesy of J. H. Wold).

CRT monitors the relative difference reaches several percent (Schanda, personal communication).

For half a century, the Judd-modified luminous efficiency curve, $V_M(\lambda)$, has been preferred over the 1924 $V(\lambda)$ as a basis for precise, vision-based photometry. Smith and Pokorny (1975) conjectured that $V_M(\lambda)$ is the sum of $L(\lambda)$- and $M(\lambda)$-cone sensitivities only, with S-cones making no contribution. This conjecture seems to hold in most experiments, and it has allowed for a simple physiological interpretation in terms of the MC pathway cells being silenced at isoluminance.

Later in the last century a third, static psychophysical method for heterochromatic photometry was developed: the minimally distinct border method (MDB; Boynton and Kaiser, 1968). The MDB method also avoids the problems with non-additivity mentioned above (Boynton and Kaiser, 1968). In the case of MDB, the matching is between two photometric half-fields, as for direct heterochromatic brightness matching. The subject is asked to ignore the color appearance of the two fields involved and to concentrate only on adjusting the relative radiance so that the border between them becomes minimally distinct. Typically, the method gives $V_M(\lambda)$ sensitivity (Valberg and Tansley, 1977) and is relatively insensitive to image defocus (Lindsey and Teller, 1989).

Luminous efficiency functions in vision research

On the physiological origin of $V_M(\lambda)$

What role does the concept of luminous efficiency play in the context of detection and discrimination? What are the neural correlates and mechanisms that can account for luminous efficiency? Its meaning is most clearly expressed by the scotopic function $V'(\lambda)$ of the dark-adapted eye. Physiologically, this function simply mirrors the spectral efficiency of rods in absorbing light quanta, thus reflecting the absorption spectrum of rhodopsin, the photopigment of rod receptors. At high luminance the rods are not modulated by light, and it is the photopic luminous efficiency function that has taken over its role in determining visual sensitivity. This sensitivity must in some way be related to the fundamental processes of quantum absorption in the cone receptors, but since there is no one cone type with a spectral sensitivity equaling this function, the relation in this case is more complex. Nonetheless, we have a good idea of what this relationship is as well.

In order to understand its physiological basis, let us first list some of the experimental situations which have been reported to yield a photometric spectral sensitivity curve close to $V_M(\lambda)$:

- determination of a flicker minimum;
- determination of minimally distinct borders (MDB);
- detection of brief and small stimuli;
- visual acuity;
- minimal motion perception.

From what we have said in previous sections one may, perhaps, have developed the impression that photometry is a field of pragmatic solutions dictated by technical requirements. However, the striking similarity of the spectral sensitivities found in the different tasks above indicates that there is a common, physiological 'photometric' mechanism at play in all of them. This may come as a surprise in view of the strong indications to the contrary (see the discussion in Sheppard, 1968). If there is no single photoreceptor type that has the spectral sensitivity of $V_M(\lambda)$, what then is this mechanism? The answer is found in recent neurophysiological experiments.

After the first neurophysiological recordings from ON- and OFF-cells in the frog retina (Hartline, 1938), only a few neurophysiologists showed interest in how the activity of these and other nerve cells was related to perceptual phenomena. The German neurologist Richard Jung (1973) was among those who repeatedly pointed to the strong correlates of 'brightness' and 'darkness' with the intensity coding of ON- and OFF-cells in the visual system (see Jung et al., 1952; Baumgartner, 1961; Spillmann, 1971). This was later followed up on by Schiller and his colleagues (see Schiller, 1992).

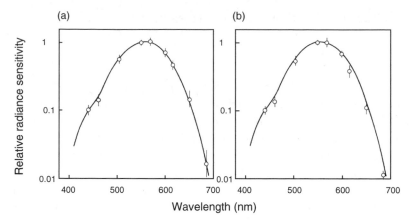

Figure 4.19 The averaged spectral sensitivity of 26 MC retinal ganglion cells of the macaque monkey is shown in (a). The data points refer to the ganglion cell sensitivities (with standard deviation) for a 4° field at 10° eccentricity while the curve is the CIE $V(\lambda)$ function for a 10° observer. In (b) the data points represent the average sensitivity of nine human subjects using the same experimental set-up as used for the cell sensitivities in (a) to minimize the flicker sensation (as in heterochromatic flicker photometry). The curve is the same as in (a). From these and other experiments, we conclude that the MC cells provide the physiological substrate for the $V(\lambda)$ curve (adapted from Lee *et al.*, 1988).

In the primate, the relevant cells have now been identified as the transient, increment and decrement (ON and OFF) magnocellular (MC) cells (Lee *et al.*, 1988; Kaiser *et al.*, 1990). Figure 4.19(a) displays the averaged sensitivity of 26 such retinal cells of macaque monkeys as measured in flicker experiments (data points), and Figure 4.19(b) shows the average sensitivity of nine human subjects who were required to minimize flicker using the same apparatus. The monkey recordings were made with a tiny microelectrode placed close to the cell body of the relatively large retinal MC cells. This electrode picked up the potential changes associated with the cell's firing of nerve pulses, the number of which were counted by the computer. The firing rate, i.e. number of impulses/s, was taken as an expression of the cell's response magnitude along the stimulus dimensions intensity, wavelength, border strength, etc. The test field was 4° in diameter and, for technical reasons, the recordings were made 10° to the side of the fovea. In the figures, the sensitivities are compared with the luminous efficiency curve of the 1964 CIE 10° observer. The agreement is excellent.

The results described above indicate that MC cells serve many different functions. Over the last years, the link between luminous efficiency and the MC pathway has been strengthened. Psychophysical tasks giving action spectra that deviated from a flicker-based spectral sensitivity in the direction of a heterochromatic brightness function (Figure 4.20) were often interpreted in terms of contributions from the opponent cells. This has, for instance, been the case for the detection and discrimination of large and long-lasting spots, revealing an irregular spectral sensitivity curve

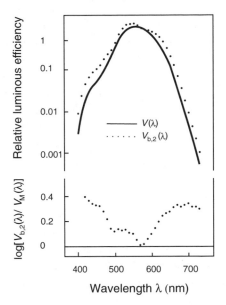

Figure 4.20 The solid curve represents the revised spectral luminous efficiency function, $V_M(\lambda)$, and the dotted curve relative spectral sensitivity for the subjective impression of equal brightness, $V_{b,2}(\lambda)$, for stationary, 2° monochromatic stimuli. The logarithm of the ratios of the sensitivities given by these two curves is shown at the bottom of the figure (courtesy of J. H. Wold). The positive difference between the two curves shows that the subjective impression of brightness of stationary color stimuli increases with colorimetric purity. This difference bears close resemblance to the relative subjective color strength of monochromatic lights, in which 570 nm yellow is perceived as much less saturated than monochromatic long-wavelength red and short-wavelength blue stimuli [see also Figure 5.23(b)].

[King-Smith and Carden, 1976; see solid curves Figure 7.6(b)] and brightness matching (Ikeda *et al.*, 1982; Ikeda and Nakano, 1986). The spectral sensitivity during saccadic eye movements also reveals this opponency (Burr *et al.*, 1994; Uchikawa and Sato, 1995; Ross *et al.*, 1996). The latter result can be taken as a sign of saccadic suppression of the MC pathway signals that have inputs to the motion centers in the brain, thus exposing an otherwise concealed opponent cell contribution. It is believed that the suppression of the MC pathway may blunt the disturbing sense of motion that saccades would otherwise elicit.

As a result of the success in referring flicker-photometric luminance to a unitary neural mechanism, the concept of isoluminance has become increasingly important in vision science. In recent years, chromatic, isoluminant stimuli have often been used in psychophysical experiments in order to silence the putative luminance system and to isolate chromatic mechanisms. The method resembles the technique commonly used to abolish the response of one type of photoreceptor (silent substitution). It has been demonstrated that several visual functions deteriorate at flicker- or MDB-based isoluminance. These include the perception of movement (Cavanagh *et al.*, 1987), illusory border and filling-in illusions such as the Kanisza triangle (Livingstone and Hubel, 1988), and depth perception. The MC system may well be the dominant neural substrate in these tasks.

However, equating the MC pathway with the luminance channel can lead to surprises because MC cells, under certain circumstances, also show cone-opponent behavior although MC cells are believed not to contribute to color vision. This opponency is between L- and M-cones (the MC cells are not likely to receive S-cone inputs), and reflected in the distinctness of the residual border in an MDB situation, where MC cells exhibit a second harmonic response to isoluminant, moving chromatic patterns (Kaiser *et al.*, 1990; Valberg *et al.*, 1992). The magnitude of the second harmonic residual response of MC cells is proportional to the absolute value of the excitation difference, $|L - M|$, of L- and M-cones across the border. Thus, phasic MC cells not only provide a signal which can mediate performance on the minimally distinct border task, but they also provide a signal which may account for residual distinctness perceived after a subject has carried out the minimization.

Brightness vs luminance

We have mentioned that, when comparing isoluminant chromatic and achromatic stimuli, the chromatic stimuli always look brighter. Color saturation contributes to brightness but not to luminance. This is known as the Helmholtz–Kohlrausch effect. The greater the colorimetric purity (or color strength or chroma) of a stimulusis, the greater its brightness (due to what Helmholtz called *color glow*). Since the early days of photometry, students, architects, artists and other consumers relying on their own judgements have been puzzled by this discrepancy between the concepts of brightness and luminance. Brightness may be more important than luminance in some everyday detection tasks, where the conspicuousness of an object is important (Venable and Hale, 1996; Schumann *et al.*, 1996).

Figure 4.20 shows the wavelength sensitivity of the subjective impression of brightness for stationary photopic, spectral stimuli, $V_{b,2}(\lambda)$, compared with $V_M(\lambda)$, and the difference between them. Both curves are for a $2°$ field, and the results are averaged over many subjects. Both curves have been normalized to 1.0 for 570 nm.

The logarithm of the ratio of the two curves is shown as a function of wavelength in the lower part of the figure. The form of this curve resembles the subjective saturation or color strength (the degree of subjective difference relative to white of the same luminance) for monochromatic stimuli. In other words, the greater the perceived saturation (or color strength) of a color stimulus relative to an isoluminant white, the brighter it looks. Yellow spectral colors around 570 nm have the least saturation, and their brightness is also not very different from that of an isoluminant white. The fact that saturation contributes to brightness is a general observation that holds for all chromatic lights (Ware and Cowan, 1983).

The dotted curve in the lower part of Figure 4.20 shows that all wavelengths appear brighter than an equiluminant yellow stimulus of 570 nm. For short-wavelength blue light and for long-wavelength red light, the difference amounts to more than 0.3 log units (more than a factor 2). Thus, compared with white, all stationary chromatic colors will appear brighter, even if their photometric luminance is the same.

For architects and artists the concept of luminance can be enigmatic, and it is not easy to understand why equal luminance does not also entail equal brightness, and this dissociation usually causes a lot of confusion, even among people who work with light and vision. However, as we have seen, the explanation lies in the formal and pragmatic nature of the flicker-definition of the photometric weighting function, V_λ, required to obtain a linear photometric system. Contrary to photometric measurements, brightness measurements do not obey the linearity laws. It is important to be aware that this fundamental difference applies also to many other perceptual and optical properties: *there is a fundamental difference between the physically defined stimulus magnitudes (if they are photometric, colorimetric or other) and the subjectively perceived qualities.*

Since the establishment of a firm link between flicker- and MDB-based luminance and the sensitivity of the MC channel, it has become common to explain the excess brightness of chromatic stimuli by additional responses of the opponent channels. For instance, it has been conjectured that a $V_M(\lambda)$-like curve can also be obtained from sustained opponent cells by a particular combination of cone opponent L–M and M–L outputs (Kandel *et al.*, 2000; Krauskopf, 1999). Such possibilities have also been expressed in models of color vision, and particularly in that of Guth *et al.* (1980). Even if spectral sensitivity of MC cells equals $V_M(\lambda)$, the physiological substrate for equal luminance, the intensity magnitude need not be correlated with responses from these cells alone. Brightness of stationary colors may be due to the combined activity of several channels (Yaguchi and Ikeda, 1983; Yaguchi *et al.*, 1993; Nakano *et al.*, 1988). However, it is not easy to conceive of a process by which the MC cells, that are only transiently active at luminance borders (and at equiluminous borders between red and green fields, but not between blue and yellow), can contribute to brightness. It is indeed more likely that brightness can be attributed entirely to the responsiveness of a combination of the sustained, opponent systems, also responding to achromatic stimuli (Valberg *et al.*, 1986a).

In this connection, it may be of interest to consider the induction of blackness by a white surround into a central field of monochromatic lights (Shinomori *et al.*, 1997). If the luminance of the white surround is held constant, the action spectrum for black perception in the center does not follow the flicker-based curve, but the same spectral sensitivity curve as that for heterochromatic brightness matching. For instance, a blue center will turn black at a lower relative luminance than red, green and yellow, and the spectral curve representing the inverse of this ratio will resemble the spectral saturation of the center stimulus. This spectral curve also parallels that obtained for chromatic thresholds from black, and for zero gray content called G_0-colors (Evans and Swenholt, 1967; Richter, 1969). Since black and zero gray are at opposing ends of the reflectance scale, this result indicates that both these achromatic percepts are compensated or neutralized in inverse proportion to the saturation (or colorimetric purity) of the center stimulus. It should be noted, however, that the effect of monochromatic surrounds in inducing blackness in an achromatic center follows the other, flicker-based curve (Werner *et al.*, 1984).

For photopic vision, surfaces with related colors elicit maximum color strength for a restricted range of relative luminance, different for every color. As relative luminance increases above this range, chromatic strength diminishes and stimuli acquire a whitish appearance, and lower luminance induces black. This dependency of color strength of stimuli of constant chromaticity on relative luminance is nonlinear, like the Bezold–Brücke effect, and can be explained entirely by nonlinear responses of the opponent cells (Valberg *et al.*, 1991b).

Conclusion

The photopic spectral luminous efficiency curve is a measure of the efficiency of electromagnetic radiation in evoking the same visual effect, and CIE 1924 $V(\lambda)$ has served this purpose for many years. Despite some shortcomings, it has provided the basis of photometry in research and in technology. This success may lead one to ask if it would be possible to obtain industrial acceptance of alternatives to the 1924 curve, even if there were relevant psychopysical and physiological arguments for a change. The history of separate usage of the 1924 and 1988 functions in industry and science is not encouraging in this respect. A change may, however, be desirable along with the introduction of a new physiologically based system for colorimetry.

At the beginning of the twentieth century, luminous efficiency did not seem to have a strong perceptual correlate – except in the sensation of minimum flicker. It appears now that the aim of photometry was not so much to do justice to lightness and brightness perceptions under different viewing conditions, but to develop a useful physical measure of light. As Le Grand (1968) writes: 'It was necessary to arrive at a solution, even if it were theoretically defective'. Luminous intensity is an adequate property of achromatic stimuli, and if it were to have a similar, although not so obvious perceptual correlate also for chromatic stimuli, the argument made sense that, after the two colors had fused in a flicker experiment, the remaining flicker perception would represent such a correlate. The early researchers in this field were not able to relate the minimization of what they regarded as an 'achromatic-intensity percept' (flicker) to the behavior of a particular physiological mechanism. It was left for later generations to show that this could indeed be done.

In retrospect, the success of photometry makes one wonder if the early advocates of flicker photometry were guided by a genuine intuition, or whether perhaps they were merely lucky. They could not have known that the flicker method and its equivalents lead to a luminous efficiency function with a firm physiological basis. The discovery that this curve is indeed the manifestation of a physiological mechanism came at the end of the last century, when it was realized that it reflects the sensitivity of transient, magnocellular cells of the primate retina (Lee *et al.*, 1988). With help of the derived concept of sensation luminance (Kaiser, 1988), it has been demonstrated that the transient MC cell population, as a putative physiological 'luminance mechanism', participates in many different visual functions.

The identification of the sensitivity of MC cells as the proper physiological basis of photometry has, perhaps, given $V_M(\lambda)$ a deceptive appearance of objectivity. The impressive work of Ikeda and co-workers (CIE, 1988) has demonstrated that it is indeed possible to make brightness matches with spectral colors, and that the difference between the luminous efficiency functions obtained by flicker photometry and heterochromatic brightness matching resembles the spectral saturation function. This does not, however, allow an immediate conclusion to be drawn as to the physiological substrate for brightness – whether it is a single qualitative entity or a sum of achromatic and chromatic parts (von Helmholtz, 1896). Several possibilities exist, and the difference of lightness and brightness, and the different percepts for related and unrelated colors, complicate the picture.

For discussions of still other aspects of luminance, interested readers are advised to consult Lennie *et al.* (1993) in a special issue of the *Journal of the Optical Society of America* (**10**, no. 6, 1993).

Contrast vision

Contrast is a word with many meanings. Here, we apply the term to physical differences in luminance and color, as well as the perception of these differences. The perception of physical contrast is fundamental for all vision. It is a prerequisite for distinguishing an object from its background, and for determining its texture, color, form, movement and depth. Contrast in a natural scene may be strongly reduced by fog, rain or snow. While these are examples of changes in physical contrast, the loss of contrast that we experience with decreasing illumination at dusk or at night reflects changes in perception, as our sensitivity to the unchanging physical contrast of the scene deteriorates at low luminance levels. The measurement of a subject's contrast sensitivity, i.e. the ability to distinguish between small differences, gives an indication of how well a person will perform on such tasks. Driving a car or flying an airplane in fog or during heavy snowfall are examples of extreme conditions that require high contrast sensitivity. Nordic skiers are well acquainted with the problem of seeing the smooth changes of a white mountain terrain on an overcast day, when shadows are virtually absent. Under such extreme conditions it does not help to have high visual acuity, i.e. the ability to see fine details. Good contrast vision, however, will be an asset.

Below we shall see how a person's contrast sensitivity and visual acuity can be measured in the same procedure. You may be surprised to learn that it is possible for a person to have quite normal vision for fine detail (normal visual resolution), and yet have reduced contrast sensitivity for larger objects. The opposite situation also occurs.

In many circles, visual capacity is still characterized by the minimum angle of resolution, i.e. by visual acuity. Resolution can be measured by means of letters of different sizes [Fig. 2.29(a)] as in the Snellen chart. This test, with black letters on a white background, requires recognition and identification of the letters, and this is more challenging than a simple detection of contrast. For children who have not yet learned to read, and for people who cannot read, other methods must be used. Observing pursuit eye movements when a high-contrast pattern drifts in front of a child is one such alternative. While acuity tests, such as the Snellen chart, identify thresholds for recognizing symbols, grating tests, for example, identify thresholds for detection of a contrast without the need for recognition. The presence of a contrast may be detected even if the form of the symbol cannot be made out. In other words, the threshold contrast for detection is lower than the threshold contrast for identification.

A person's *contrast sensitivity* is the inverse of his contrast threshold. In the luminance domain it tells us how good a person is at seeing small luminance differences, ΔL, between a test field and some comparison, or reference field with luminance L (Figure 4.6; the smaller this difference, the better the contrast sensitivity). In this case, we are talking about *luminance contrast*. In the color domain, the threshold contrast is a measure of the ability to see color differences. In dealing with pure color contrast, without any luminance difference, we shall use the term *chrominance contrast*. Examples of contrast sensitivity functions are, for instance, plots of contrast sensitivity as a function of absolute background luminance, as in Figure 4.7, or sensitivity vs size of the stimulus, or the contrast necessary to detect flicker as a function temporal frequency (see below).

As indicated above, the perception of contrast depends on the size of the object, its form and its temporal variation. In good illumination, sensitivity for static contrasts with sharp borders is greatest for objects that are larger than 0.2° in visual angle. If the borders are less distinct (as is often the case with shadows), sensitivity for larger areas than this is reduced. In the case of a diffuse edge zone of a large field spanning about 12 min arc (or 0.2°), sensitivity decreases to about 70 percent of the value with sharp edges (Schober, 1958). The sensitivity for pure chrominance contrasts, on the other hand, still increases for objects larger than 0.2°. Surprisingly, this sensitivity is not affected by the blurring of sharp borders.

For theoretical reasons (see the discussion of Figures 2.33 and 2.34), contrast sensitivity is often measured using sinusoidal gratings with different spatial frequencies (different bar widths). In such gratings the luminance varies as a sine-function, forming a regular pattern of blurred bars. Even when the optics of the eye is not optimal, the bars of the grating keep their sinusoidal form after being imaged on the retina. A grating with sharp edges, i.e. with square-wave luminance distribution, no longer has sharp edges after being imaged on the retina; the edges are rounded off. A square-wave grating offers no real advantage in testing visual functions since it, and many other stimuli, can be described as a linear sum of sinusoidal gratings of different spatial frequencies (see Figure 2.34). Having determined the threshold sensitivity for sinusoidal gratings, it is then possible to derive the sensitivity to other stimuli. For example, knowing the sensitivity to sinusoidal gratings of different frequencies makes it possible to calculate the sensitivity to a square-wave grating. The reverse calculation is not always possible.

Luminance contrast

For non-periodic stimuli, such as letters, viewed against a specific background, the physical contrast is often expressed as the Weber ratio between figure and background,

$$(L - L_b)/L_b = \Delta L/L_b$$

This definition of contrast is the most common one in lighting engineering. As shown in Figure 4.21, the Weber ratio can be either negative (black letters on a white or gray background), or positive (white letters on a black or gray background).

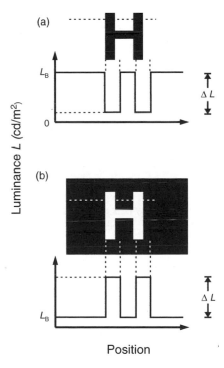

Figure 4.21 The luminance distribution of the black letter H on a light background. (b) The luminance distribution of the bright letter H on a dark background. In (a) the black letter has a negative Weber contrast, $\Delta L/L_b$ between 0 and -1, and in (b) the white letter has a positive Weber contrast ($\Delta L = L - L_b$) that is much higher than 1.

For printed optotypes, such as for Hyvärinen's (1992) vision test charts, Weber contrast can either be expressed in terms of luminance, as above, or by means of the reflection factors β:

$$C_{\text{Weber}} = (\beta - \beta_b)/\beta_b$$

For diffusely reflecting materials, and for optimal geometry of the illumination and viewing conditions (diffuse illumination, no specular reflection, etc.), this contrast is independent of the illuminance and absolute luminance level. Physical contrast is the same for low light levels at night as in daylight. Our ability to see a particular small contrast, however, changes dramatically from scotopic to photopic luminance levels (Figure 4.7).

Now let us take a closer look at sensitivity for luminance contrast measured with a sinusoidal grating as in Figure 4.22. The luminance, L, of parallel bars varies as a sine function across the bars (along a horizontal x-axis for vertical bars):

$$L(x) = L_m(1 + C_M \sin \omega x)$$

where L_m is the mean luminance (see Figure 2.33), C_M is the *Michelson contrast*, ω is the angular frequency, and the x-axis is scaled in degrees of visual angle. Michelson contrast, C_M, is defined as

$$C_M = (L_{\max} - L_{\min})/(L_{\max} + L_{\min}) = (L_{\max} - L_{\min})/2L_m$$

This is the same as the ratio a/b of *amplitude/mean luminance* in Figure 2.33. For reflecting materials, for instance printed test charts with gratings (Teller, 1990), the Michelson contrast C_M can be written:

$$C_M = (\beta_{\max} - \beta_{\min})/(\beta_{\max} + \beta_{\min})$$

where β is the reflection factor of the printed pattern or of its background. From this expression it is clear that the physical contrast of the pattern does not change with illumination, although contrast perception undergoes a drastic change in the transition from light to dark.

Spatial contrast sensitivity

Figure 4.22(a) shows a sinusoidal grating with three different contrasts, and in Figure 4.22(b) the contrast is the same but spatial frequencies (number of white/black cycles per degree) are different.

A person's contrast sensitivity (CS) determines the contrast at which a grating is just visible. The inverse of this threshold contrast, as determined for several spatial

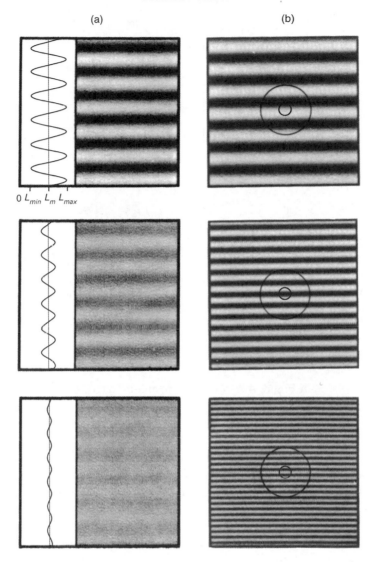

Figure 4.22   (a) A sinusoidal grating of the same spatial frequency, but with different contrasts. The luminance across each grating varies as a sinusoidal function. The spatial frequency is the number of whole periods (one white and one black bar) within a visual angle of $1°$ ($1° = 1$ cm at a distance of 57.3 cm from the eye). Contrast is defined as $C = (L_{max} - L_{min})/(L_{max} + L_{min})$. (b) Three sinusoidal gratings with the same contrast but with different spatial frequencies. Circle and annulus illustrate the relation between spatial frequency and receptive field size (adapted from Frisby, 1979).

frequencies, gives rise to a curve of contrast sensitivity as a function of spatial frequency. Figure 4.23(a) and (b) shows the mean contrast sensitivity curve for sinusoidal gratings of varying spatial frequency for 10 normally sighted subjects (mean age 46 ± 5 years). It is common to present spatial CS curves in double

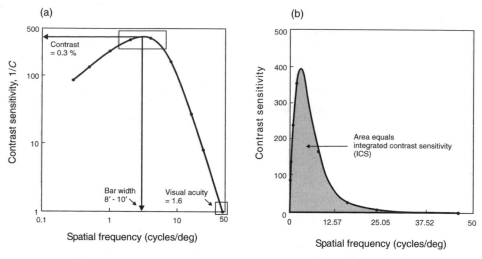

Figure 4.23 (a) A double-logarithmic plot of the mean spatial contrast sensitivity of 10 subjects (average age 46 years) plotted as a function of spatial frequency. In (b) the same curve is plotted using linear scales. The gray area, called the integrated contrast sensitivity, can be used in evaluating visual performance along with maximum contrast sensitivity and visual acuity shown in (a).

logarithmic plots, as in Figure 4.23(a). For a mean luminance of 40 cd/m², maximum contrast sensitivity was found for a spatial frequency of 3 cycles/deg (bar width of 6 min arc). At this spatial frequency, the minimum threshold Michelson contrast was 0.3 percent. Visual resolution corresponded to a grating acuity of 1.6/deg. In Figure 4.23(b) the same curve as in (a) is redrawn using linear axes. The area under the latter plot, called integrated contrast sensitivity (ICS), may be interpreted as a measure of the excitation of all hypothetical spatial frequency channels (see the next section), analogous to the way excitation is calculated for photorecepetors on a wavelength scale. This area under the CS curve is usually taken to represent the amount of spatial information dealt with by the visual system. We shall return to this measure when dealing with low vision.

Figure 4.24 shows many such CS curves, each for a different adaptation luminance. For photopic luminances, the curves have band-pass characteristics (an inverse U-form), because spatial frequencies in the mid-range, between 3 and 5 cycles/deg, are more easily detected than higher and lower spatial frequencies. The curves of Figure 4.24 indicate low-pass filter characteristics for the lowest luminance and a band pass filter at high luminance. At the highest luminance, the sensitivity to a grating of 0.3 cycles/deg is only 10 percent of the maximum sensitivity around 3 cycles/deg. The fall-off at low frequencies seen in Figure 4.24 is not an effect of the optics of the eye; it is related to lateral inhibition in the nervous system.

Except at low spatial frequencies, contrast sensitivity increases with increasing photopic luminance. The Weber law, stating that threshold contrast is constant and

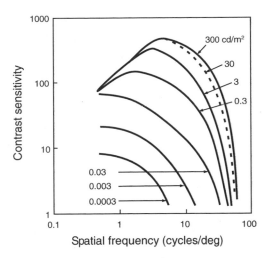

Figure 4.24 Spatial contrast sensitivity and maximum resolution (grating acuity) generally improve with increasing luminance. Maximum grating acuity (far right) corresponds to the resolution of a 60 cycles/deg grating at 100% contrast. Note that the Weber law fails at low spatial frequencies and low luminance levels.

independent of luminance, thus holds over a broader luminance range for low spatial frequencies than for high.

In humans, maximum contrast sensitivity is usually found at spatial frequencies between 3 and 5 cycles/deg and in young eyes it typically has a value of 500, corresponding to the detection of 0.2 percent contrast (Watson *et al.*, 1983). The position where a curve crosses the *x*-axis shifts towards higher spatial frequencies (narrower stripes) when luminance increases, indicating that resolution also improves with luminance (see Figure 2.30). Under optimal conditions, around 300 cd/m², the curve crosses the *x*-axis at 60 cycles/deg. At this spatial frequency, the distance between two black stripes is 0.5 min arc, which corresponds to a visual grating acuity of $1/0.5 = 2.0$. In conclusion, when the adaptation luminance increases, we are generally able to discriminate lower contrasts and resolve finer details.

For static gratings, photopic contrast sensitivity and resolution deteriorate as one moves away from the fovea, and the change can be described as a downward and leftward shift of the curve in Figure 4.24. Contrast sensitivity also depends on the temporal rate of change. The effect of temporal frequency is seen for frequencies beyond a few Hz, and is strongest for low spatial frequencies (Kelly, 1994).

Contrast sensitivity curves have been determined for different species using a behavioral criterion. Similar curves are found for macaque monkeys and humans, and Figure 4.24 can serve as an example for both. For cats and goldfish, on the other hand, the curves are shifted towards lower spatial frequencies, to the left on the *x*-axis. Their maximum contrast sensitivity is found around 0.5 and 0.3 cycles/deg, with a correspondingly lower grating acuity than humans. Some birds, for instance falcons, have curves that are shifted towards higher spatial frequencies. They have better visual resolution than humans, and a sensitivity maximum around 20–30 cycles/deg.

Spatial channels

In the 1970s Fergus W. Campbell and John G. Robson of the University of Cambridge put forward the hypothesis that the visual system has several separate mechanisms, or 'channels', sensitive to different spatial contrasts. Each of these channels respond within a relatively narrow frequency band, and together they span our visual spatial frequency range. The sensitivity of each channel is limited to about one octave bandwidth (one octave corresponds to a doubling or halving of the frequency; Campbell and Robson, 1968; Blakemore and Campbell, 1969). This idea was inspired by the discovery of the organization of receptive fields in excitatory and inhibitory areas. Receptive fields and spatial frequencies are complementary concepts in spatial image analysis and Fourier theory. On several occasions, the predictions of this hypothesis have been found to hold true, especially at threshold. For instance, by selective adaptation to a grating of a particular spatial frequency, a temporary sensitivity loss is found for that and adjacent frequencies. This is interpreted as 'fatigue' in strongly activated channels, which are most probably cortical cells.

Psychophysical studies have concluded that there are six cortical frequency channels or cell types, with maximum contrast sensitivity centered around six different spatial frequencies (e.g. Wilson and Gelb, 1984; Wilson et al., 1990). This number has later subsequently increased. The CS curve of Figure 4.23(a) and (b), and each curve in Figure 4.24, can be regarded as an envelope around the sensitivities of these individual mechanisms. The sensitivities of the postulated mechanisms are illustrated in Figure 4.25(a). Figure 4.25(b) shows the spatial contrast sensitivity for single cells in the macaque monkey, and how they together may determine the primate psychophysical spatial contrast sensitivity curve. Probably more than six units would be required to obtain a smooth envelope. In particular, there seems to be one unit missing at high spatial frequency in order to account for the limit of resolution.

For a given stimulus size, the most sensitive of these component mechanisms is thought to determine the psychophysical threshold, but it has also been considered possible that every mechanism contributes its share to the response. If the latter is correct, the resulting threshold response could, for instance, be a vector sum of the responses from several mechanisms. In addition to being selective for spatial frequency, these cells often have a preferred angular orientation. This is ascribed to each cell having an elongated receptive field, giving a higher response when a contrast edge is aligned with the long axis of the receptive field than when it is at right angles to it.

In summary, the above hypothesis states that detection of contrast relies on a group of specialized brain cells, and it is assumed that there are several distinct sets of such

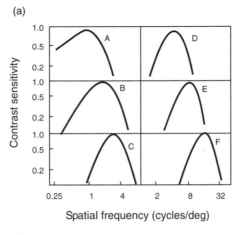

(a)

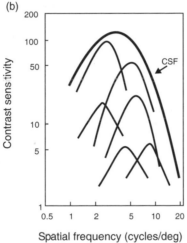

(b)

Figure 4.25 (a) One theory postulates that a normal spatial contrast sensitivity curve (Figure 4.23) reflects the contribution of at least six separate neural mechanisms. Each mechanism is sensitive to a narrow spatial frequency band, with maximum sensitivity at 0.8, 1.2, 3, 4, 8 and 16 cycles/deg. These mechanisms can be regarded as spatial frequency filters with a certain bandwidth. (b) The psychophysical contrast sensitivity function, CSF, may be regarded as the envelope of the sensitivities of the separate mechanisms shown in (a).

cells, constituting separate, parallel spatial channels. The cells in each set analyze a particular position in the retina (the position of its receptive field), and they respond to a narrow range of spatial frequencies, each with a preference for a relatively small sector of angular orientations.

Contrast sensitivity changes with increasing distance from the fovea, and persons who are forced to use their peripheral vision due to a foveal defect, have atypical sensitivity curves. This applies, for instance, to people with macular degeneration, a disorder in which central vision is compromised by the destruction of receptors in the macula. A person with macular degeneration will therefore have drastically reduced visual acuity. This affliction is more frequent among the elderly, and about 10 percent of people above 60 years have AMD. AMD subjects typically have reduced contrast sensitivity and resolution (see Figure 4.33).

Curves of spatial contrast sensitivity can provide valuable insights into a person's visual capacity that cannot be expressed by visual acuity (De Lange, 1958). It has been found (Hess and Woo, 1978) that persons with reduced sensitivity at lower spatial frequencies have the greater visual handicap in daily life. For instance, a reduced ability to see contrasts of large objects leads to problems in navigating in an unfamiliar environment, and impairment in the mid-frequency range makes it difficult to recognize faces. This latter problem can be understood in light of the fact that contrasts within 2–4 cycles per face are particularly important for face recognition, whereas visual acuity is of less importance. It has been reported that some people with Parkinson's disease have sensitivity curves with two maxima, with a saddle in between. Following Campbell and Robson's hypothesis, this has been taken to imply that one, or a few, of the spatial frequency channels are affected.

Some individuals have reduced contrast sensitivity for high spatial frequencies, whereas the low frequencies are less affected. Spatial contrast sensitivity curves for elderly with normal vision for their age have this character, as illustrated in Figure 4.26. For this group, contrast reduction is greatest for fine-detail vision. This can be explained partly by increasing opacities of the eye media and a smaller pupil size, leading to lower effective retinal illuminance in the elderly. Typically, young children have reduced sensitivity for all frequencies relative to adults.

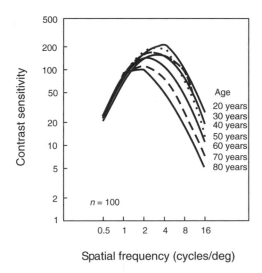

Figure 4.26 Spatial contrast sensitivity for different age groups. Increasing age leads to reduced sensitivity for the higher spatial frequencies, while sensitivity to lower frequencies is unchanged.

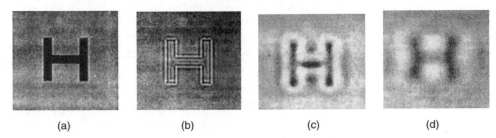

(a)   (b)   (c)   (d)

Figure 4.27 Spatial filtering of the letter H in (a) corresponding roughly to what one would expect for some of the mechanisms in Figure 4.25. When these images are seen at a distance of about 1 m, the letter height will be about 1° and the filtering in (b) will correspond roughly to that of a mechanism with the filter characteristics of F in Figure 4.25. The H in (c) corresponds to filtering by mechanism B, and (d) to A. Owing to limitations of the printing process, the rendering of the filtered letters cannot be expected to be accurate.

An object, such as a face or a letter would give rise to different responses in each of the frequency channels. Figure 4.27 is an attempt to simulate what the letter H would look like to a person who has only one of the postulated spatial mechanisms intact. The examples in Figure 4.27 are digitally filtered on a computer, with the frequency band in each picture emulating the filtering resulting from the sensitivities in Figure 4.25. For a viewing distance of 95 cm, Figure 4.27(b) would approximate the reproduction by mechanism (F) in Figure 4.25(a). Figure 4.27(c) would correspond to Figure (B) and Figure 4.27(a) to (A) in Figure 4.25(a). The printing process will have distorted the original pictures somewhat, so these figures can only be seen as an illustration of the principle that a combination of all frequency bands will result in the normal, unfiltered image of Figure 4.27(a).

We have mentioned that, to some extent, the visual system corrects for bad optical imaging (e.g. border enhancement). Therefore, one may ask if a person with vision loss would recognize images more easily if they were selectively enhanced in his 'weak' frequency domain. Experiments by Peli *et al.* (1991) suggest that this is at least partly the case. They showed how image manipulations with digital frequency filtering and selective amplification may transform an image diffused by cataract, for example, in a way that renders it more easily recognizable.

Temporal sensitivity

Our eyes are constantly in motion, a phenomenon that combines slow and fast components. Without these movements, the visual image would fade within a few seconds. Eye movements refresh the physiological processing of contrasts, preventing local adaptation. Thus, receptors and other cells are continuously exposed to temporal modulation. There is no evidence indicating that it makes much difference to retinal cells whether this modulation is caused by stimulus motion or by eye movements (Kelly, 1994), although higher level processing may give rise to saccadic suppression, a temporary reduction of contrast sensitivity during saccades. Without stabilizing natural eye movements, it is not possible to isolate pure spatial effects of a stimulus from temporal ones in psychophysical experiments. Many, if not most, psychophysical measurements of spatial contrast sensitivity have been made without such stabilization, and the results are commonly contaminated by temporal effects, just as in normal viewing situations.

Figure 4.28 shows how contrast sensitivity for a grating depends on the frequency of temporal variation. Threshold sensitivity increases with luminance up to about

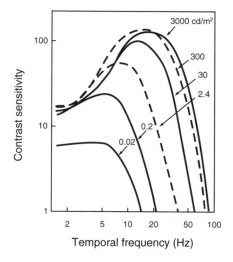

Figure 4.28 Contrast sensitivity (modulation sensitivity) plotted as a function of temporal frequency. Except at the lowest frequency, sensitivity and temporal resolution increase when photopic luminance increases. Maximum temporal resolution, i.e. the fastest flicker that can be detected under optimal conditions, is about 80 Hz (right-most curve).

300 cd/m^2, where flicker sensitivity is best between 10 and 20 Hz. The highest frequency that can be perceived as flicker under optimal conditions and maximum contrast is close to 80 Hz. The frequency at which consecutive phases blend and no flicker is detected for a given set of conditions, is called *critical fusion frequency* (the frequency at which the curves cross the *x*-axis). Critical fusion frequency rises with increasing luminance.

In fluorescent light bulbs that are driven by 50 Hz alternating current, the light flux varies at 100 Hz (because fluorescence is triggered in both phases of the current).

However, fluorescent coating on the inside of the tube has a certain afterglow so that the lamp is not totally dark when the current is zero. Different kinds of coating give different light modulation. Modern fluorescent lighting often has a high frequency light modulation, in the kHz range. Incandescent light bulbs have a relatively thick metal filament that does not stop glowing when the alternating current is zero. The modulation of the light flux is therefore less than for fluorescent lighting.

In the photopic luminance range, flicker sensitivity does not change with luminance in the low-frequency range (the left part of Figure 4.28). For slow temporal change, there is complete adaptation, and Weber's law holds. This is not the case for higher temporal frequencies. This means that adaptation does not contribute in the same way when the stimulus is modulated at high temporal frequencies. Then it is the amplitude of modulation, ΔL, and not the contrast, $\Delta L/L$, that is constant at threshold.

When a grating moves across the retina, the conversion from spatial frequency, f_s, and speed, v, in deg/s, to a temporal frequency, f_t, is simple:

$$f_t = vf_s \, (\text{Hz})$$

A sinusoidal grating with spatial frequency of 3 cycles/deg drifting with a speed of 2 deg/s, has a temporal frequency of 6 cycles/s.

When a red and a green light are interchanged in the same field at a low temporal frequency, we see the color change clearly. If the interchange is faster, at a temporal frequency of about 15 Hz or more, we see a steady and uniform yellow color mixture. When we change the relative luminance of the red and green stimuli, e.g. by increasing or decreasing the relative luminance of the red stimulus, the impression of flicker will change and reach a minimum or disappear at a certain red/green luminance ratio. This is exploited in the previously mentioned method of hetero-chromatic flicker photometry for determining equal luminance (see Figure 4.17). Several phenomena associated with flicker have been traced back to the behavior of magnocellular cells. These correlates will be discussed later.

Does the visual system perform spatial frequency analysis?

To be noticed, everyday stimuli must have a rather high contrast, far above threshold. Nonlinear intensity–response functions, like those in Figures 4.10 and 4.11, make it difficult to accept that the brain interprets visual information of high contrast as separate responses to constituent sinusoidal components in the stimulus, as in Fourier analysis of an image. In addition, cells in the retina and at later stages of visual processing have relatively small classical receptive fields, whereas Fourier analysis requires that the analyzing fields are very large, extending over the whole retina. The classical receptive fields of brain cells may have different size, orientation and resolution, but they are nevertheless confined to a small area. The analysis of size and form in these cortical cells is therefore most likely to be performed locally. However, in recent years evidence has accumulated that the classical receptive fields are, indeed, influenced by a larger area surrounding them (Valberg *et al.*, 1985a), indicating that there is a global influence on these cells as well. One other problem in accepting the idea of Fourier analysis, however, is the assumption of discrete frequency channels. It seems more likely that the size distribution of receptive fields at a particular retinal location is continuous, and that spatial frequency analyzers (channels) have a more or less continuous spatial frequency distribution.

Whether or not such discrete mechanisms are in fact a feature of the visual system, the underlying assumptions of Fourier analysis have proven valuable in image (stimulus) analysis. It is probably fair to say that these ideas have led to valuable contrast sensitivity tests (e.g. sinusoidal grating tests), even if one no longer believes that Fourier analysis is directly implemented in the primate visual system. The analogies are striking, however, and for a restricted linear domain of the intensity–response relationship of a cell, not far above threshold, one may well find the use of Fourier principles and methods of analysis instructive.

As we have seen, receptive fields can function as spatial analyzing units which, with the limitations mentioned above, can be represented by frequency channels. Taken together, these channels would be sensitive to about 20 different orientations, with a resolution of about 10°. There are several thousand such overlapping receptive fields at any location in the retina, and for every point in the field of view. Receptive field sizes range from a few minutes of arc in the fovea, to several degrees outside the fovea. Analyzing units for size must exist side by side with analysis mechanisms for all other features that are transmitted in parallel for each position of the retina, such as color, depth, movement, contour, etc.

Chromatic contrasts

The visual system deals with chromatic contrasts and luminance contrasts in distinctly different ways, indicating fundamental difference in the neural mechanisms for chromatic and luminance processing. One difference is the way in which diffusion of an image, for example a pattern of wide, sharp-edged bars, affects sensitivity to luminance contrast and chromatic contrast between areas of equal luminance (chrominance). Whether the image is deliberately transformed into a low-frequency sinusoidal grating pattern or unavoidably diffused by faulty refraction or cataract, sensitivity to equiluminant chromatic contrasts will be less affected than sensitivity to pure luminance contrast. For elements larger than about 0.5°, chromatic contrasts are more easily detected than achromatic ones (the cone modulation required for detection is smaller), and particularly so when the retinal image is blurred. In an experiment with subjects having unilateral cataract, Seim and Valberg (1988, 1993) confirmed this by demonstrating that chromatic contrasts were reduced far less than achromatic contrasts in the affected eye.

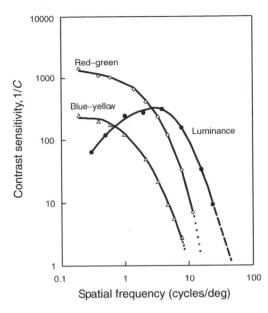

Figure 4.29 Spatial contrast sensitivity for luminance and isoluminant chrominance gratings. The sensitivity curves for pure color differences resemble low-pass filters, while the curve for luminance contrast sensitivity corresponds to a band-pass filter. When using a common cone contrast measure for luminance and chrominance, the results can be compared, and we see that contrast sensitivity is best for a red–green sinusoidal grating of low spatial frequency. Resolution is best for luminance contrast. The data are averages of 10 subjects.

Figure 4.29 compares red–green and yellow–blue chrominance contrast sensitivities with luminance contrast sensitivity for sinusoidal gratings. The curves represent the average values of the same group of 10 persons who participated in the experiments determining luminance contrast sensitivity of Figure 4.23. Here, the subjects were required to discriminate the bars of an equiluminant sinusoidal grating by means of color differences. For luminance contrasts, sensitivity peaked at 3–4 cycles/deg [0.17–0.13° bar width, Figure 4.23(a)], while for chrominance contrast, sensitivity increased monotonically with increasing bar width. The extrapolated contrast sensitivity curves for chrominance cross the x-axis in different places: at about 9–10 cycles/deg for yellow–blue gratings, and at about 15 cycles/deg for the red–green combination. This means that the spatial resolution is different for these color combinations, in both cases much smaller than for luminance contrast. However, when a detection criterion is used instead of chromatic discrimination, the two curves both show a break around 3 cycles/deg (Granger and Heurtley, 1973; Valberg et al., 1997) and a high frequency cut-off at about 46 cycles/deg.

It needs to be pointed out that there is no generally accepted method for comparing luminance contrast with red–green and yellow–blue chrominance contrasts quantitatively. When plotting their contrast sensitivity curves together in the same diagram, as in Figure 4.29, the common sensitivity measure is derived from cone modulations, combining the Michelson contrast of L-, M- and S-cones. For the M-cone we have

$$C_M = (M_{max} - M_{min})/(M_{max} + M_{min})$$

The letter M represents light absorption, or excitation, in the M-cones. The combined cone contrast is defined as:

$$C = [1/3(C_L^2 + C_M^2 + C_S^2)]^{1/2}$$

This contrast definition has the advantage that, for pure luminance changes, it reduces to the commonly used Michelson contrast. This linear, physical definition is the closest one can come to a physiologically relevant definition without introducing nonlinear responses and weightings of opponent mechanisms. These latter parameters and corrections are not fixed, dependent as they would be on stimulus parameters such as size and temporal modulation. This speaks in favor of the simple combination of cone contrasts given above (see also Brainard, 1996).

When the same quantitative measure is used for both luminance and chrominance contrasts, the respective sensitivity curves can be compared directly. Before the spectral sensitivities of the cones were firmly established, this was a uncertain method, but now, after the spectral sensitivities of the cones have become known with a relative high degree of certainty (Smith and Pokorny, 1975; Stockman et al., 1993; Stockman and Sharpe, 2000), the contrast definition above makes it possible to compare luminance with different color dimensions.

Weber's law and Weber–Fechner's law

As mentioned earlier, the more intense a stimulus background, L, the stronger the physical increment or decrement, ΔL, must be in order for us to detect a difference. In the photopic range, the ratio, $\Delta L/L$, is constant at threshold [Figures 4.6(c) and 4.7]. This empirical fact can be expressed mathematically in the form of the Weber law:

$$\Delta L/(L + L_o) = \text{constant}$$

where ΔL is the increment or decrement in L that gives a 'criterion threshold response'. L_o is a constant that often is regarded as equivalent to 'neural noise' or 'dark light'. L_o is small, and is only important for low values of background luminance, L. At photopic luminance levels the expression approximates $\Delta L/L$. An example of psychophysically measured values of the inverse ratio, the contrast sensitivity, $L/\Delta L$, as a function of L, is given in Figure 4.7.

The numerical value of the Weber ratio at threshold depends on the size of the object and its shape. The Weber ratio is smallest, and contrast sensitivity is highest, for fields larger than 0.2–0.5°. When the contrast borders are sharp, contrast sensitivity does not change much for larger objects, contrary to the case of sinusoidal gratings, where sensitivity is significantly reduced for wide bars (Figure 4.23). Weber law behavior does not apply to small details, as can be seen from Figure 4.24. Here, it is the threshold increment or decrement, ΔL, alone, that is constant at threshold, and not the ratio, $\Delta L/L$.

Scales of lightness contrast

The German physicist Gustav Theodor Fechner (1801–1887) took advantage of the fact that the Weber ratio is constant at threshold to establish a mathematical relationship between physical luminance and the subjective impression of *lightness*. The idea, which according to Fechner occurred one morning while still in bed, was that the sensory magnitude, Δr, corresponding to the smallest noticeable relative difference, $\Delta L/L$, could be chosen as a constant. Fechner assumed that this 'sensory difference', Δr, which he proposed as a unit for sensory difference, was proportional to the Weber ratio. If we assume this to hold independently of background luminance or color, we can write:

$$\Delta r = c\,\Delta L/L$$

where the constant of proportionality, c, determines a suitable unit for the sensory magnitude. This equation is often called the *Weber–Fechner law*. For high values of luminance, where Weber's law holds, we omit L_o in the general expression of the Weber ratio. Fechner treated Δr and ΔL as differentials, dr and dL, and integrated the equation above. This integration is based on the assumption, later considered rather doubtful, that one can add the sensory threshold values, dr, to a finite sensory magnitude R. The integral is:

$$R = \int dr = \int dL/L$$
$$R = c\,\ln L + A$$

where A is a constant, and ln is the natural logarithm. We can derive an expression for the finite integral between the values L_t and L, where L_t is the smallest detectable value of the luminance L. If we set $R = 0$ for $L = L_t$, the equation will be simplified to *Fechner's law*:

$$R = c\,\log L/L_t$$

In this way the sensory magnitude, R, which represents the magnitude of the subjective experience of lightness (or brightness) of a stimulus with luminance L, obeys the following expression: $R \approx \log I - \log L_t$ (where 'log' means ^{10}log).

It is important to notice that the sensory magnitude, R, is not measurable in the physical sense: according to Fechner's theory, it is the subjective impression of intensity or lightness magnitude. It may be expressed by a subjective scale of lightness or brightness, i.e. by assigning numerical values to these impressions. For instance, the German DIN system uses the logarithm of reflectance to construct a gray scale with perceptually equal step sizes. While this logarithmic relationship is better than a linear scaling of reflectance, or of luminance ratio, it must nonetheless be considered as an approximation that has validity within a limited range of stimulus intensities.

A similar logarithmic relationship is used to establish a scale of physical sound intensity, measured in W/m^2. The perceived loudness measured in decibels (dB) is:

$$D\,(\text{dB}) = 10\,\log(I/I_t).$$

Here, I_t is the sound intensity reference value of 10^{-12} W/m^2, which corresponds to the lowest detectable sound intensity (threshold).

Taking neural intensity–response curves as the starting point, Seim and Valberg (1980, 1986) developed a different formula for the scaling of lightness, V_Y, in perceptually equal gray steps:

$$V_Y = 40(Y - 0.43)^{0.51}/[(Y - 0.43)^{0.51} + 31.75]$$

where Y stands for the luminance ratio between the reflecting probe and a white surface in percent,

$$Y = 100\,L/L_w.$$

As shown in Figure 4.30, this formula reproduces the value scale, V, for grays in the Munsell system with a high degree of accuracy. The value scale was arrived at in an experiment using a medium-gray background, with a reflection factor of 0.2 surrounding the target. By extrapolation, this formula can also be used to calculate subjective brightness of a range of self-luminous surfaces (unrelated colors) with values of the luminance ratio Y greater than 100 percent.

We have seen many examples of how the contrast and lightness scaling of surfaces depend on the surroundings. Lightness scales for probes embedded in surrounds of different reflectance values are shown in Figure 4.31, where $Y_b = 1$ percent implies a completely black surround. In this case, small changes in the reflectance, β, of a probe will lead to large changes in the perceived lightness (see Figure 4.32). For high reflectance values of the probe, however, relatively large changes in reflectance fail to induce significant brightness changes. A medium-gray surround, $Y_b = 20$ percent, gives a more shallow curve, as in Figure 4.30. For $Y_b = 100$ percent, which indicates a white surround, one approaches, without quite reaching, a linear relationship (dashed curve) between reflectance and lightness.

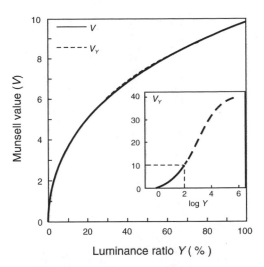

Figure 4.30 The Munsell value scale, V, for the lightness of related colors (solid curve) compared with a mathematical response function V_Y (dashed curve; see text). The curves are practically identical. V_Y suggests the extrapolated relationship, shown as the dashed curve in the inset, between the luminance ratio, Y, and the brightness of stimuli brighter than the adaptation background.

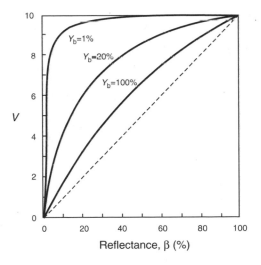

Figure 4.31 The scaling of lightness, V, for probes embedded in backgrounds with different reflection factors, Y_b (given in percent). Against a black background ($Y_b = 1\%$), a probe surface with a small reflection factor, β, will look relatively light, while the same object will be much darker when seen against a brighter background ($Y_b = 20$ or 100%). All curves are normalized to a maximum value of $V = 10$ for 100% reflectance of the probe.

Figure 4.32 Demonstration of simultaneous lightness contrast and the influence of different surrounds (see Figure 4.31). Four identical series of gray squares appear different when embedded in light and dark surrounds (see also Figures 1.5 and 1.6).

In the equation above, Y is the ratio, expressed in percent, between reflection factor, β, of the probe and that of a white reference of pressed magnesium oxide with a reflection factor of 97.5 percent. This means that $Y = 102.568$ for value, $V = 10.0$.

As shown in the inset to Figure 4.30, the formula describes a sigmoid curve with the same form as the curves in Figure 4.15(b). This means that the curve has a large range in the middle where the subjective lightness is proportional to the logarithm of the luminance ratio Y. A logarithmic relationship was also used in old astronomical systems for classifying stars into different magnitude classes according to their light intensity. The magnitude m of a star depends on the illumination I on the retina according to the formula: $m = a - 2.5 \log I$, where a is a constant. The weakest star we can see in the sky has a magnitude of 8.5, whereas the sun has magnitude -26.7. From the formula we find that the ratio between the retinal illuminance from the star and the sun has a logarithm of 14.1. This means that, for a given pupil size, the sun leads to 100 000 000 000 000 times higher illumination on the retina than the faintest star that we can see with the naked eye.

Stevens' law

According to empirical works of Miescher *et al.* (1982), a gray scale with 10 equal steps from black to white on a white background can be approximated by the power function:

$$V \sim 10\,\beta^{0.5}$$

The symbol '$\sim$' means 'corresponds to'. As we have seen in Figure 4.31, the curves for V become steeper when using a middle gray background ($\beta = 20$ percent), and the formula then changes to

$$V \sim 10\,\beta^{0.33}$$

As before, V represents the subjective estimate of lightness, while β is the reflection factor. Figure 4.32 gives an idea of how simultaneous contrast changes the appearance of physically identical squares when they are embedded in different backgrounds. Further examples are given in the section 'Visual illusions' in Chapter 1 (p. 18).

The American psychologist S. Stevens developed a power equation relating the magnitude of a subjective attribute to the physical stimulus, implying that the perceptual magnitude, R, is proportional to the stimulus magnitude, I, raised to a power n:

$$R \sim \text{const.}\,I^{n}$$

For instance, the subjective perception of the weight of an object (whether it is light or heavy) will be a power function of its mass in kilograms. Stevens found that a power law was valid within several sensory modalities, such as the perception of weight, heat, pressure, sound, light, etc. At threshold, Stevens's power law has the following form:

$$\Delta r / R \sim n\Delta I / I$$

According to this law then, it is not the subjective difference at threshold, Δr, that is proportional to the Weber ratio, as in the Weber–Fechner law. Rather it is the *relative response difference*, $\Delta r / R$, which is proportional to the Weber ratio. The power law above is the result of integrating this relationship to get an expression for the perceptual magnitude R.

This idea was not new. The Belgian physicist J. Plateu and the German philosopher F. Brentano both anticipated such relationships between subjective impressions and physical magnitudes. This can be illustrated with a widely shared attitude to the value of money: if you receive 10 dollars to add to the 100 you had before, your sense of value will be similar to what you might feel upon receiving 100 dollars when you already have 1000. The value of what you receive depends on the value you attach to what you already have.

The logarithmic form of the Weber–Fechner law is similar to Stevens's law for $n = 0.3$, and this is a power found in many of Stevens' experiments, including the loudness of sound. Therefore, the two laws are not as different as they might have seemed, if we keep in mind that relatively small changes in the stimulus parameters and in the viewing conditions influence the results of scaling experiments (Stevens, 1961).

Vision loss

The term 'vision loss' implies a reduction of a person's visual capabilities to such an extent that it impedes visual activities. It is a general term, including both total loss (blindness) and partial loss (*low vision*). A person is considered to have a severe vision loss when he is unable to read newspaper print with standard optical correction, or when his peripheral visual field is severely reduced. This applies to a heterogeneous group of people, from the totally blind to those with quite useful visual capabilities. Some are born with a visual problem and others have lost their vision as a result of injury or disease. The expression 'low vision' is recommended to be used for lesser degrees of vision loss, where individuals can be helped by vision enhancement aids and devices (WHO, 2001).

Impaired vision is commonly divided into two separate categories, depending on whether the primary site of the defect is in the optic media or in the retina and higher brain centers. Optical problems give poor image quality despite normal transmission of signals from the retina to the visual centers of the brain. For people with a cortical defect, comprehension of the visual image might be the larger problem.

Which type of visual loss renders a person most disabled? Is it functions attributed to the eye media, such as acuity and contrast sensitivity, or functions associated with the higher brain processes, such as detection, recognition, orientation, localization, memory, etc.? Problems associated with imaging can, to a large extent, be compensated for by the use of special optics. Strong positive lenses, for example, provide high magnification and allow a person to see fine details even when acuity is strongly reduced, and telescope systems can help in detecting small details at a distance. Visual resolution or clarity is important, but so is an intact visual field. Imagine the difficulty you would have in moving about if your visual field were restricted to only a few degrees. This is, in fact, what happens in a disease called retinitis pigmentosa (RP). Failure of depth perception does not have the same effect, and while color vision contributes to discrimination and object recognition and to the quality of life, it is not crucial for coping with the activities of daily life. Cortical damage can lead to a variety of visual problems, some of which will be discussed later under the heading cerebral visual impairment (CVI).

The severity of a disability resulting from reduced visual ability must be judged in relation to the tasks at hand. Poor acuity is a greater problem for a person who needs to read than for one who does not. Thus, one may conclude that it is the requirements of one's profession or of society at large – its adoption of some standards related to normality – that makes a person disabled, more than the visual defect itself. Such standards may, for instance, be the print sizes commonly used in newspapers, periodicals and on the internet, acuity limits required to possess a driver's license, or the illumination conditions in shops, streets, official buildings, homes, etc. It would therefore in many cases be possible to make environmental improvements that would lower the threshold for activity and participation for a person with low vision.

Young people with normal vision for their age group need not think about visual problems at all. In fact, it is hard to make them understand the ways in which poor vision can affect daily life. Some examples are the failure to detect low-contrast objects, poor color discrimination, not being able to see fine details, feeling at loss in a new environment, the problems of interpreting the environment and predicting the movements of people, cars and bicycles, etc. Normal vision is defined as a decimal acuity equal or better than 1.0. This corresponds to resolving a detail of about 0.17 mm at an arm's length. Most young, normally sighted people perform much better than this. They are capable of resolving details half this size and of detecting luminance contrasts as small as 0.2–0.3%. The former corresponds to about 500 steps on a gray scale from black to white. They have what we might call a substantial visual reserve, and thus they encounter few visual problems in every day life. However, in extreme situations, such as in some sports, where speed and reaction time are important, this visual reserve may still decide between a winner and a loser.

As one gets older, say entering the forties and fifties, the depletion of this visual reserve may start to become noticeable. The most common development is a reduced ability to accommodate and to discriminate between small contrasts. Progressive glasses usually take care of the accommodation problem. Even if this normal development continues until vision enters the lower normal range of acuity, one can still cope pretty well because society's standard for normality has been adapted to this low range. However, if the deterioration of functional vision progresses much further, either because of some ailment like cataract, which is common among the elderly, or for other reasons, one may eventually be left with a visual problem. Vision loss is usually defined by the failure to pass certain tests. It is a restriction that cannot be corrected by regular spectacles or contact lenses. According to the World Health Organization (WHO, 1973), it is the visual resolving power and restrictions of the visual field that determine whether or not one belongs to the group that is entitled to social benefits. The upper limit is a decimal Snellen visual acuity of 0.33 after correction, while a visual field that is restricted to equal to or less than 10° in radius around the central fixation qualifies the person as blind even if his or her acuity is good.

Let us take a brief look at some functional aspects of a few frequently encountered causes of vision loss.

Age-related macular degeneration

Age-related macular degeneration (AMD) is an eye disease that is most frequently encountered in elderly people over 65 years. It is the most common cause of legal blindness in the western world. AMD includes a whole range of progressive and degenerative changes in the human macula in which the normal functioning of the photoreceptors of both eyes is disrupted. It leads to areas of the retina developing low resolution and poor contrast sensitivity (the so-called 'scotoma' or blind spot in the visual field). Since the macular region of the retina is where we have sharp, high-acuity vision, central scotoma are usually devastating for the individual. He or she is unable to focus normally or to read newspaper print, and has problems coordinating the images of the two eyes. Some AMD subjects therefore need to learn to use another, eccentric area during fixation and reading, the so-called preferred retinal locus of fixation (PRL). Although color vision and stereoscopic vision are much reduced, peripheral vision is seldom affected. Therefore the visibility of larger objects in the peripheral field of view is nearly normal, and orientation and mobility are relatively good. AMD subjects need more time than normal to recover from the glare effects of bright lights, and this interferes with vision at night. In addition to a distorted vision, about 10 percent of AMD subjects experience visual hallucinations (flashes, lines, dots, grids, brickwork, etc.) within the scotoma at some stage of the disease (Charles Bonnet Syndrome). Currently, there is no effective treatment for AMD, but laser treatments in the exudative (wet) form of AMD seem to slow down the progression of the illness.

Functionally, AMD is a rather unspecific diagnosis, and the diagnosis alone does not permit a reliable prediction of a person's visual performance, for instance in reading. However, several reports state that tests of contrast sensitivity provide valuable information as to monocular and binocular visual performance. For AMD subjects who are able to use their right visual field during reading, maximum contrast sensitivity of the best eye shows good correlation with reading speed. A proper adjustment of the level of illumination can often help AMD subjects to improve their reading performance (Fosse and Valberg, 2001).

Figure 4.33(a) gives examples of individual contrast sensitivity cures for AMD subjects (Valberg and Fosse, 1997; Fosse and Valberg 2001), and Figure 4.33(b) shows the sensitivity range for 15 AMD subjects with no other ailment affecting visual function. In Figure 4.33(a), the curves all cross the x-axis in the same place, despite differing significantly from each other at medium and low spatial frequencies. While visual acuity measures might lead us to think that these persons see equally well, their contrast sensitivity tells another story. These examples demonstrate that, even if sensitivity to fine detail is the same for two people, their contrast sensitivity for large objects can nevertheless be very different, and thus lead to problems with orientation in an unfamiliar environment.

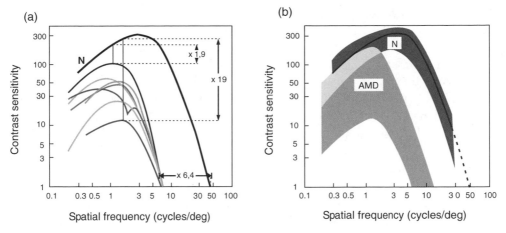

Figure 4.33 (a) The colored curves show spatial contrast sensitivity for seven subjects with AMD, an eye disease that attacks the receptors in the macula. N is the normal curve. This figure demonstrates that, while grating acuity (point of intersection with the *x*-axis) was severely reduced and rather similar for these subjects, their best contrast sensitivity varied by a factor 10. (b) The green area (AMD) shows the range of contrast sensitivities for a group of 15 subjects with AMD. The corresponding range for 11 normally sighted persons between 40 and 60 years is also shown in red (N). The curve N in (a) represents their geometric mean. The difference between the normal and AMD groups is greater at high spatial frequencies, while there is some overlap for low spatial frequencies (shown in yellow). See also Color Plate Section.

Cataract

Cataract, too, is a common ailment in elderly people. It affects the clarity of the eye media (crystalline lens and vitreous) and causes blurred vision that cannot be corrected by optical means. Consequently, visual acuity and contrast sensitivity are reduced. As in AMD, difference in contrast sensitivity among subjects, despite similar acuity, is typical for patients with cataract. The cataract eye is particularly susceptible to glare from bright light sources. The light from a glare source is diffused by the lens and covers the retinal image like a veil that reduces all contrasts.

It is assumed that about half of the population over the age of 60 will develop cataract. This age group is currently growing larger, and in Scandinavia it currently accounts for about 25 percent of the population. This high number means that cataract is a significant problem before intervention. When cataract has reached an advanced stage, the lens can be replaced surgically by an artificial interocular lens. This is now considered to be a fairly straightforward procedure, and it normally restores vision to an acceptable level.

Binocular inhibition

When corresponding points in the two retinas are stimulated simultaneously, binocular summation normally results in improved acuity and better contrast sensitivity. In normal vision, contrast sensitivity of both eyes together is about 40 percent better than for monocular vision. Failure of binocular summation may occur when the two eyes are unequally illuminated (Fechner's paradox; Fechner, 1966), and for some subjects the result is inhibition. The inhibition is stronger the greater the difference in illumination of the two eyes. In many AMD subjects and in others with unilateral cataract, for example, the poorer eye disturbs vision in the better eye (Pardham and Elliott, 1991; Valberg and Fosse, 2002). This sometimes results in having a poorer binocular contrast sensitivity function than that for the better eye alone. In other words, binocular inhibition leads to a subject seeing better with one eye than with two.

When comparing monocular and binocular contrast sensitivity and acuity for a group of 13 subjects with AMD, we found that most of them showed greater reduction in binocular functions than expected. For a range of spatial frequencies, we found a better performance with one eye than with both eyes in eight subjects. This may have implications for reading and orientation and should be considered in the rehabilitation practice. Figure 4.34 shows an example of spatial contrast sensitivity functions for monocular and binocular vision for our reference group of normal subjects (a), and for an AMD subject with strong binocular inhibition (b). In the latter case, binocular inhibition was severe for all spatial frequencies. When contrast vision was measured in terms of integrated contrast sensitivity, ICS (see p. 186), only two out of the 13 AMD subjects had a normal monocular to binocular sensitivity ratio (M/B ratio in the figure). The current conjecture is that binocular inhibition and failure of summation occur when non-corresponding retinal areas are used for binocular viewing. It remains to be seen if this also explains Fechner's paradox.

Glaucoma

Glaucoma is a disease of the optic nerve that is usually caused by an elevated pressure within the eyeball. This causes visual field loss, usually starting in the periphery, which can lead to severe disability. The disease does not cause blindness unless acute, but it may result in blindness if it is not controlled at an early stage. If detected early, the pressure can be controlled by means of medication, usually by eye drops.

Diabetic retinopathy

Diabetes may, after some years, lead to damage to the small blood vessels in the retina and this may in turn lead to leakage of serum or blood into the fluid filling

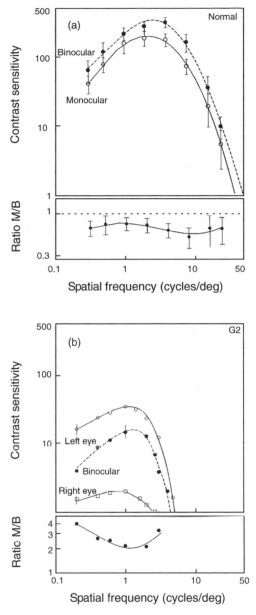

Figure 4.34 (a) Spatial contrast sensitivity for monocular and binocular vision for our reference group of normal subjects. The lower panel shows the ratio M/B of monocular to binocular sensitivity of the upper panel. (b) The results for a subject with AMD. For normal subjects, binocular vision is better than monocular vision by about 40%, but many AMD subjects have a higher monocular sensitivity, demonstrating substantial inhibition of binocular vision by the poorer eye (from Valberg and Fosse, 2002).

the eyeball. This causes macular edema (swelling of the retina) which again leads to blurred vision. The visual loss ranges from moderate to severe. At an advanced stage, internal bleeding usually causes immediate and severe visual loss. Control of sugar level in the blood and a proper diet may postpone the onset of the disease.

Retinitis pigmentosa

Classical retinitis pigmentosa (RP) is a progressive, inherited disease, in which the rods in the peripheral retina degenerate, the visual field becomes narrower and night vision deteriorates. The more advanced the illness, the more restricted the field of view. The subject develops tunnel vision, where all peripheral vision is destroyed, leaving only the central macula functioning. Mosaic RP is another form where the affected retina appears patchy. A person with RP has difficulty moving about and detecting obstacles and probably uses a white cane, but may nevertheless be able to read newspaper print – an odd combination for those not familiar with the disease. In the later stages of the disease, RP may cause blindness.

5 | Color

Color order systems

Let us look at how the wealth of color experience can be given a simple geometric order based on particular perceptual dimensions and color attributes, more or less independent of their physical properties. Commonly, the perceived color of a reflecting surface is described by means of the three dimensions *hue*, *saturation* or *color strength*, and *lightness*, and this gives rise to three-dimensional *object color solids* and color atlases [see Figures 5.1 and 5.4(a)]. The hue dimension separates, for instance, a reddish orange from a yellowish orange, and all hues can be represented on a circle, as in Figures 5.2(a) and 5.3. A color's saturation, or color strength, indicates how much the chromatic component departs from achromatic, white or gray. Saturation corresponds to a fraction of the radius of the full color circle. Lightness, or value, is an intensity dimension that for surface colors increases with increasing reflectance factor. It is low for blackish colors (brown, olive) and high for whitish ones (pastel colors). Willhelm Ostwald (1921) used additive mixtures on rotating disks to specify colors within the geometry of a triangle. This principle was later taken over by the Natural Color System (NCS) and developed into an analogy where the perceived amounts of chromaticness (*c*), blackness (*b*) and whiteness (*w*) should sum to unity:

$$c + b + w = 1$$

Thus, the unidirectional lightness dimension was here divided into something composed of chromaticness, as well as whiteness and blackness.

Even if most color systems are constructed on the basis of hue, saturation and lightness, or on closely related attributes, they differ in their scaling of these properties (Derefeldt, 1991). Different applications of colors, in the graphics and

Light Vision Color. Arne Valberg
© 2005 John Wiley & Sons Ltd

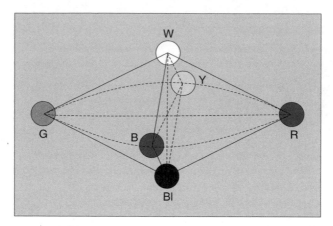

Figure 5.1   A three-dimensional color solid in the form of a double cone. The hue circle forms the common base of the two cones, and the neutral grays are situated on the central axis. Object colors find their natural position within this double cone. Bl = black; W = white. (See also color plate section.)

the paper and textile industries, in interior design and painting, in lighting engineering etc., have given rise to different practical solutions of the three-dimensional ordering of color stimuli. Many of the color systems are hybrid in that they have been constructed on compromises between perceptive arrangements and technical require-ments. Equal perceptual scaling of visual differences is one important requirement for a system that is to be used for product control. This has been attempted in the widely used *Munsell System.*

The best known color-ordering systems today, where color atlases are available, are the Munsell System from the USA, the Swedish NCS (*Natural Color System*), and until recently the German DIN system (Deutsche Industrienormen). The structure of the Munsell system is shown in Figure 5.4(a). A vertical cut through the color solid is displayed in Figure 5.4(b), with the coordinates *chroma* (color strength) and *value* (lightness). Ideally, in this system all chroma steps and all value steps should have the same perceptual difference (which is, however, only approximately fulfilled). These differences were determined empirically by over 70 subjects participating in scaling experiments. Because of the established use of the Munsell system in determining color tolerances in industrial processes, i.e. in the dying of ceramic tiles and textiles, this system has served as a reference standard for quantitative color scaling and discrimination, and also for color vision models.

The NCS is an attempt to realize a system based on Leonardo da Vinci's simple colors and Hering's unique colors by using them as reference points under well-defined viewing conditions (Johansson, 1952). The hue circle has been scaled in terms of relative contributions of elementary hues (Figures 5.1–5.3).

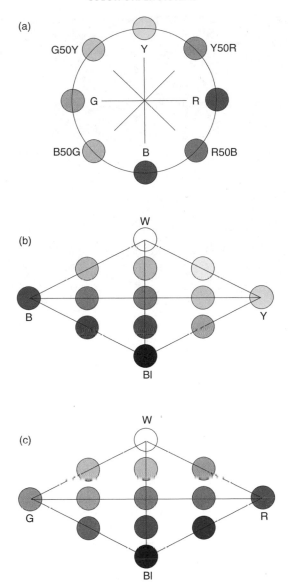

Figure 5.2 (a) A hue circle with the elementary hues Y (yellow), R (red), B (blue) and G (green) on the axes. The hues in between contain proportions of the two nearest elementary hues. (b, c) Examples of opposing hue triangles from vertical cuts through the color solid of Figure 5.1. The horizontal and vertical dimensions are chromatic strength and lightness, respectively. (See also color plate section.)

Unique yellow is characterized by it being 'neither reddish nor greenish'. It is thus determined purely subjectively by means of the other, neighboring unique hues on the hue circle. Unique blue satisfies the same criterion. The yellow–blue pair is opponent in that the two color percepts mutually exclude one another. No object color is seen as

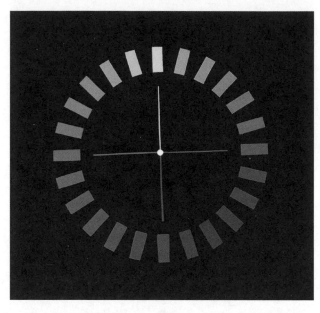

Figure 5.3 A hue circle with 24 steps (after Miescher *et al.*, 1961), here reproduced from the original (the printing process will have reduced chromatic strength and has introduced hue distortions as well). Opposing unique hues on the axes. (See also color plate section.)

both yellowish and bluish at the same time, in the way that purple can be said to be perceptually composed of blue and red. The same reasoning applies to the 'neither yellowish nor bluish' unique red or unique green.

All other hues are experienced as being a transition between two unique, or *elementary hues* (e.g. orange is a transition between yellow and red). This leads to an arrangement of surface colors in a three-dimensional color solid, e.g. like the double cone in Figure 5.1. The vertical central axis represents all neutral, *achromatic* colors between black and white, whereas the chromatic colors of maximum color strength are situated on a hue circle, in a common plane of the double cone [Figures 5.2(a) and 5.3]. Figure 5.3 reproduces a 24-step hue circle made by the Swiss chemist and color scientist Karl Miescher (1892–1974). In an experiment with 28 subjects, Miescher used Hering's principle to construct a hue circle for daylight illumination (Miescher, 1948; Miescher *et al.*, 1961). A vertical cut through the solid of Figure 5.1 displays schematically the dimensions of color strength and relative lightness (Figure 5.2(b) and (c)]. A similar schematic order of colors was described in 1611 by the Swede A.S. Forsius in his book *Physica*.

Before we present other, psychophysical color systems, we need to take a closer look at the physical properties of light and matter that are of importance for color perception and for color technology.

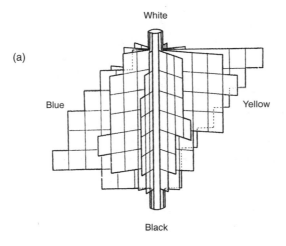

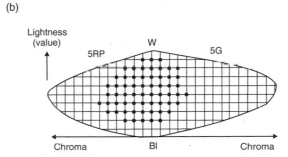

Figure 5.4   (a) The three-dimensional arrangement of color chips in the Munsell system. (b) The colors in a hue plane of the Munsell system have the coordinates lightness (value) and color strength (chroma). Of all colors that are theoretically possible, the black dots represent the color chips that are realized in the atlas. The outer boundary is determined by the optimal colors. RP = red–purple; G = green; W = white; Bl = black.

The physics of color stimuli

Spectral distributions

Under normal lighting conditions, the color of a surface depends on several physical factors, such as its structure (if it is matte, polished, etc.), its spectral reflectance, and the geometry and spectral distribution of the illumination. Yet the physiological adaptation of the eye due to successive contrast and the surroundings (simultaneous contrast and color induction) is equally important. Let us first take a closer look at the effect of the spectral distribution of radiant power reflected or transmitted by a color stimulus.

Surfaces can be characterized by spectral reflection curves that quantify how much of the incident light is reflected back for each wavelength of the visible spectrum relative to an ideal white surface. Some examples of typical reflection curves for surfaces are shown in Figure 2.25. In the same way, transparent colored glass and gelatin filters possess spectral transmission curves for the light that is transmitted. The sum of reflected, transmitted and absorbed intensities of an object must equal the incident intensity. For most objects in nature, the spectral distribution curves have a relatively slow and smooth variation across the visible spectrum.

It has been demonstrated that the spectral reflection curves, β_λ, of natural objects can be described, with good approximation, as the sum of as few as three basis spectral curves. For instance, three curves ($s_{1\lambda}$, $s_{2\lambda}$, and $s_{3\lambda}$) with different weights (a, b and c) add up to the reflectance curves, β_λ, of leaves, fruit, tomato, skin, flowers, etc.:

$$\beta_\lambda = a\,s_{1\lambda} + b\,s_{2\lambda} + c\,s_{3\lambda}$$

It is important to note, however, that the eye does not function well as a spectral analyzer; it cannot retrieve the spectral distribution of light reflected from a surface. The popular view that surfaces which reflect long-wavelength light are always red, and that surfaces which look blue always reflect short-wavelength light, is correct for objects that are viewed in neutral or dark surroundings, like in the laboratory. Under more normal viewing conditions, however, in daylight or in broad band artificial light, the reflection properties of the surroundings of an object tend to contribute more to the object's color than does the illumination (see later sections 'Induced colors', p. 263 and 'Chromatic adaptation' p. 265).

There exists an infinite number of different spectral power distributions that for the same viewing conditions give rise to the same color. For instance, sunlight shining on a white wall has a continuous spectrum that includes all wavelengths [see Figure 2.23(c)]. This color can be matched in the laboratory by a mixture of two monochromatic wavelengths, one blue of 480 nm and the other one yellow of about 580 nm. Another example of different spectra of lights with the same color appearance is given in Figure 5.5. Stimuli that have the same color appearance, despite different spectral distributions, are called *metameric colors*.

Metameric surface colors are usually equal only under one type of illumination. Therefore, the car paint that matches your old car color in the artificial illumination of the garage may not match it in daylight.

Subtractive mixtures

A so-called 'subtractive color mixture' is not a real *color* mixture. When a yellow and a blue filter are placed one after the other in the light path of a projector, the resulting color on the screen could equally be purple instead of the expected green. It is not unusual to find such filter combinations in an assortment of gelatin filters. Thus, you

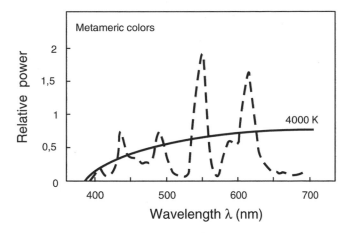

Figure 5.5 Two lights with metameric spectral distributions, meaning that the color appearance of the two lights will be the same (and in this case equal to white light with a color temperature of 4000 K).

cannot trust filters, or paints from different companies, that appear to have the same color to give the same result in a subtractive mixture. In this case it is not the perceived color of the filters that determines the mixture, but their spectral transmission.

Given that the spectral transmission factors of two filters are $\tau_{1\lambda}$ and $\tau_{2\lambda}$ and that the spectral power distribution of the (white) light is P_λ, then the spectral distribution, Φ_λ, of the color that results on a white screen from a subtractive mixture will be:

$$\Phi_\lambda = P_\lambda \tau_{1\lambda} \tau_{2\lambda}$$

This relationship is illustrated in Figure 5.6, and we may conclude from this figure that it would have been better to call these two mixtures *multiplicative* instead of subtractive. When light from the light source is transmitted through one filter, some of the light is removed (absorbed and transformed into heat), and the light loses some of its luminance. In the example of Figure 5.6(a), white light first passes through a blue filter, and the bluish light then passes through a yellow filter. As we can see from the dashed curve, describing the spectral transmission of both filters taken together, there is a maximum transmission in the middle of the spectrum. Therefore, the resulting color will normally appear greenish. However, as mentioned above, it is possible to find another blue and another yellow filter, leading to a mixture color that is not green, but purple. This is illustrated in Figure 5.6(b). Consequently, it is the spectral transmission curves of the component filters that are the most important here, and not their color.

Many of the great old painters used a glazing technique of putting a colored transparent layer over an opaque background. The light reflected from the opaque

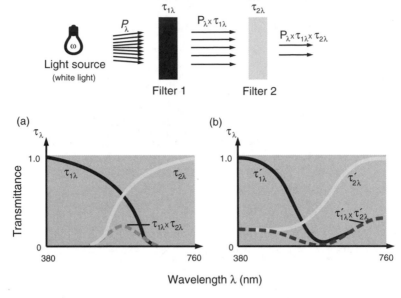

Figure 5.6 (a, b) Examples of multiplicative (subtractive) color mixtures. Individually, filters 1 and 1′, and filters 2 and 2′ in (a) and (b) give rise to pairwise metameric colors when transmitting light from the same source (color 1 equals color 1′ while color 2 equals color 2′), despite the fact that the transmission curves $\tau_{1\lambda}$ and $\tau'_{1\lambda}$, and $\tau_{2\lambda}$ and $\tau'_{2\lambda}$ are different. Because the spectral transmission curves are different, the spectral distribution of the resulting color mixture resulting from combining filters 1 and 2 and filters 1′ and 2′ will be different. The dashed green curve in (a) illustrates that the mixture is greenish and the dashed purple curve in (b) that the mixture is purple. The different mixture colors demonstrate that multiplicative (subtractive) color mixtures depend more on the spectral distributions of the component colors that on their color appearance. See also Color Plate Section.

background would be filtered by the transparent layer and acquire a deep, lustrous color effect. The colors produced by the glazing technique follow the subtractive principle. A yellow glaze over an opaque blue layer usually results in a green color.

Additive color mixtures

When two colored lights are projected onto the same region on a screen, they are also superimposed on the retina. The resulting additive color mixture is brighter than each of the components. Figure 5.7 demonstrates an additive mixture of yellow and blue, derived, for instance, from the same color filters as in Figure 5.6. The resulting color in this additive mixture is white and, in contrast to the multiplicative mixture of Figure 5.6, the spectral distributions of the yellow and blue colors do not matter for the outcome of the mixture. All yellow colors that have the same appearance regardless their spectral distributions (the stimuli being metamers), will yield the same white in an additive mixture with metameric blue stimuli.

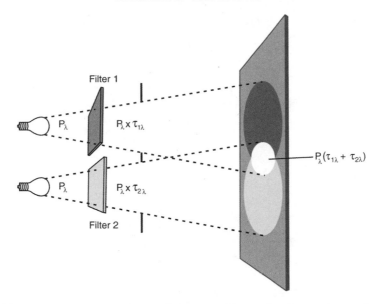

Figure 5.7 An additive color mixture of the same blue and yellow colors as in Figure 5.6 produces a white mixture on the screen, independently of the spectral distribution of the contributing colors. In an additive mixture of colored lights, only the colors matter; their spectral distribution is irrelevant. Colors of the same appearance will always give rise to the same additive mixture color, regardless of their spectral distributions. See also Color Plate Section.

The principle of equality and color matches

The laws of color mixtures have been important for our understanding of color vision, and they have provided the theoretical basis for a flourishing color technology. Experiments with additive light mixtures have shown that a subject with normal color vision can match a given color with a mixture of only three *base colors*, often called *primary colors*. However, there are a few conditions for the best choice of these three base colors: neither of them must be simply a mixture of the two others, and 'negative' mixtures are allowed. Let us assume that, in field A of Figure 5.8(a), light with the color **F** is projected on a white screen, and that its color is matched by an additive light mixture of a red, a green and a blue primary in field B. When the mixture is just right, the halves A and B merge to form a homogenous disk. To achieve this, it is sometimes necessary to add one of the primary colors to the left side, in field A. This is what is called a negative mixture and, as we shall see later, it has some consequences for the arithmetic of color matches.

We have already mentioned that, when two metameric lights, for instance two yellow stimuli with different spectral power distributions, are added to two other metameric stimuli, for instance two spectrally different stimuli of the same blue colors, the mixture color of the yellows and the blues will be equal (in this case white). This is the same as to say that if we have two metameric colors A and B, such that 'color A = color B' (the yellows), and we add two other metameric colors C and

(a)

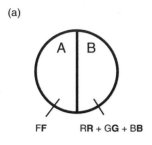

FF   RR + GG + BB

(b)

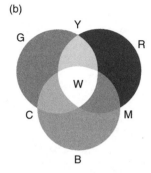

Figure 5.8 (a) In a colorimetric experiment, a bipartite field is used to compare the color F in the left half-field (A) with an additive mixture of red (R), green (G), and blue in the other half. (b) The mixture colors are R + G = Y (yellow), G + B = C (cyan), B + R = M (magenta). Opposing colors (blue and yellow, red and cyan, green and magenta) all add to white and are therefore complementary colors. See also Color Plate Section.

D (the blues where C = D) to A and B, then for the resulting mixture colors (which are white) we have:

$$A + C = B + D$$

This additivity law applies to color appearance and holds independently of the spectral composition of the colored lights. The colors, and not their spectral power distributions, decide if the equality holds true or not. Thus *for stable and equal viewing conditions, the color of a subtractive color mixture is determined by the spectral power distributions of the mixture components, whereas in an additive mixture only the color appearances matter.*

Moreover, when the intensities of two lights that are equal (A and B in Figure 5.8) are increased in the same proportion, the two lights are still equal (the color appearance of both stimuli undergoes the same change). This is implicit in the law of proportionality (if A = B, then $\alpha A = \alpha B$, for $\alpha > 1$). This law holds only for photopic conditions, where the rods do not influence the result. When we reduce the intensity of two metameric lights, the relationship is more complex, because the rods complicate matters for mesopic and scotopic luminance. Transitivity, too, holds for additive light mixtures; if $A = B$, and $B = C$, then $A = C$. These linearity laws were implicit in the German mathematician H. Grassmann's (1853) three laws for additive color mixtures. To keep things simple, we assume that symmetry, $A = B$ and $B = A$, applies with regard to interchanging the retinal positions of fields A and B.

So far we have only dealt with phenomena that do not change the equality between two metameric color stimuli. We have used the normal eye as an equalizing instrument, and in such experiments the only task for the subject is to decide if two lights match or not. All other observations on their appearances, whether they are light or dark, saturated or whitish, red or green, etc., are of no interest in this context. The irrelevance of the qualitative appearance of colors in a color match seems to make matches of little use for specifying colors. However, despite this restriction, the equality principle has been sufficient as a basis for the development of modern color technologies.

The principle of equality is basic to all physical systems for the measurement of *color stimuli*. This principle together with the additivity laws led to the development of an international system for objective color measurements in 1931 by the International Commission on Illumination (CIE). The system makes it possible to specify, and to communicate, colors by three numbers without having color chips to look at. In such systems, the coordinates of a given color stimulus are the relative amounts of three specific primary stimuli (for instance R, G and B) that match the color in question. In principle, no particular triplet of primary stimuli is better suited than another set, and every color reproduction technique has its own base colors chosen from practical considerations (TV, liquid crystal display LCD, and digital light projectors DLP, the printing industry, etc.). In color reproduction and in printing, both additive and subtractive techniques are often used in the same picture.

Colors on monitors and TV

On the screen of a color monitor or a color TV the color mixtures are made by red, green and blue primaries. The electron beam excites three different phosphors that produce the base colors as small dots on the screen. One can see them through a strong magnifying glass as dots of light about 0.3 mm in diameter, or as short lines. They are so densely packed that they cannot be resolved by the naked eye; they melt together and mix by averaging. Since the additivity laws above are valid for such spatial averages too, the color and brightness of an image detail is determined by the relative intensity of the colored dots. The three phosphor primaries set the limit on how well natural colors can be rendered on TV. Quite often colors in nature are too saturated to be correctly reproduced on the screen, but as we seldom notice the difference we seem to be rather tolerant to such color distortions.

In the art of painting this same mixture technique, with many small dots densely packed, is called 'pointillism'. The method, which was used by the neo-impressionists, was well represented by the French artists Georges P. Seurat (1859–1891) and Paul Signac (1863–1935).

Cone color space

Color matches are readily explained by the light absorption and the excitations of the three types of cone receptors. When the two half-fields A and B in Figure 5.8 have the same color, it means that the light from both halves leads to equal excitation of the three types of cone receptors in the retina. As viewed by the cones, the light of one half-field has simply been extended to the other half. If the relative magnitudes of the excitations of the three cone types were known, these magnitudes could themselves directly serve as coordinates in a physiological color specification system. One could use as color coordinates the relative excitations, represented by the lengths of three vectors in a three-dimensional vector space, each vector representing the excitation of one type of cone. However, there are also other alternatives. One could, for instance, use the relative intensities of the three primaries (e.g. R, G and B of Figure 5.9) in an additive color mixture. Still another possibility would be to use the readings of three photodetectors, each with a defined spectral response. Erwin Schrödinger (1925) showed that the laws that apply to additive color mixtures, in the way these laws were formulated by Grassmann (1853), are the same as those that apply to linear vector spaces. Thus, all the alternatives mentioned above are equivalent, and one could exchange one system for another by a simple mathematical transformation. The international CIE 1931 system represents one such vector space. Before we go into details about the CIE system, let us look at the vector representation of color mixtures.

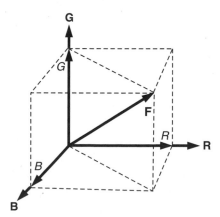

Figure 5.9 An additive color mixture, as in Figure 5.8, represented by addition of the three basic color vectors, **R**, **G** and **B**. Suitably chosen magnitudes R, G and B of these colors will lead to a color represented by the vector **F**.

In the following we shall describe the mathematical laws that apply for additive color mixtures – the same laws as those that hold for linear, three-dimensional vector spaces. The fundamental laws are the same, regardless of whether the mixture takes place on the TV screen or in the theatre. We shall illustrate the principles using three arbitrary primary colors, **R** (red), **G** (green) and **B** (blue), that might, for example, be lights from three projectors illuminating a white screen. With a fourth projector, we project a color F on the screen, as in Figure 5.8, and this color F is matched by a mixture of **R**, **G** and **B**:

$$F = R\mathbf{R} + G\mathbf{G} + B\mathbf{B}$$

where the length of the vector F is

$$\| F \| = [R^2 + G^2 + B^2]^{1/2}$$

R, G, and B can be called 'tristimulus values', and they signify the magnitudes of the unit vectors **R**, **G** and **B** in a particular mixture. In a linear, three-dimensional vector space the relation between the base vectors **R**, **G** and **B** and the unknown color vector F is as shown in Figure 5.9. The length $\| F \|$ is proportional to the intensity of the corresponding light as defined in some arbitrary system.

If it should be necessary to add one of the primaries (for example red) to the left side of the bipartite field of Figure 5.8(a) to obtain a match, we write

$$F + R\mathbf{R} = G\mathbf{G} + B\mathbf{B}$$

or

$$F = -R\mathbf{R} + G\mathbf{G} + B\mathbf{B}$$

Here, we have an additive color mixture with one negative component (red) because the color F was too saturated a green to be matched by adding the three primaries on the right side.

In the RGB-vector space, we can define a plane (Figure 5.10):

$$R + G + B = \text{constant}$$

In such a plane, where the intensity information is lost, it is possible to characterize a color by means of two coordinates only, for example r and g, as in Figure 5.11:

$$r = R/(R + G + B)$$
$$g = G/(R + G + B)$$
$$b = B/(R + G + B)$$

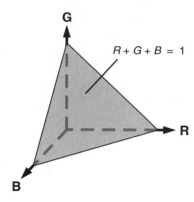

Figure 5.10   The unit color triangle, $R + G + B = 1$, in a color space constructed from the basic colors (the color vectors) **R**, **G** and **B**.

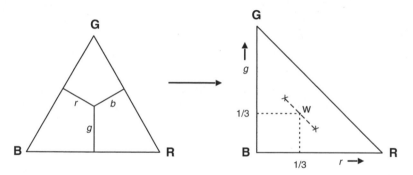

Figure 5.11   The chromaticity diagram (r, g) resulting from the unit color diagram of Figure 5.10, using the color vectors, **R**, **G** and **B**. White has the coordinates $(r, g) = \left(\frac{1}{3}, \frac{1}{3}\right)$. Additive color mixtures of two colors are situated on a straight line connecting their chromaticities.

and

$$r + g + b = 1$$

The triangle shown in Figures 5.10 and 5.11 is called a 'chromaticity triangle', and the coordinates r, g and b are the 'chromaticity coordinates'. In this two-dimensional space, the meaning of a negative coordinate is that the color to be matched is situated outside the triangle but by mixing a proper amount of one of the primaries to it, the chromaticity of the mixture color is brought to one of the triangle's sides, or even inside it.

As we have mentioned earlier, the choice of the primaries **R**, **G** and **B** in an additive mixture need not necessarily be a red, green and a blue light. Theoretically, the base colors can be any combination of three-color stimuli provided that they are mutually independent, meaning that none of them can be matched by a mixture of the other two. If enough light is available, it is in practice an advantage to use stimuli of the highest *spectral purity*. This provides a greater color range, and it is normally achieved with primaries that are *monochromatic*. The international CIE 1931 system for color measurement is special only in its choice of base color stimuli, all other principles being as described above. It is important to remember that *this method of measuring colors, by specifying their tristimulus values or chromaticity coordinates, only tells us which light stimuli will match under identical viewing conditions, and that it gives little information about their color appearance in a normal viewing situation.*

The color appearance of stimuli positioned in a certain location in a chromaticity diagram (when using a particular set of primaries) can only be predicted for standardized viewing conditions. The reason for this is that the color appearance of a surface is, besides its chromaticity, dependent on several factors, such as the color of the illumination, simultaneous contrast and the adaptation state of the eye. Fortunately, however, a color match made under one physiological adaptation condition will also be a match for another state of adaptation, or for another surround, even though the appearance of the stimulus may change.

It is often practical to normalize a system with the primaries $\mathbf{R}'$, $\mathbf{G}'$ and $\mathbf{B}'$ to equal magnitudes for a white (W) diffusely reflecting surface:

$$R'_\mathrm{w}\mathbf{R}' + G'_\mathrm{w}\mathbf{G}' + B'_\mathrm{w}\mathbf{B}' = \mathbf{W}$$

such that

$$R_\mathrm{w} = G_\mathrm{w} = B_\mathrm{w} = 100$$

This is achieved by letting the ratios

$$R = 100R'/R'_\mathrm{w}$$
$$G = 100G'/G'_\mathrm{w}$$
$$B = 100B'/B'_\mathrm{w}$$

become the new tristimulus values. In this way, the *white point* (in practise representing the color of the illuminant) will take the position in the center of the diagram with the chromaticity coordinates $(r_\mathrm{w}, g_\mathrm{w}) = (1/3, 1/3)$. Thus the colorimetric purity of a color stimulus will increase the further its chromaticity is from the white point. For instance, a plot of the chromaticities of all monochromatic lights will form a curve circumscribing the chromaticity triangle [R', G', B' in Figure 5.13(a)]. Additive color mixtures between two chromatic stimuli will always be situated somewhere on the straight line drawn between their chromaticity coordinates in the diagram. Therefore, two colors that in the diagram are positioned on opposite sides of

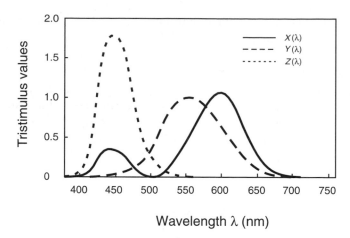

Wavelength λ (nm)

Figure 5.12 The spectral tristimulus values $X(\lambda)$, $Y(\lambda)$ and $Z(\lambda)$ that form the basis of the CIE 1931 system for color measurement for a 2° field. $Y(\lambda)$ corresponds to the relative luminous efficiency function $V(\lambda)$ of 1924. These curves can be regarded as the spectral sensitivities of three detectors (photocells) X, Y and Z that are used for the measurement of light and colors. For historical reasons these curves are usually written as $\bar{x}(\lambda)$, $\bar{y}(\lambda)$ and $\bar{z}(\lambda)$.

the white point, on a line through it (Figure 5.11), will give white when added in suitable proportions. Such stimuli are called *complementary colors*. Thus, as one can imagine, there is an abundance of complementary colors that will add up to the same white (metameric whites).

Color differences do not depend directly on differences in cone excitations. Equal distances in the chromaticity diagram, or in the three-dimensional color space of Figure 5.9, are not related to equal color differences. Perceived color differences are related to differences in tristimulus values in a complex way that is not yet fully determined. Most likely, color differences are a consequence of the nonlinear responses of color coding cells at later stages in the visual pathway.

In areas where quantitative color specifications are important, such as in the graphics industry, in color photography, for TV and video-projectors, in the food industry, in lighting and signaling technology, etc., the international CIE *XYZ* system has been used successfully for many years. This system will be described in the next section.

The CIE *XYZ* system for color measurements

In 1931 the International Comission on Illumination (Commission Internationale de l'Eclairage, CIE) introduced a system for color measurements (colorimetry) for a standard field size of 2°. The coordinates of this system can be regarded as the responses of three light-sensitive devices (e.g. photocells or diodes) for measuring

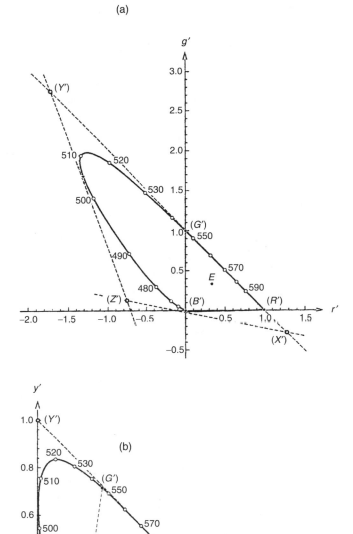

Figure 5.13 The transformation from a color triangle (r', g') in (a), based on additive color matches using the spectral colors R' (700 nm), G' (546.1 nm), and B' (435.8 nm), to an (x', y') chromaticity diagram in (b), in this case the Judd-Vos modified CIE (x', y') chromaticity diagram (Courtesy of J.H. Wold).

light intensity: three light detectors X, Y and Z with different spectral responses, where X has a maximum response for long-, Y for middle- and Z for short-wavelength lights. The spectral sensitivity and response of the Y detector was chosen to be the same as for the human 2° luminous efficiency function $V(\lambda)$. Since $V(\lambda)$ is the basis of all photometric definitions of light intensity (see the section on 'Photometry' in Chapter 4, p. 165), this property of Y as one of the sensors means that the value Y can be normalized to become equal to luminance, and therefore a colorimetric measurement is at the same time a photometric measurement. This is convenient, eliminating the need for two separate systems, one for the measurement of the intensity of a light (such as luminance, light flux, illuminanance, etc.) and another for its color. Today, the CIE 1931 system is widely used whenever a technical specification of color stimuli is required.

The simplest way to understand how color measurements are done, is probably to imagine, as we did above, three light detectors with three different spectral sensitivities $X(\lambda)$, $Y(\lambda)$, and $Z(\lambda)$, called 'spectral tristimulus values'. The particular spectral sensitivities that are used by the CIE are shown in Figure 5.12. Let us further assume that each of the detectors is connected to an ammeter, an instrument for measuring electric current. When the detectors are illuminated by light, let us say by greenish-blue light, they will be excited differently according to their spectral sensitivities. The ammeters show the respective currents X, Y and Z (e.g. $X = 12$, $Y = 55$ and $Z = 65$ for a greenish-blue light, only the relative magnitudes being of interest). These values are the three-dimensional color coordinates of the light stimulus, represented by a point in the CIE system's XYZ color space. The color stimulus can thus be represented as a vector, as in Figure 5.9.

The excitation of a receptor, or of an optical detector, is equivalent to performing a mathematical summation or integration of excitations over all wavelengths to which the detector responds. If the detector's spectral sensitivity is $V(\lambda)$, and the incident light has a spectral distribution $\Phi(\lambda)$, then the excitation is proportional to the product $V(\lambda)\Phi(\lambda)$, integrated or summed over the whole spectrum [$\int V(\lambda)\Phi(\lambda)d\lambda$ or $\sum V_\lambda\Phi_\lambda\Delta\lambda$; $\Delta\lambda$ or being, for instance, an interval of 5 nm]. If the light that reaches the eye has been reflected from a surface with a *spectral reflection curve* $\beta(\lambda)$, illuminated by a light source with the spectral power distribution $P(\lambda)$, then $\Phi(\lambda) = \beta(\lambda)P(\lambda)$ (see Figure 2.24). The following integrations will yield the excitation X, Y and Z of the three photodetectors:

$$X = \int X(\lambda)\beta(\lambda)P(\lambda)d\lambda$$

$$Y = \int Y(\lambda)\beta(\lambda)P(\lambda)d\lambda$$

$$Z = \int Z(\lambda)\beta(\lambda)P(\lambda)d\lambda$$

These three numbers, X, Y and Z, determine a color's coordinates in the CIE color space. As mentioned earlier, $Y(\lambda)$ is the spectral luminous efficiency function $V(\lambda)$, and Y can therefore be normalized to a photometric unit, such as cd/m^2 for luminance, with X and Z being normalized accordingly. With this normalization, the Y value is a photometric magnitude.

When all magnitudes that enter the equations above are known, the XYZ values can be calculated mathematically as shown. Alternatively, the values can be measured directly by three detectors. The calculation requires physical measurements of the spectral radiance or power distribution, $P(\lambda)$, of the light source and the spectral reflectance factor, $\beta(\lambda)$, of the probe, all of which can be done in a physics laboratory. Direct measurement of color coordinates by means of three detectors requires that the detectors have exactly the spectral sensitivities $X(\lambda)$, $Y(\lambda)$, and $Z(\lambda)$. It is technically difficult to manufacture detectors with precisely these sensitivities and therefore such instruments are rare and expensive. When computational power is available, it is in many cases preferable to do the calculations above step by step.

Ideally, the tristimulus values X, Y and Z are related to the absorption of light in three types of cone in the human eye, measured relative to the light that enters the cornea. For historical reasons, this is not exactly the case for the CIE 1931 XYZ system. The CIE system was developed long before the sensitivities of the three cone types were known, and it later became clear that the CIE spectral tristimulus values X, Y and Z were not quite representative for the average human eye, particularly at the shorter wavelengths. The spectral tristimulus values were therefore corrected, and replaced by X', Y' and Z' by the American colorimetrist D.B. Judd (1951a). Among vision scientists, Judd's system has been regarded as better than the CIE 1931 system for precise psychophysical color specifications. Smith and Pokorny (1975) adopted this system as the colorimetric basis in the development of the cone sensitivities $L(\lambda)$, $M(\lambda)$ and $S(\lambda)$. Normalized to equal energy white, these sensitivities are exact linear transformations of X', Y' and Z'.

$$L(\lambda) = 0.2331\, X'(\lambda) + 0.8159\, Y'(\lambda) - 0.0494\, Z'(\lambda)$$
$$M(\lambda) = -0.4640\, X'(\lambda) + 1.3665\, Y'(\lambda) + 0.0983\, Z'(\lambda)$$
$$S(\lambda) = 1.0414\, Z'(\lambda)$$

In recent years the CIE has attempted to reach international agreement on a physiologically based color measuring system. After much discussion, the responsible committee has agreed to establish a new system based on the cone sensitivities derived from the work of Stockman and Sharpe (2000). We shall return to this system later after a brief summary of the physics of CIE color specifications (see also Appendix on p. 415).

The CIE (x,y) chromaticity diagram

Three coordinates are required to completely specify a color stimulus. In technical measurements, the most common combination is the chromaticity coordinates (x, y), together with the tristimulus value Y (normalized either to the reflection factor β in percent or to give luminance in cd/m^2, or some other photometric unit). The chromaticity coordinates (x, y) are:

$$x = X/(X + Y + Z)$$
$$y = Y/(X + Y + Z)$$
$$z = Z/(X + Y + Z)$$
$$x + y + z = 1$$

As for (r, g) in Figure 5.11, the coordinates x and y can be plotted in a coordinate system with axes at right angles, called the chromaticity diagram (x, y). Figure 5.13(a) shows the spectral stimuli plotted in the (r', g') chromaticity diagram of the Judd–Vos modified 2° observer using the color primaries **R'**, **G'** and **B'**. The $X'Y'Z'$-color space and the (x', y') diagram in Figure 5.13(b) shows how the circumscribing triangle of the X', Y', Z'-coordinate axes yields only positive color coordinates.

Figure 5.14 shows the spectral stimuli plotted in the traditional CIE 1931 (x, y)-chromaticity diagram. The inner white triangle in Figure 5.14(a) shows the range of colors that can be produced using the common R, G and B colors of a color monitor (see also Figure 5.16). The spectrum locus forms a curve resembling a horses shoe, and the circumscribing triangle in Figure 5.14(b) applies for the special XYZ primaries of the CIE 1931 system. The curve within the diagram of Figure 5.14(b) shows the chromaticities of a black body radiator following Planck's radiation law, with temperatures given in degrees Kelvin between 24 000 and 1000 K ($0 K = -273$ °C). This curve is close to the color coordinates of the tungsten filament at different temperatures, as in a light bulb run at different voltages. A common 40 W tungsten bulb radiates with a temperature of about 2500 K ($= 2227$ °C). The color temperature curve is close to that for the different phases of daylight from a blue sky at mid-day to evening sunset. On this curve are represented the daylight colors C and D, corresponding to two types of normalized daylight from the overcast northern sky. Equal-energy white light E (with the spectral power distribution being a constant for all wavelengths) has the chromaticity coordinates $(x, y) = (\frac{1}{3}, \frac{1}{3})$.

In the diagram, the end points of the visible spectrum, 380 and 760 nm, are connected by a straight line, the so-called 'purple line'. The colors represented by the purple line are not present in the Newtonian spectrum, but are mixtures of the red and violet colors at the spectral endpoints. A color's colorimetric purity, p_c, increases with distance from the white point. White light has $p_c = 0$, and for all monochromatic stimuli the value is $p_c = 1.0$. Generally, colorimetric purity depends on the luminance, L_λ, of monochromatic light in an additive color mixture with white (L_w) and it is defined as:

$$p_c = L_\lambda/(L_\lambda + L_w)$$

This is the same as

$$p_c = (y_b/y)[(y - y_w)/(y_b - y_w)]$$

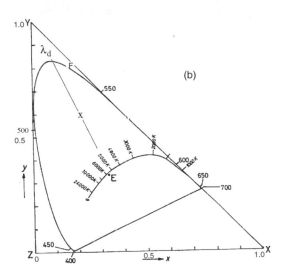

Figure 5.14 (a) The chromaticity diagram (x, y) of the 1931 CIE 2° standard observer with the spectrum locus (coordinates of monochromatic lights). The colors illustrate their approximate positions within the diagram. The 'display gamut' represents the colors that can be reproduced on a CRT monitor or on a TV. 'Printer gamut' gives the range of colors that can be reproduced by six printing colors. The colors near the spectrum locus cannot be correctly reproduced by these techniques. (b) The (x, y) chromaticity diagram giving the locations of the spectral colors (with some of the wavelengths indicated), of equal energy white, E, of a color F (marked as x) with its dominant wavelength, λ_d, and a curve showing the chromaticity of black-body radiation at various temperatures (Planck's locus). See also Color Plate Section.

or

$$p_c = (y_b/y)[(x - x_w)/(x_b - x_w)]$$

the choice of equation depending on which coordinates give the more precise calculation. x_b and y_b are the chromaticity coordinates of the stimulus' dominant wavelength, λ_d (see below).

The extension of the straight line from the chromaticity point of equal energy white, E, through the chromaticity (x, y) for a given color, F, cuts the spectral locus in a point representing the *dominant wavelength* λ_d of F. Purple colors do not have a dominant wavelength, but are characterized by their *complementary wavelength* λ_c found by extending the straight line until it cuts the spectral locus at the other side of equal energy white.

Object colors – surface colors – related colors

We have already mentioned that, for a reflecting surface, the spectral distribution of the light that enters the eye, $\Phi(\lambda)$, is determined by the spectral distribution of the light source, $P(\lambda)$, and the spectral reflection factor, $\beta(\lambda)$, of the surface (Figure 2.24):

$$\Phi(\lambda) = \beta(\lambda)P(\lambda)$$

For most surfaces the shift in color appearance is very small as the color of daylight changes during the course of the day. The shift is small even for a transition from indoor incandescent light to daylight, despite the drastic change in spectral distribution of the illumination, and hence of the reflected light (Figure 2.25). A sheet of white paper looks white for a wide range of illuminations, even though its CIE coordinates are very different, corresponding to color temperatures from 2500 K in the case of indoor light to above 20 000 K for outdoor light if the sky is clear. The ratio of relative excitations of the three cone types will change accordingly. However, since the color of the white paper does not change much (a phenomenon called color constancy), it is as if the visual system evaluates the reflectance of the surface more or less independently of the color of the illuminant. This demonstrates that the color appearance of a surface cannot be unequivocally deduced from its chromaticity coordinates, or from the relative excitations of the three cone types, except under strictly standardized viewing conditions. It was this phenomenon of color constancy that made von Helmholtz think that the eye actually discards the illuminant.

The physiological mechanisms responsible for color constancy are still largely unknown. A better understanding of the spatial and temporal integration of color-coding cells is necessary to account for the underlying mechanisms. Many theories do well in approximating the color shifts that occur for small changes in the color of the illuminant. However, with larger changes, like those often occurring between natural and artificial illuminations, they all fail. This applies to the coefficient theory of von Kries (1905) as well as to the retinex theory of Land (1983). We shall return to this theme later on; here we only want to point to the advantage of this phenomenon of surfaces 'resisting' a shift in their color appearance with changes of the illumination. It makes it easier to recognize objects by means of their color even though the illumination undergoes changes in color and spectral composition. The visual system strives to achieve color constancy, but it never reaches a state of absolute constancy; the more saturated the color of the light the larger the departure from absolute constancy. Therefore, in everyday life, it is of importance to characterize a light source by its failure to render surfaces in the same color as some chosen standard. A von Kries 'centering transformation' of the tristimulus values to those of the light source mimics to some degree the physiological processes of adaptation in that it leads to approximate color constancy for near whites, especially in cone excitation space. This centering operation is performed by dividing the standard tristimulus values (centered to Illuminant E) by those of the actual light source. In CIE *XYZ* space 'centering' of the tristimulus values *X*, *Y* and *Z* to, for example, illuminant A (incandescent light) leads to new coordinates, $X' = X/X_A$, $Y' = Y/Y_A$, $Z' = Z/Z_A$, and to an ideal white, diffuse reflecting surface having chromaticity coordinates $(x', y') = (\frac{1}{3}, \frac{1}{3})$ also under the new illumination.

The CIE color space can be normalized to become a well-defined color solid for object colors. This is simply done by dividing the tristimulus values by the value Y_W of the ideal white surface:

$$X = 100 \int X(\lambda)\beta(\lambda)P(\lambda)\mathrm{d}\lambda / Y_W$$

$$Y = 100 \int Y(\lambda)\beta(\lambda)P(\lambda)\mathrm{d}\lambda / Y_W$$

$$Z = 100 \int Z(\lambda)\beta(\lambda)P(\lambda)\mathrm{d}\lambda / Y_W$$

Because $\beta(\lambda)$, the *spectral reflectance* or *luminance factor* for an ideal white surface is 1.0 for all wavelengths,

$$Y_W = \int Y(\lambda)P(\lambda)\mathrm{d}\lambda$$

Here, $P(\lambda)$ is the relative power distribution of the light source. For a diffusely reflecting ideal white surface, *Y* becomes equal to 100, and all other surfaces will then have tristimulus values smaller than 100, and *Y* will represent the reflection factor β in percent.

The *XYZ* vector space of surface colors resulting from this latter normalization is limited by a well-defined boundary and is called an *object color solid*. The outer boundary of the color solid is defined by Schrödinger's (1920a) *optimal colors*, which are surface color stimuli with an idealized spectral reflection curve (see below and Figure 5.15). In contrast, a color space for light sources is usually normalized to *Y*, representing some absolute photometric unit, and it therefore has no such boundary.

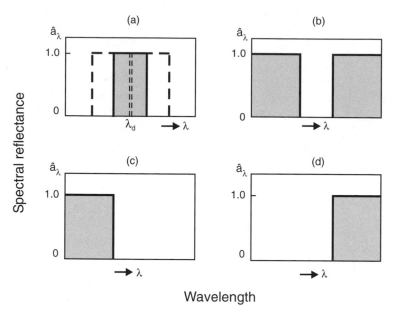

Figure 5.15   Optimal colors have one, or at most two transitions between 0 and 1.0 in the spectral reflection curve. (a) The spectral reflection for three optimal color stimuli with the same dominant wavelength; (b) the complementary distribution to the fully drawn curve in (a). Spectral reflection of a bluish and a reddish optimal border color are shown in (c) and (d).

Schrödinger's optimal colors

Optimal colors are of special interest because they represent idealized cases of spectral reflection and transmission. An optimal color has a spectral distribution with one, at most two positions in the spectrum where the spectral reflection (or transmission) factor changes abruptly from zero to 1.0 (or from 1.0 to 0). Such surface colors are not common in nature, but they can be produced optically. The special feature of the optimal colors is that, for a given chromaticity, the optimal color has the highest luminance factor of all possible spectral distributions. Therefore, they form the outer surface (boundary) of the object color solid.

Figure 5.15 shows some examples of the spectral distribution of optimal colors. In Figure 5.15(a) the spectral distributions are centered on a constant wavelength. The shaded distributions of Figure 5.15(a) and (b) give complementary optimal colors. The spectral distributions of Figure 5.15(c) and (d) are for *border colors*, which are observed when sharp black and white edges are imaged through a prism. Interested readers are referred to Bouma (1946) for a thorough treatment of optimal colors.

Color reproduction

As mentioned above, the colors on a conventional CRT (cathode ray tube) type of color monitor are generated by modulating the light emittance of three dot phosphors, R, G and B. In order to produce a particular color on the screen, with given chromaticity coordinates (x, y), one needs to know the relationship between the voltage of the three electron guns that excite the phosphors and the luminance of the corresponding dots. An example of such a relationship is shown in Figure 5.16(a) for each of the phosophors R, G and B. These nonlinear curves are called the γ-curves of the monitor. The DAC values (digital to analog convertion) determine the voltage of each gun. As is apparent from the figure, for a DAC value of 1000, the green phosphor has the highest luminance ($130 \, cd/m^2$), the red a lower value ($60 \, cd/m^2$) and the blue phosphor produces only $20 \, cd/m^2$. The reason for these different values is the unequal effectiveness of the three spectral distributions in evoking a light sensation, which is determined by the spectral luminous efficiency function $V(\lambda)$. With a DAC value of 1000 for all guns, the color is white with a luminance of $210 \, cd/m^2$.

The manufacturer of color monitors has the problem of achieving the highest possible colorimetric purity of the phosphor colors, without reducing their luminance. Narrow-band, pure spectral colors and high luminance are conflicting requirements, and the phosphors that are in use in most monitors today are a compromise. The phosphors commonly used on CRT monitors and TV screens have CIE chromaticity coordinates corresponding to the vertices of the triangle in Figure 5.16(b). Figure 5.16(b) shows the typical differences between the target colors of a calibrated monitor ($\bigcirc$) with the chromaticities actually produced ($+$). The calibration curves (the γ-curves) are those of Figure 5.16(a).

None of the colors outside the white (R,G,B) triangle of Figure 5.14(a) [the same as in Figure 5.16(a)] can be correctly reproduced by conventional TV monitors. Colors beyond the monitor's range can be reproduced in different ways. The out-of-range-colors can be 'clipped' and reproduced with lesser saturation (closer to the white point). To maintain photographic realism, however, it is usually better to squeeze the color space of the whole image into the smaller color space of the monitor, thus preserving relative differences between colors instead of absolute differences. This method works satisfactorily provided it is not possible to compare the reproduced color with the original. Available software can transform the true color coordinates of an image pixel into the best perceivable color within the monitor's smaller color space. In this process, the CIE space is the reference space, independent of the color spaces of a given input scanner and a given output printer. This reference allows the color data to be stored independently of any device, and the devices themselves to be characterized separately without reference to any other device.

However, conventional file formats have only eight bits of information or 256 levels in each of the R, G and B channels, and this is not enough to ensure a smooth transition in an image for all colors without noticeable steps. The file format must

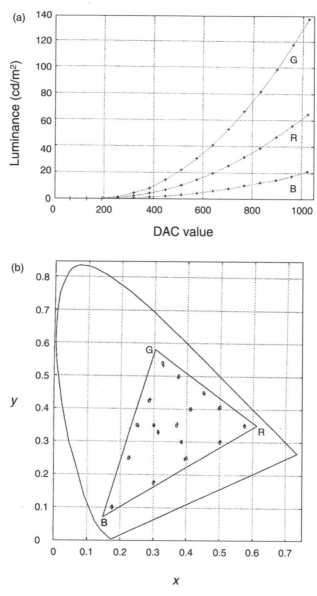

Figure 5.16 (a) γ-Curves for calibration of the *R*, *G* and *B* electron guns of a CRT monitor. The curves show the nonlinear relationship between the 1024 voltage values (DAC = digital-to-analog converter) for the three guns and the resulting luminance on the monitor screen. The *G*-gun gives the highest luminance, *R* somewhat less and *B* the lowest. The sum of the three highest luminance values gives the luminance for white (about 220 cd/m² in this case). Accurate reproduction of colors on the screen, within the triangle in (b), requires precise knowledge of these three curves for the monitor's intensity/contrast setting. (b) An example of the accuracy in the reproduction of colors on a CRT monitor screen using the calibration curves in (a). (○) A target value read by the software; (+) the rendered value as measured by a precision colorimeter. Correspondence is quite good (from Björnevik, 1992).

contain enough bits to give a seamless transition in the most sensitive areas. To save bits, other color spaces with a smaller color gamut have been designed by industry standard committees. One example is the AdobeRGB space, which is frequently used in photography.

Modifications of the CIE 1931 system

We have already mentioned that the CIE 1931 system for the 2° observer uses the spectral luminous efficiency function, the 1924 $V(\lambda)$ function, for the *spectral tristimulus value* $Y(\lambda)$. Although this is economical in that it combines photometric and colorimetric measurements in one, in other respects it has been a disadvantage. The curve from 1924 has been demonstrated to be erratic for wavelengths below 460 nm, and therefore it is not a suitable luminous efficiency function for blue light. In 1951, Judd introduced an alternative curve, the so-called 'Judd-$V(\lambda)$', that, after some minor revision by Vos became an additional standard of the CIE and named $V_M(\lambda)$. These two $V(\lambda)$ curves are compared in Figure 4.18. The difference between them increases for shorter wavelengths below 460 nm, and at the lower end of the spectrum the $V_M(\lambda)$ curve is about 10 times more sensitive than the 1924 $V(\lambda)$. Therefore, the international standard for photometric measurements, the 2° $V(\lambda)$ curve, gives values that are too small for all photometric magnitudes at short wavelengths, be it light flux, illuminance or luminance. The same errors occur in the spectral sensitivities of cone fundamentals that are derived from 1931 CIE spectral tristimulus values.

The CIE has for a long time been planning to introduce a new, physiologically based system for color measurements, which is likely to co-exist with the 1931 system. In color science, an alternative to the 1931 system has long been awaited, and a committee has been working on this task since the early 1970s. One major problem has been to agree upon a common set of human cone spectral sensitivities. The 1931 system is based on the color matching data of Wright (1928/1929) and Guild (1931), and the cone sensitivities of Smith and Pokorny (1975) correspond to the Judd–Vos modification of the data of Wright and Guild. However, the new committee has adopted the cone sensitivities of Stockmann and Sharpe (2000) that are based on the color-matching data of Stiles and Burch (1955, 1959). An earlier version was developed by MacLeod and Boynton (1979) based on Smith and Pokorny cone excitations and stimuli of equal luminance. The committee is trying to make a smooth transition from a 2° system to a color measurement system for 10° fields, the latter currently being recommended for stimulus sizes larger than 4°.

A physiological system for color measurements

Because the laws of additive color mixtures are consequences of the linear and independent absorption processes in three types of cone, one would expect to find

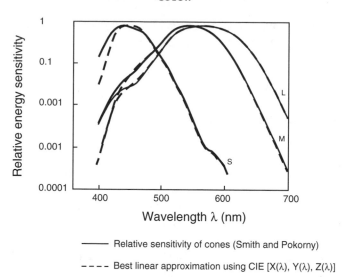

Figure 5.17 A comparison of the set of three cone spectral sensitivities for the L-, M- and S-cones by Smith and Pokorny (1975) with the sensitivities obtained from a transformation of the 1931 CIE tristimulus values $X(\lambda)$, $Y(\lambda)$ and $Z(\lambda)$. The difference is caused mainly by errors in the data for the 1931 CIE standard observer (from Wold, 1998).

linear transformations between X, Y and Z of the CIE system and the excitations L, M and S of three cone types. Figure 5.17 shows how close one can come to the cone sensitivities of Smith and Pokorny by linear transformations of X, Y and Z. As we have already mentioned, the reason for the difference is mainly the errors in the CIE spectral tristimulus values at short wavelengths. For Judd/Vos's modified spectral tristimulus values $X'(\lambda)$, $Y'(\lambda)$ and $Z'(\lambda)$, however, there is an accurate linear transformation to Smith and Pokorny's cone sensitivity curves $L(\lambda)$, $M(\lambda)$, and $S(\lambda)$.

Figure 5.18 shows a three-dimensional cone excitation (color) space that is spanned by three vectors corresponding to the fundamental cone mechanisms. The curve connecting the symbols represents the relative excitations of the cones by mono-chromatic light, called the 'spectrum locus'. The basis vectors **L**, **M** and **S** are normalized such that the vector for an equal energy white stimulus passes through the point $(\frac{1}{3}, \frac{1}{3}, \frac{1}{3})$. The projection of the locus down into the (L,M) plane is also shown. In Figure 5.19(a) and (b) the central projection of the spectrum locus into the unit plane

$$L + M + S = 1$$

is shown. Figure 5.19(b) shows the (l, m) diagram corresponding to the parallel projection into the (L,M) plane. This (l, m) chromaticity diagram plays the same role as the (x, y) chromaticity diagram of the CIE color space. A larger version of this

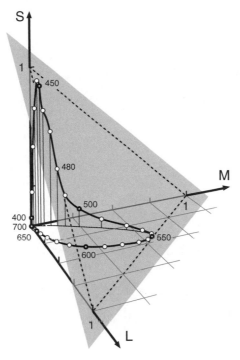

Figure 5.18 A three-dimensional vector space based on the relative excitations of the human L-, M- and S-cones of Stockman and Sharpe (2000). The spectrum locus is given by the curve connecting the data points. The projection down into the L, M-plane corresponds to the reduced color space of a tritanope, lacking S-cones. (Courtesy of J.H. Wold, 2002.)

diagram is found in Appendix I, and the tabulated cone spectral sensitivities can be found at http://cvision.ucsd.edu/database/data/cones/linss2_10e_1.txt

Figure 5.20(a) shows schematically a plane in color space where

$$L + M = \text{constant}$$

For a particular normalization of the L- and M-cone sensitivity curves, this plane is an isoluminant plane, i.e. a plane for which luminance is constant. This is a consequence of the luminous efficiency function being a sum of L- and M-cone spectral sensitivities, with no contribution of the S-cone. In Figure 5.20(b) an isoluminant diagram has been constructed from linear opponent transformations of the cone sensitivities of Stockmann and Sharpe (2000).

The linear cone excitation vector space of Figures 5.18 and 5.19 gives a direct explanation of the laws of color matching without introducing virtual color vectors like **X**, **Y** and **Z**. This simplification is an advantage for the understanding of colorimetry and in the teaching of color measurements.

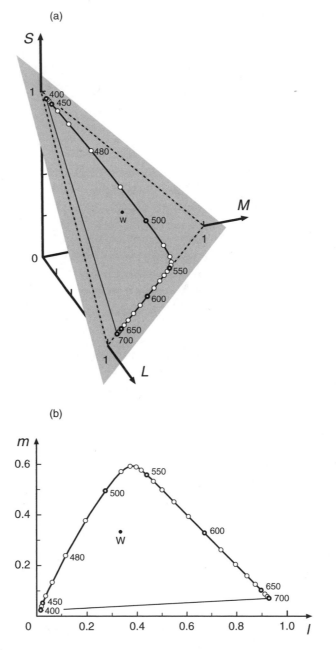

Figure 5.19 (a) A chromaticity diagram, $L + M + S = 1$, in an LMS-cone excitation space based on the fundamentals of Stockman and Sharpe (2000). (b) The (l, m)-chromaticity diagram from (a), analogous to the CIE (x, y) chromaticity diagram. See also Figure A1 in the Appendix.

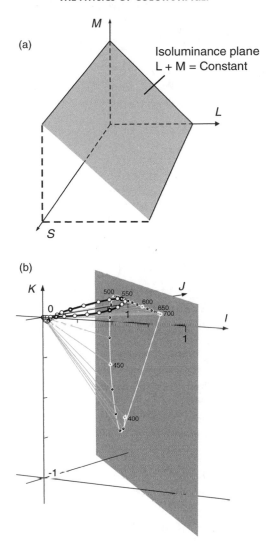

Figure 5.20 (a) An isoluminant plane is a plane in which the sum of L- and M-cone excitations is constant. S-cones are not likely to contribute to luminance as defined by heterochromatic flicker photometry. (b) The spectrum locus in an isoluminant plane where I, J, and K are opponent transformations of the human cone fundamentals L, M, S of the previous figures (Courtesy of J.H. Wold).

An XYZ representation for different sets of color-matching data

Because of the extensive use of the *XYZ* color space, it might be of some interest to look into the consequences for this representation of using different color matching data. In Figure 5.21 we see the results of a transformation that minimizes the differences in a two-dimensional (x, y) chromaticity diagram when using the Stiles

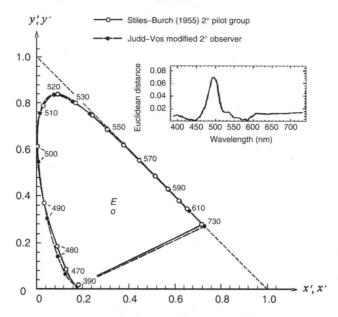

Figure 5.21 The chromatcicity diagram of the Judd–Vos modified 2° observer (long-dashed spectrum locus) compared with the chromaticity diagram of the Stiles–Burch 1955 2° pilot group (solid spectrum locus). The result is obtained by minimizing the difference between the spectrum loci in the new and the old (x, y)-diagrams. The inset shows the Euclidian distance between corresponding wavelengths in the two diagrams. This distance is largest around 500 nm. (Reproduced from Wold and Valberg, *J. Opt. Soc. Am.*, 1999, by permission of The Optical Society of America.)

and Burch (1955) 2° pilot color matching data (which are compatible with the Stockmann and Sharpe 2000 fundamentals) and the reference diagram of the modified Judd–Vos 2° observer. The inset shows the distance between the points on the spectrum locus in the Judd–Vos diagram and the corresponding points in the diagram of Stilesn and Burch (Wold and Valberg, 2001). Similar differences are to be expected when using the Stockmann–Sharpe 2000 fundamentals to develop an *XYZ* color space instead of the Stiles–Burch data.

Luther–Nyberg color space and the Hurvich–Jameson opponent model

The CIE *XYZ* color space and the (x, y) chromaticity diagram may appear rather abstract with little resemblance to perceptual aspects of color vision. This has made the teaching of colorimetry difficult. We shall therefore describe a particular simple object color space, derived from *XYZ*, but based on the work of Luther (1927) and Nyberg (1928) and combined with the opponent color vision theory of. Hurvich and Jameson (1956; see Hurvich, 1981). Hurvich and Jameson did not directly use cone excitations to quantify their opponent theory, because these were not sufficiently known at the time. Instead, they used transformations of the CIE 2° tristimulus values X, Y and Z to illustrate their ideas. The resulting color space is limited by the coordinates of optimal colors, and the outlines of the object color solid are shown in Figures 5.26 and 5.27. Object colors are by far the most common natural colors in our environment (according to Wilhelm Ostwald they make up 99 percent of the colors seen during the day). Below we shall demonstrate how this object color solid leads to a more transparent relationship between color attributes and their colorimetric representations.

The most important coordinates in the aforementioned opponent theory are the opponent chromatic magnitudes M_1 and M_2, which we will call *chromatic moments*, and the luminance factor Y (with X and Z being accordingly normalized, see p. 232):

$$M_1 = X - Y \ (\text{`red–green moment'})$$
$$M_2 = 0.4(Y - Z) \ (\text{`yellow–blue moment'})$$
$$Y = \beta \text{ in percent (luminance factor)}$$

The resulting moment (or metric color strength),

$$M = [M_1^2 + M_2^2]^{1/2}$$

In the model of Hurvich and Jameson, the spectral opponent moments $M_1(\lambda) = X(\lambda) - Y(\lambda)$ and $M_2(\lambda) = 0.4[Y(\lambda) - Z(\lambda)]$ were linear approximations of Hering's physiological opponent processes. These model curves are shown in Figure 5.22(a) (a similar transformations of X, Y and Z to form two component color signals and a third luminance signal is used in color TV). The corresponding resultant spectral moment, $M(\lambda)$, shown in Figure 5.22(b) is

$$M(\lambda) = [M_1(\lambda)^2 + M_2(\lambda)^2]^{1/2}$$

These magnitudes have proved to be useful approximations of physiological opponent processing. Hue cancellation experiments, and several other psychophysical studies, have been inspired by this quantitative formulation to investigate properties of opponent channels and the individual variations of the opponent magnitudes.

Two other important coordinates, the *opponent purities* p_1 and p_2, can be derived from the moments and defined in a plane of constant luminance ($Y = $ constant):

$$p_1 = M_1/Y$$
$$p_2 = M_2/Y$$

The resulting purity, p, and the spectral opponent purity $p(\lambda)$ are defined as:

$$p = [p_1^2 + p_2^2]^{1/2}$$
$$p(\lambda) = [p_1(\lambda)^2 + p_2(\lambda)^2]^{1/2}$$

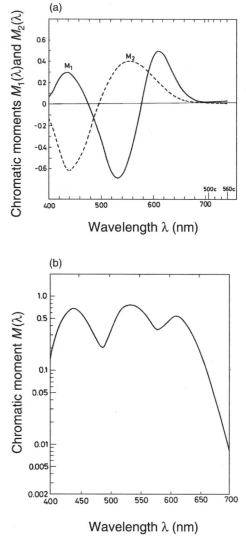

Figure 5.22 (a) The opponent spectral moments $M_1(\lambda)$ and $M_2(\lambda)$ in the opponent theory of Hurvich and Jameson, based on linear transformations of the Judd (1951a) modified spectral tristimulus values $X'(\lambda)$, $Y'(\lambda)$ and $Z'(\lambda)$. (b) The resulting chromatic moment $M(\lambda)$, based on the partial chromatic moments shown in (a).

The absolute value of the opponent spectral purities $p_1(\lambda)$ and $p_2(\lambda)$ are shown in Figure 5.23(a), and the resultant spectral purity $p(\lambda)$ in Figure 5.23(b). The colorimertic purity, p_c, of an arbitrary chromaticity is related to the opponent purity p as follows:

$$p_c = p/p(\lambda)$$

(a)

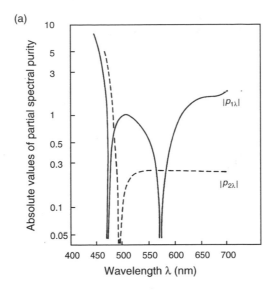

(b)

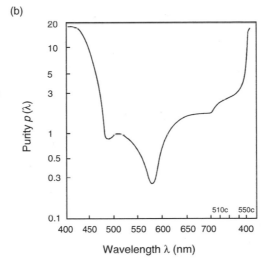

Figure 5.23 (a) Absolute values of the two partial opponent spectral purities $p_1(\lambda)$ and $p_2(\lambda)$, derived by dividing the partial moments of Figure 5.22(a) by the spectral luminance factor $Y(\lambda)$. (b) The resulting opponent spectral purity $p(\lambda)$.

where $p(\lambda)$ is the spectral opponent purity of the corresponding dominant wavelength. As is shown in Figures 5.24 [see also Figure 5.26(a)], the purity diagram (p_1, p_2) results from central projection of the spectrum locus onto the equal luminance plane $Y = $ constant.

This symmetric Luther–Nyberg color solid resulting from the coordinates M_1, M_2 and Y incorporates important features of color stimuli and simplifies important

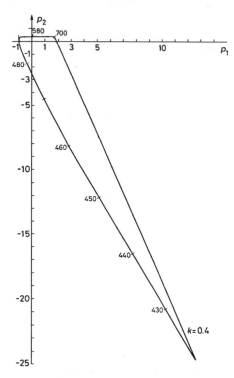

Figure 5.24 An isoluminant color plane (p_1, p_2) with the locus of the opponent spectral purity $p(\lambda)$. The distance from the white point $(p_1, p_2) = (0, 0)$ to the spectrum locus roughly corresponds to the saturation of spectral colors as determined by psychophysical experiments (see Figures 4.20 and 5.23(b)) k represents the relative weight of p_2 relative to p_1.

relationships that are more or less hidden in the CIE 1931 system. This applies to correlates to perceived attributes such as color saturation, color strength, color differences, etc. For instance $p(\lambda)$ in Figure 5.23(b) corresponds closely to the subjective saturation of isoluminant spectral colors (MacAdam, 1942; Chapanis, 1944). When compared with the lower curve of Figure 4.20, we see that $p(\lambda)$ also resembles the contribution of equiluminant spectral stimuli to brightness (except for a scaling factor; see also Evans and Swenholt, 1967, 1968). $M(\lambda)$ in Figure 5.22(b) shows a great resemblance to the chromatic threshold sensitivity for spectral colors when they are projected on a white background (King-Smith, 1975). Equal color differences are also rendered better in this opponent space than in the CIE system (Holtsmark and Valberg, 1960). In Figure 5.25, for Munsell chroma up to chroma 6, the equal chroma contours signifying a constant chromatic difference from white approximate circles in the (p_1, p_2) diagram.

The symmetric object color solid is visualized in Figure 5.26(a, b) and 5.27. These figures, taken from Valberg (1981), show in Figure 5.26(a) the interrelation in vector space between the horizontal moment and purity planes, and the vertical (M, Y)

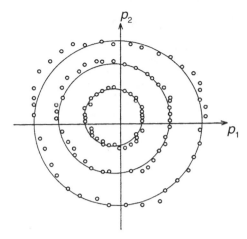

Figure 5.25 The loci of Munsell color chips of all hues of *Chroma* 2, 4 and 6, and value 5 approximate circles in the (p_1, p_2)-diagram.

projections. Vertical projections of color coordinates end in the moment diagram (M_1, M_2) for which $Y = 0$, and central projections from the black point (Bl) end in the purity diagram of constant luminance $(Y = 1.0)$. Along the latter lines, the ratio $p = M/Y$ is constant, thus defining a line of constant purity. If we take the spectral distributions of optimal colors in Figure 5.15(a) for color stimuli of constant dominant wavelength, and start out with a narrow spectral distribution of monochromatic light [λ in Figure 5.26(b)] and gradually broaden the wavelength range, the coordinates of the stimuli will be in a plane and follow the bold curved line in Figure 5.26(b), which represents the loci for optimal colors of the same dominant wavelength on the surface of the object color solid. Purity is greatest for the (dark) monochromatic light and decreases as the spectrum widens. As the stimulus gets lighter, its chromatic moment increases, and at some point a color of maximum moment is reached, after which the color becomes more and more whitish. For the complementary wavelength, to the left in the diagram of Figure 5.26(b), the coordinates of the complementary colors will follow a curve, which, relative to the curve on the right, is symmetrical about the mid-point of the achromatic axis. (C) and (C′) are complementary color vectors adding to white (W). Figure 5.27 shows a photograph of a three-dimensional model of the object color solid with the vertical (M, Y) planes (based on Luther, 1927).

Some other correlates of this linear representation to perceptive attributes are indicated. For instance, a luminance ratio, Y, or a reflection factor, β, between 0 and 1 corresponds to reflecting surfaces of varying lightness (or Munsell value), whereas a luminance ratio higher than 1 corresponds to the brightness domain. Resultant chromatic moment correlates with perceived chromatic strength (or Munsell chroma). Analogous to the triangular coordinates of the Ostwald and NCS systems, being chromatic content, whiteness and blackness can easily be recognized in the (M, Y)

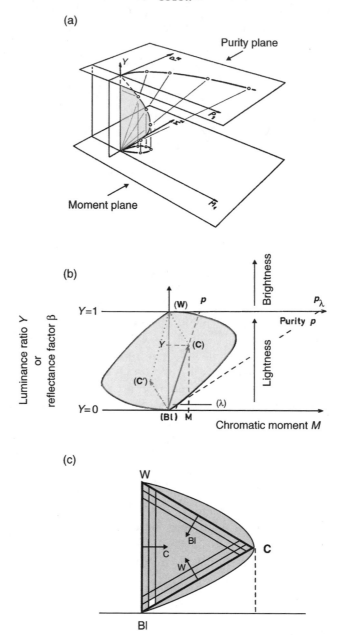

Figure 5.26 Schematic representation of an object color space for related colors based on opponent transformations of 1931 CIE tristimulus values. (a) A plane of constant hue with the moment (M_1, M_2) and purity (p_1, p_2) projections. (b) Opponent coordinates (M, Y) in a plane of constant dominant wavelength, with purity defined as $p = M/Y$. Monochromatic light λ has the highest purity $p(\lambda)$. The vectors **(C)** and **(C′)** add to white **(W)** and therefore represent complementary colors. Object colors (related color stimuli) have a reflection factor $(Y = \beta < 1)$, and unrelated colors have $\beta > 1$. (c) Colors can also be characterized by the triangular coordinates $C + W + Bl = 1$.

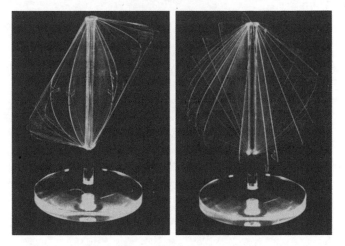

Figure 5.27 A photograph of a plexiglass model of the object color solid.

plane [Figure 5.26(c)]. Some advantages of this color solid over the CIE space in representing psychophysical correlates have been discussed elsewehere (Valberg, 1981).

Color differences

Hue and saturation in the (x, y) diagram

Figure 5.28 displays the CIE chromaticities of the basic chromatic attributes hue and chroma of equal luminance factor of the American Munsell color ordering system. The slightly curved lines radiating from the white point represent the chromaticities of colors of constant hue, varying in chroma. Since a constant hue follows a curved, rather than a straight line, this is an indication that the process determining hue, a perceived attribute, is nonlinearly related to cone excitations. The ellipse-like forms encircling the white point characterize stimuli with the same chroma, or the same color difference from white.

Let us take the hue 5R, represented by a solid curve in Figure 5.29. This hue in the Munsell system comes close to what most people regard as an elementary, or unique red, being 'neither yellowish nor bluish'. Additive mixtures between a color of maximum chroma and white are in the diagram situated on the straight dashed line connecting these chromaticities, whereas constant perceived hue follows the solid curve. Thus the hue of the mixture of red with white becomes bluer as more white is added, until it reaches a maximum blueness somewhere between the red and white, after which the blueness decreases. This change of perceived hue of an additive

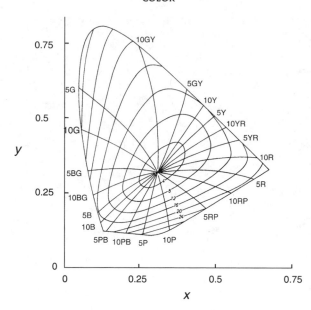

Figure 5.28 Constant hue and chromatic strength (Munsell hue and chroma) for medium reflectance (value 5) plotted in the CIE chromaticity diagram. Color stimuli that have the same hue, but different chroma, are situated on curves radiating from the white point. Equal chromatic differences from white give ellipse-like loci about the white point. The chroma steps are 4, 8, 12, etc.

mixture with white is called the 'Abney effect', and it is common for all hues (except maybe for hues close to 5GY and 5P; see also Figure 5.28). For instance, a yellow–green spectral color becomes greener in an additive mixture with white. Later we shall associate such peculiarities with nonlinear responses of the opponent cells beyond the stage of receptor excitations and receptor potentials.

Color scales

When comparing color differences, it is difficult for two persons to reach agreement on what are equal differences. In color science the specification of color differences is a recurrent topic of discussion. Take two clearly different nuances of orange, for example. What is the same color difference for two purple nuances, or between two different greens? Problems of comparison and a simple quantification of color differences (in terms of larger than, equal to or smaller than) occur whenever the color of a product must be reproduced consistently over a period of time in the manufacture of paints, textiles, ceramic tiles, etc. These are examples of cases where the tolerances for deviations from a standard must be quantified. Most of us find it particularly hard to compare color differences if differences in hues are involved, and

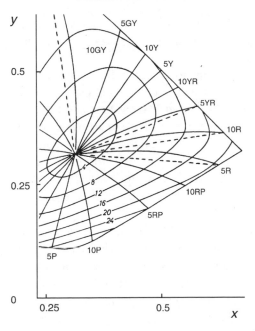

Figure 5.29 A section of the chromaticity diagram of Figure 5.28. A solid curve starting at the white point end ending at the spectrum locus represents chromaticities that all have the same hue. The hue curves deviate from the dashed straight lines which represent additive color mixtures of the spectrum colors with white. Let us take the red elementary color 5R of maximum purity as an example: in a mixture with white, the color will become increasingly bluish, depending on how much white is added. Yellow–green colors will become greener when white is added to them. Only the yellow color 10Y and violet 5P seem to retain their hues when mixed with white.

the difficulty increases with increasing difference. Who is able to say what difference between a red and a green is equal to a difference between a certain yellow and a certain blue? Isn't this a bit like comparing apples and oranges? Nor is it easy to compare even a small step of grays or saturations with a difference in hue. Nevertheless, industry needs objective measures of color differences and quantitative specifications of tolerances across the different color dimensions. This is of particular importance in color reproduction, and it is also required in the quality control of food, in color TV, for color displays in general, lighting engineering, etc.

It has been suggested that psychophysical thresholds, i.e. the just noticeable difference of a repeated color match, can be used as a unit for color difference, and that larger color differences can be characterized by the number of threshold steps they contain. This hypothesis stems from Weber and Fechner's laws (see the section on 'Contrast Vision' in Chapter 4, p. 196), and it is analogous to saying that a distance of 1 cm always equals 10 mm. The problem with this idea is nonlinearity, as with scales of lightness: 10 thresholds do not lead to the same perceived difference everywhere in color space. This is an empirical fact contradicting the conjecture that it is possible to summate thresholds to get a predictable finite step. However, in some cases, and because of the lack of an alternative, summation can be a useful approximation.

Based on empirical judgements of color differences by many individuals, atlases have been constructed that visualize a color scaling system for the entire color space. The Munsell system (USA) is probably the best known. Although far from perfect, this system is based on extensive scaling experiments with many observers. In the past it has often been used as a 'yard-stick' in quality control. It has also served as a basis for the development of quantitative formulae of color differences. Such formulae are the CIE-Lab and CIE-Luv formulae (Wyszecki and Stiles, 1982; CIE, 1970). A review of old and new formulae for color scaling can be found in Richter (1996).

The need for fast and reproducible control procedures has led to the development of standardized instruments to specify color differences. However, such instruments often lack the desired precision. This has two main causes: (a) the failure of the three light sensors to match the standardized CIE spectral sensitivity curves $X(\lambda)$, $Y(\lambda)$, and $Z(\lambda)$, and (b) the use of questionable mathematical model approximations to convert differences in tristimulus values to perceived color differences.

Color difference discrimination

MacAdam (1942) made a study of the precision of color matches, and showed that, for a number of repeated matches, the chromaticities of the matching stimuli were distributed on an ellipse around the targeted chromaticity in the (x, y) diagram, These so-called MacAdam ellipses varied in size across the diagram, as shown in Figure 5.30 (here magnified 10 times). This means that equally large color differences are reproduced in the diagram as different lengths, depending on the target chromaticity and on the direction from the target. The ellipses are rather small in the blue and violet corners of the diagram and large in the green region. Later works demonstrated that these ellipses were cross-sections through three-dimensional ellipsoids in XYZ-space.

Because of its importance in practical applications, many attempts have been made over the years to develop good empirical formulae for color differences. They have, however, met with little success. One requirement for a good formula has been that it describes the scaling of color differences in the Munsell system. Another system against which to test color difference formulations is the American Optical Society Uniform Color Scale (OSA-UCS; see Wyszecki and Stiles, 1982).

Most descriptions of small and large color differences have been based directly on mathematical manipulations of cone excitations, or of the XYZ tristimulus values. Relatively few attempts have been made to utilize knowledge about the physiological processing of color information later in the nervous system. One of the earliest attempts in this direction was that of Adams (1942). In want of something better, he

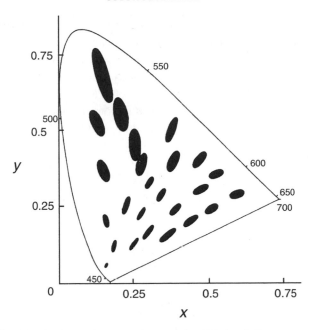

Figure 5.30 MacAdam's discrimination ellipses in the CIE (x, y) diagram, magnified $10\times$. The ellipses represent the uncertainty (standard deviation) of repeated matches of the center color. They are therefore taken to represent the same sensory difference from the center color in all directions.

used nonlinear transformations of X, Y and Z in an opponent formulation of color processing. A more recent attempt in this direction was the SVF formula developed by Seim and Valberg (1986). The magnitudes S, V and F represent the three most significant steps in the process; the linear excitations S of the three cone types, the nonlinear receptor potentials V of the same cones, and the responses F of opponent cells and their linear combinations. The introduction of nonlinear intensity–response curves for each of the three cone types was the most important feature of the SVF model. All later steps in the model's signal transmission were assumed to be linear. These assumptions were later confirmed by a mathematical modeling of the responses of opponent ganglion cells in the retina and in the lateral geniculate nucleus (LGN) of the macaque monkey (Valberg *et al.*, 1986a; see Chapter 6).

These responses were obtained from recordings where color stimuli were exchanged with a neutral, white background. This viewing situation resembles the way in which object colors are seen in nature, related as they are to near and far surrounds that determine the adaptation level of the eye, and with eye movements introducing a temporal component. These cell recordings also showed that the opponent ON- and OFF-cells (here called Increment and Decrement cells; I- and D-cells) divided the luminance range between them, with the opponent D-cells responding to darker colors than the I-cells. In fact, opponent D-cells are so strongly inhibited that they are silenced for a luminance just slightly higher than the adaptation luminance of a white stimulus, and therefore they cannot code for color of such stimuli. The obvious next step was to develop a model adding the responses of I- and D-cells with the same opponent cone input (in other parallel channels, however, these cells should maintain their independence). As we shall see later, a vector treatment of cells with different opponency, for instance of 'M–L' and 'S–L' cells, provides reasonably good correlates to such properties of colors as the Abney effect, the Bezold–Brücke phenomenon, and equidistant color scales.

A comparison of the early SVF formula with the CIE-Lab formula showed a clear advantage of the former in describing the data of the Munsell system. In Figure 5.31

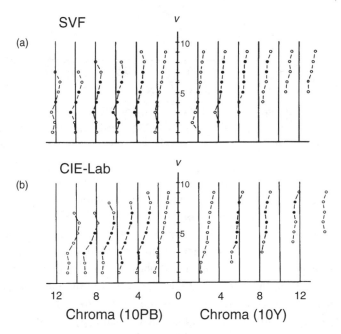

Figure 5.31 A plane defined by the achromatic axis of the Munsell system and the line through the opposing hues 10PB and 10Y. The parallel vertical lines represent an ideal scaling of chroma in steps of two chroma units up to chroma 12. The SVF formula, using nonlinear intensity–response functions for the cone signals, comes closer to modeling an ideal spacing than does the empirical CIE formula. Solid symbols are for the experimental data and open circles for interpolated and extrapolated data.

one sees the chroma scaling (abscissa) of the two representations for two opposite hues 10Y and 10PB and value V 1–9. The chroma scales in these two opposite hue directions are more equidistant for SVF than when using CIE-Lab (the vertical bars indicate the targeted Munsell scaling). In Figures 5.32(a) and (b) the coordinates of the color chips of the Munsell system (for value 5) are compared in CIE-Lab space

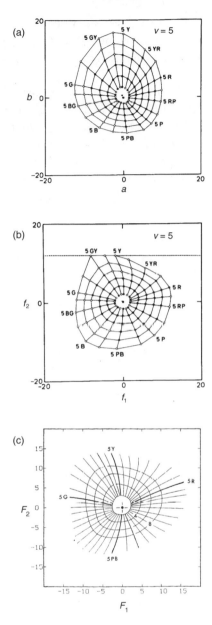

Figure 5.32 Munsell hue and chroma for value 5, as modeled by (a) the CIE formula, (b) the SVF formula, and (c) a neural model of color vision to be presented later. The chroma steps are 2, 4, 6, ..., 12. Solid symbols represent the original data of Munsell scaling, whereas the open circles are for inerpolated and extrapolated data. Compared with (a), the scaling of hue and chroma is more uniform in (b) and (c). In the neural model used to obtain (c), the scaling of Munsell hue and chroma is based upon a mathematical transformation (F_1, F_2) of a combination of six opponent cell types found in recordings from the LGN of the macaque monkey. In this model, the hue steps around the hue circle are more evenly spaced and the contours for equal chroma are more radial symmetric than for the empirical CIE formula in (a).

and in SVF-space. Figure 5.32(c) shows the result of the physiological model mentioned above, which was based on our recordings from macaque opponent cells. In this model, to be described later, representative opponent cells are the actual response units.

A calculation of deviations between an 'ideal' Munsell representation and the coordinates of the CIE-Lab, the SVF-system and the cell response coordinates (F_1, F_2) of the physiological model, demonstrates that the two latter representations of Figures 5.32(b) and (c) are both better than the empirical CIE-Lab representation in Figure 5.32(a). Thus, major improvements of the quantitative formulation of color differences can be achieved by taking into account the nonlinear response curves of cone receptors and their subtraction in the cone opponent cells. The result of Figure 5.33,

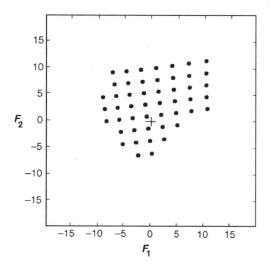

Figure 5.33 A representation of the chromaticities of the Uniform Color Scale of the Optical Society of America (OSA-UCS) in terms of the coordinates (F_1, F_2) of the neural color vision model also used in Figure 5.32(c). The original data of OSA-UCS have equal perceptive steps sizes horizontally and vertically, and this is very well rendered by the neural model. The reflection factor for these colors were near $Y = 30\%$ (from Valberg *et al.*, 1986a).

showing how the physiological model yields a close to perfect scaling of a plane in the OSA-UCS system, proves the usefulness of this approach. These results lend strong support to the idea that cone nonlinearity and a simple subtractive opponent processing is a better model for color scaling than using Weber ratios of cone receptors and combining them to form line elements (see below).

However, the theory of line elements has played an important role in developing a physical measure of color difference and contrast, and in the next section we shall give a brief description of this approach.

Line elements

Although it has not been very successful, the theory of line elements has played an important historical role in attempts to describe small color differences. Von Helmholtz was the first to refer the Weber law to the relative excitation of the three cone receptors, stating that the Weber fraction was constant at threshold for each of them and independent of cone excitation.

In a three-dimensional vector space (u_1, u_2, u_3), a small distance, ds, the length of a line element, is found by (see Figure 5.34):

$$(ds)^2 = (du_1)^2 + (du_2)^2 + (du_3)^2$$

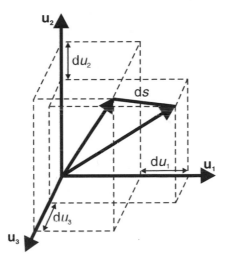

Figure 5.34   The definition of a line element as the distance 'ds' between two vectors in the three-dimensional vector space with the basic vectors $\mathbf{u}_1$, $\mathbf{u}_2$ and $\mathbf{u}_3$.

The three vector segments du_1, du_2 and du_3 may be regarded as increments or decrements of the cone excitations L, M and S, but this is not the only possibility. One might also imagine them as the response differences of three independent channels at a later stage of visual processing, for example of one luminance channel and two opponent chromatic channels. If we imagine the three vectors $(\mathbf{u}_1, \mathbf{u}_2, \mathbf{u}_3)$ in Figure 5.34 as those of cone fundamentals $\mathbf{L}$, $\mathbf{M}$ and $\mathbf{S}$, we can write

$$(ds)^2 = (dL)^2 + (dM)^2 + (dS)^2$$

However, according to Weber's law, it is the differential excitation, dL divided by excitation, L, and not dL alone, that is constant at threshold. If the excitation is increased from L to $2L$, the increment dL must be increased to $2dL$ to maintain threshold visibility (but as we have seen, this is restricted to low spatial and temporal frequencies of the stimulus; see Figures 4.24 and 4.28). Such considerations led von

Helmholtz (1892, 1911) to propose a line element, ds, based on the three-receptor hypothesis of color vision and the Weber law:

$$(ds)^2 = (dL/L)^2 + (dM/M)^2 + (dS/S)^2$$

Here, 'ds' is meant to represent a small perceptual difference of the same magnitude everywhere in cone excitation space, independent of the magnitudes of L, M and S.

This equation describes a three-dimensional ellipsoid with the main axes along the directions of **L**, **M** and **S**. In a plane, for example the (L, M) plane, the contours will be ellipses, as MacAdam found to be the case in his measurements of the standard deviations of color matches. However, the equation above does not account for MacAdam's empirical data with respect to the size and the orientation of the ellipses. We know that the Weber ratio is not the same for all three receptor types (Wyszecki and Stiles, 1982), and the weight of each cone type in the equation (their relative contributions to the line element) depends on a number of parameters, such as the size of the fields that are compared, their temporal modulation, the subject, etc. Nevertheless, this latter expression is a useful physical measure of contrast that can be applied for all dimensions of color space, regardless of whether the difference is in luminance or chrominance, or both.

Schrödinger (1920b) pointed to some weaknesses in the line element of von Helmholtz and developed a new line element that tried to take some of them into account:

$$(ds)^2 = LUM^{-1}[a(dL/\sqrt{L})^2 + b(dM/\sqrt{M})^2 + c\,(dS/\sqrt{S})^2]$$

This equation represents an ellipsoid in the three-dimensional space spanned by the axes $\sqrt{L}$, $\sqrt{M}$ and $\sqrt{S}$. LUM stands for luminance, and being a denominator in the equation it takes into account the fact that thresholds decrease when luminance increases. a, b and c are weighting factors that depend on experimental parameters such as test field size, temporal frequency, the test subject, etc. Schrödinger's concept was relatively modern in that the ratio

$$(dQ/\sqrt{Q}) = \text{signal}/\text{noise}$$

applies for a Poisson process where Q is the mean number of quanta that are absorbed in a cone per second. For equal detectability of a signal against the background, the signal dQ must always be proportional to $\sqrt{Q}$, i.e. the signal–noise ratio must be constant. This corresponds to an adaptation of the cone to a background activity $\sqrt{Q}$. Vos and Walraven (1972) utilized this ratio in a complicated line element, where the first step resembles that of Schrödinger.

Opponent transformations

Psychophysical experiments have shown that the color discrimination ellipses tend to have two main axes in the two-dimensional (L,M) color space; the luminance direction

(L + M) and the chrominance direction (L − M). Noorlander and Koenderink (1983) determined the difference thresholds for combinations of red–green contrast and luminance contrast when the spatial and temporal frequencies were changed. Figure 5.35 shows a schematic example for a large field of how the discrimination ellipses depend on temporal frequency. For a large stationary stimulus and for slowly varying stimuli, the discrimination ellipse is oriented with the long axis along the luminance

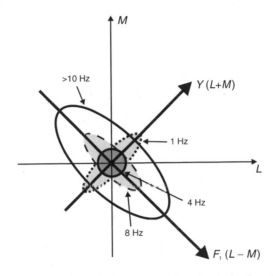

Figure 5.35 A schematic drawing of the contours for discrimination thresholds in different directions of color space with respect to white and for different temporal frequencies. The major axes for the threshold ellipses represent directions of pure luminance change (*Y*) or isoluminant red–green changes (*F*₁). Stationary and slowly changing stimuli (∼1 Hz) give best discrimination (lowest threshold) for color (chrominance) and least discrimination (highest threshold) for luminance. This causes the long axis of the ellipse to be orientated along the luminance direction $Y = L + M$. As frequency increases, the luminance threshold decreases, whereas the chromatic threshold for isoluminant stimuli increases. This results in a circular threshold contour around 4 Hz. From 5 Hz and up to about 7–8 Hz, the luminance threshold changes little, whereas chrominance threshold increases significantly, resulting in a major axis (low sensitivity) oriented along the chrominance direction L–M. For still higher frequencies (>10 Hz), both luminance and chrominance thresholds increase (after Noorlander and Koendrink, 1983).

direction $Y = L + M$ and the short axis along the chrominance direction $F_1 = L − M$. Thus, discriminability is less for luminance than for chrominance. As the temporal frequency increases, the long axis becomes shorter, and around 4-5 Hz the ellipse resembles a circle. If the temporal frequency is increased further, chrominance discrimination becomes poorer and the axis increases in the red–green chrominance direction $F_1 = L − M$. The transition from a circle to having a long axis in the chrominance direction takes place between 5 and 8 Hz for large fields. Above 8 Hz, discrimination of both luminance and chrominance becomes increasingly poorer.

If the same holds true for the yellow–blue opponent direction, one would expect to obtain a better line element by introducing orthogonal luminance and chrominance channels:

$$(ds)^2 = a(dY/\sigma_Y)^2 + b(dF_1/\sigma_{F1})^2 + c(dF_2/\sigma_{F2})^2$$

where F_1 is a red–green opponent process and a function of $L - M$, and F_2 is a yellow–blue process that may be a function of the cone combination $(L + M) - S$. σ is the uncertainty in Y, F_1 and F_2. For high luminances, σ is proportional to Y, F_1 and F_2 as in the Weber ratio. a, b and c are weighting factors that depend on experimental parameters.

In the experiment described in Figure 5.35, the discrimination was made along a line where F_2 (yellow–blue) was constant. We will come back to Figure 5.35 later when we compare the responses of opponent cells with psychophysical results.

Combined luminance and chrominance contrast

Line elements were early attempts to find a common, combined measure for contrasts that included differences in luminance as well as in chrominance. However, experiments have shown that there is no simple way to design a psychophysical 'yard stick' for contrast that can be used anywhere in color space. No known formula is able to give a satisfactory account of equivalent luminance and chromatic differences. For a given stimulus size, perceptive color differences can be empirically characterized in the CIE space, as MacAdam did by plotting discrimination ellipses and ellipsoids. However, there is no obvious and direct way to model these differences mathematically, and there is no measure that can directly translate a given luminance contrast into an equivalent chromatic difference, or the other way around. There is no simple way to combine photometric and colorimetric magnitudes to get something that resembles perceptual scaling. Moreover, color and luminance discriminations are differently affected by spatial and temporal parameters, making the task even more complex.

It is, however, possible to use a simple measure of contrast, for example one that is linearly related to the physical processes of excitation. The nonlinear CIE Lab formula and the other empirical scaling formulae are not what we are thinking of, developed as they were from a series of *ad hoc* concepts and hypotheses. It appears that the closest one can come to a physical measure is to reconsider the excitations of the three types of cone. Threshold values of relative cone excitations are expressions of the effectiveness of a stimulus in evoking physiological and psychophysical response differences.

One simple possibility is the Helmholtzian approach – to combine the Weber ratios or the Michelson contrasts of the three cone excitations, with equal weights on all three cone types. This combination of cone contrasts should not be attached to a particular color vision model; it simply represents the last linear stimulus stage before the generation of non linear signals in the receptors and in the neural pathways. A common measure for luminance and chrominance contrast would thus be to sum three quadratic contrast expressions, one for each cone type, divide by 3, and calculate the square root:

$$C^2_{\text{Weber}} = 1/3[(\Delta L/L)^2 + (\Delta M/M)^2 + (\Delta S/S)^2]$$

If the stimulus is periodic, as is the case for a sinusoidal grating, it is natural to replace the Weber ratio with the Michelson contrast for each cone type; for example $C_M = (M_{\max} - M_{\min})/(M_{\max} + M_{\min})$ for the M-cone and similarly for the L- and S-cone:

$$C^2_{\text{Mich}} = 1/3[C^2_L + C^2_M + C^2_S]$$

For a pure luminance increment or decrement, the contrast magnitude is the same in each of the three cone types and, since we divide by 3, the resulting combined contrast will be the same as if we used either the Weber or the Michelson contrast definitions directly on photometric luminance. Thus we have here a formula that leaves the luminance contrast definition unchanged, while allowing us to compute a combined contrast based on linear cone excitations. These two formulae treat the contrast, if it is in luminance or in chrominance, or a combination of both, as a distance in a three-dimensional cone vector space. However, nothing is said about the relationship between the contrast measures and perceived differences. We know already that identical contrast values, calculated by these equations for different cone excitations (and hence for different locations in LMS color space), do not correspond to a judgement of equal differences, even if spatial and temporal parameters are left unchanged. Therefore, as for the relationship between luminance and lightness, the relationships between cone excitations and the magnitude of a perceptual color difference must be worked out separately for different experimental conditions. Examples are shown in Figure 5.35 for temporal modulations, and in Figures 4.29 and 5.36 for changes of spatial frequency (size). Only for stationary sinusoidal gratings with a spatial frequency of about 2.5 c/deg, where the curves of Figure 5.36 cross, is the threshold, in terms of combined cone contrasts, the same for luminance as for red–green chrominance. For yellow–blue color differences, the chrominance and luminance thresholds are equal for a stimulus of about 0.6 c/deg (see Figure 4.29).

Color is what the eye sees best

For stationary sinusoidal gratings, Figure 5.36 displays human cone contrast sensitivities ($1/C_{\text{Mich}}$) as a function of spatial frequency for luminance and red–green chrominance gratings. The curves are the mean values of 11 and 10 subjects, respectively. Compared with a similar result for yellow–blue discrimination in

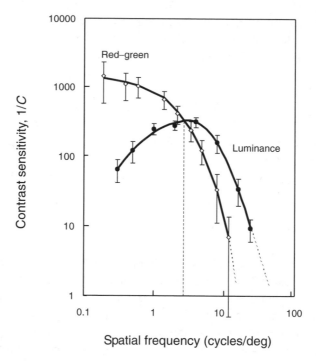

Figure 5.36 Spatial contrast sensitivity for the detection of a static sinusoidal grating modulated either in luminance or in red–green chromaticity at isoluminance. Contrast sensitivity is given as the inverse of combined cone contrasts at detection threshold. At low spatial frequencies, chromatic red–green contrast sensitivity is more than 10 times higher than pure luminance sensitivity (see text).

Figure 4.29, we see that cone contrast sensitivity is higher for red–green discrimination than for yellow–blue. At 0.2 c/deg, threshold sensitivity along the red–green dimension is about 1500. This is about five times higher than the maximum cone contrast sensitivity for pure luminance contrast (which has a maximum at 3.5 c/deg), and about 50 times higher than the luminance contrast sensitivity at 0.2 c/deg. The red–green sensitivity is about six times higher than for yellow–blue at 0.2 c/deg. However, since the curves do not seem to have yet reached a plateau at low spatial frequencies, we cannot exclude the possibility that the maximum sensitivities are even larger for still lower spatial frequencies.

These results confirm the contention that one needs less cone modulation to see pure chromatic differences than to see pure luminance differences (Mullen, 1985; Chaparro *et al.*, 1993). At the cone level, the sensitivity for chrominance contrast is significantly greater than for luminance, and thus isoluminant chromatic stimuli are the more effective in creating a perceived difference. The dependence of discriminability on spatial and temporal parameters is related to the processing of cone signals in nerve cells at later stages.

In an experiment performed by Chaparro *et al.* (1993), it was shown that, when expressed in terms of cone modulation, the threshold for detecting an isoluminant color change was much smaller than the threshold for detecting a pure luminance change. For luminance, the smallest thresholds for single stimuli were found for stimuli between 5 and 10′ in size (corresponding to the bar width of a 2–3 c/deg grating). For chrominance the optimal size was closer to 15′. Optimal presentation duration (the stimuli were turned on and off) was 50 ms for luminance and 140 ms for chrominance. Using these optimal parameters they found that optimal luminance stimuli needed between five and nine times larger cone contrast at threshold than optimal chrominance stimuli, in accordance with the results of Figures 4.29 and 5.36.

Color induction and adaptation

Simultaneous contrast

The phenomenon of lightness contrast (see for instance Figures 1.5 and 4.32) is also known as simultaneous contrast or lightness induction. A bright surround induces grayness or blackness in a central field of lower luminance. If the surround were replaced by a chromatic color, the central square would appear to be tinted in a complementary hue (see Figure 1.11). Green surrounds induce a tint of purple, and reddish surrounds induce blue–green. Blue surrounds would induce orange, whereas yellow induces blue–violet. Figure 5.37 shows some striking and more complex

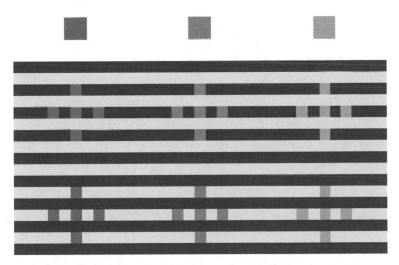

Figure 5.37 Simultaneous contrast. The larger squares on top of the figure are printed in the same colors as the smaller squares below, and they are therefore physically equal. If you look at the figure at a greater distance or tilt the book, you will see a significant distortion in the colors of the small squares. (See also color plate section.)

examples of combinations of simultaneous lightness and chromatic contrasts. The small squares on the top of the figure are printed with the same pigments and are thus physically the same as those imbedded in the blue and yellow stripes. The color shifts of the latter are enhanced even more if one tilts the book or views the figure at a greater distance, demonstrating that relative size contributes significantly to the effect. These phenomena are poorly understood, and we shall offer no explanation here.

Colored shadows are much discussed phenomena caused by color induction. You can see them in a viewing condition like that of Figure 5.38. An opaque object is

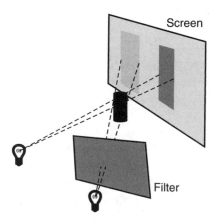

Figure 5.38 Colored shadows. When a blue and a white light illuminate the same object, the object casts two shadows on the wall. The shadow that is illuminated by the blue light looks blue, as expected, but the one that is illuminated by the white light appears yellow instead of white. The additive color mixture in the surround is nearly white. Other colored lights will also produce a contrast color in the shadow (red light will give a greenish shadow, and green light a reddish shadow, etc.). This surprising contrast effect will be reproduced in a color photograph of the scene. (See also color plate section.)

illuminated by two light sources, one emitting white light and the other chromatic light (bluish in the example). The object casts two shadows on an opposing screen. Either shadow will be illuminated by the other source. The shadow that is illuminated by the blue light looks bluish as expected, whereas the shadow that is illuminated by white light does not look white, but yellow. If the color-inducing light changes from blue to purple, the unchanged white shadow now turns greenish. In some cases, when the color surrounding the shadows is so unsaturated that it can be mistaken for being white, the chromatic shadow effect is still present and surprisingly strong.

It is easy to be convinced that the physically unchanged shadow − the one that is mostly affected by simultaneous contrast and induction − is reflecting the same light in all cases when the color of the chromatic light changes. One can, for instance, isolate the shadow by looking at it through a black tube. If someone changes the color of the chromatic light while you are looking through the tube, you do not to see any change at all in the achromatic shadow. This can be confirmed by physical color measurements. Consequently, the phenomenon is not physical, meaning that there is no direct physical correlate in the shadow itself that would indicate a change in its colored appearance. The correlate lies in the surround inducing a contrast color into the shadow.

The various forms of simultaneous contrast and induction effects in vision have often been regarded as failures of the visual system to adjust to the actual stimulus and lighting conditions. Von Helmholtz's psychological interpretation of these phenomena as optical illusions or errors of judgement has strengthened this view. The physiological explanations of Hering and Mach on the other hand, relied less on cognitive, top-down processes than von Helmoltz's 'unconscious judgement'. They pointed to induction as a normal neural function, and they regarded the mutual interaction of adjacent areas in the visual field as most important for the enhancement of small physical differences. Since color induction effects appear strong close to threshold contrast (Valberg and Seim, 1983), this may be taken as a confirmation of the latter view.

After one has become aware of the existence and the importance of the many contrast and induction phenomena, one tends to find them everywhere. Indeed, under normal viewing conditions, they are so pronounced that one may well ask if normal vision would be possible without them.

Induced colors

Induced colors are obviously related to a normal neural activity of the visual system, and their dependency on spatial and chromatic parameters provides information about the functioning of the system. For instance, how does an induced color depend on the inducing stimulus? In older literature it was believed that these two colors are complementary, but accurate measurements have revealed that this is correct only for rather unsaturated inducing stimuli. For inducing stimuli that are strongly chromatic, the induced colors deviate more from complementarity the more saturated they become. This is shown in the (x,y)-diagram in Figure 5.39 for an experimental situation as in Figure 5.40. A solid curve in Figure 5.39 shows how the induced color (that was matched by a separate comparison field viewed by the other eye) changes in chromaticity when the inducing color increases in purity (as shown by the two arrows in opposite directions from the white point W). The inducing surround colors are indicated by short dashed line segments ending at a '+' symbol. One sees that the induced colors follow a curve even though the inducing stimuli are all situated on a straight line of constant dominant wavelength (forming a tangent to the induced hue loci at the white point W). It is worth noting that the curved lines resemble the loci of constant hues in Figure 5.28 (Valberg, 1974; Valberg and Seim, 1983).

The induction process seems to be particularly effective for an unsaturated surround. A weakly colored surround is apparently capable of inducing a contrast

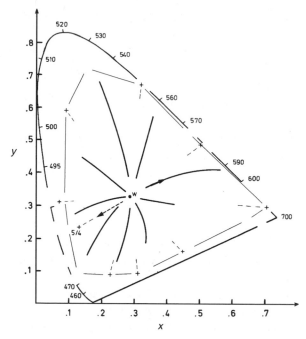

Figure 5.39 The curves radiating from the white point (W) towards the periphery of the CIE (x, y) diagram represent the chromaticities of the colors induced in a white center field by a concentric surround (see Figure 5.40). The saturation of the induced color increases as purity of the inducing surround changes from from white (W) to maximum purity (+). When, for instance, the blue surround 5/4 increases in purity, shifting the inducing stimulus color in the direction of the arrow along the dashed line, the saturation of the induced orange color also increases, shifting the color away from white along the fully drawn curve, as indicated by the arrow. A surround color of low purity induces an approximately complementary contrast color, of low saturation, in the center field. Saturated surround colors induce saturated contrast colors but the departure from complementarity is greater than for unsaturated colors. The resulting curves for induced colors bear a strong resemblance to the plots for constant hue with increasing color strength, as seen in Figure 5.28 and 5.29, suggesting a common physiological basis. See Figure 5.40 for experimental setup.

color with a greater chromatic content (chroma) than that of the surround itself (Valberg and Seim, 1983). Chromatic simultaneous contrast is thus amplifying small color differences, making them more visible. Near threshold, this effect is very strong. For instance, when two white colorimetric half-fields match (like the fields A and B in Figure 5.8), and one of them is altered by adding, for instance, a little green to it, it is the other, the physically unchanged field, that becomes redder before you notice the green.

Color induction also contributes to *color constancy* in complex scenes, a phenomenon that we shall deal with in greater depth in the next section. Consider what happens when, for instance, the illumination becomes redder. A reflecting surface,

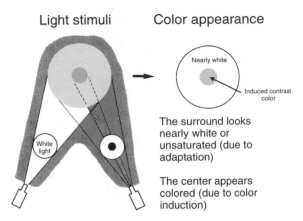

Figure 5.40   This figure shows the experimental arrangement used to obtain the results shown in Figure 5.39. An achromatic central field is surrounded by a mixture of white and a chromatic color of somewhat higher luminance. The induced color (viewed by one eye only) is matched by a physical adjustable color mixture (viewed by the other eye). This is accomplished by using slide projectors. Projector 1 projects white light onto a screen, and projector 2 adds a chromatic annulus. The image from projector 2 alone is a black disk surrounded by a chromatic ring, while projector 1 illuminates the whole screen evenly, superimposing white light on the black disk and the chromatic ring. When both projectors are on, measurements with a spectrophotometer will indicate that the disk at the center is white while the surround is an additive mixture of white and the chromatic light. However, the eye sees something else: it sees a chromatic, colored disk in the center (the induced color), surrounded by a nearly white annulus. The induced center color can be matched by an isolated, adjustable color mixture, viewed independently by the other eye. The coordinates of the matching mixture plotted in the CIE diagram can be seen in Figure 5.39. See also Color Plate Section.

and particularly a neutral gray or white one, will reflect the spectral composition of the illumination and become redder too. The brighter reddish areas in the surroundings will induce the opposite color, a bluish green, on the darker surfaces in their vicinity. Since this induced bluish green is able to neutralize some of the redness of the illuminant, the net result comes close to color constancy, the better the less saturated the color of the illuminant.

Chromatic adaptation and color constancy

My dark gray sweater looks gray in the evening twilight as well as in the sunshine at noon, despite the fact that it reflects 10 000 times more light at noon than in the evening. Neither does its color change much when I walk from outdoor daylight into a room with artificial lighting (although in other cases metamerism may cause problems; if buttons have the same color as the textile in daylight, they will probably look quite different in fluorescent lighting). How is it that the appearance of my sweater does not change as expected from the large physical changes of its luminance and chromaticity?

In conventional photography you need to use different types of film in artificial light and in daylight. If you don't, the color of the illuminant will give serious color distortions (for example yellow pictures when using daylight film indoors). Since the same distortions do not occur in vision (a banana looks yellow in bluish daylight as well as in the reddish evening sunset), the visual system must have the ability to neutralize the color of the illuminant, at least within certain limits. The eye adjusts to (adapt to) the light level and to the color of the illuminant so that both the color and the lightness of an object appear largely unchanged.

The color temperature of the illumination also affects the photograph taken with a digital camera. These cameras usually have some kind of automatic adjustment of the white point, using algorithms that seek to infer the color of the ambient light. This may cause problems when one is transferring the image to a computer in an unknown color space.

The color of an object is much less dependent on the spectral distribution of the light coming from it than one tends to believe. It appears that the reflection properties of surfaces relative to their surround are more important for color vision than the actual spectral distributions reaching the eyes. In yellow incandescent light, for instance, one may expect all surfaces to shift their color towards yellow, and slightly bluish surfaces to turn grayish. Yet the white paper still appears white, and blue is still blue. To find a satisfactory quantitative description of this adaptive neutralization of the color of the illuminant is still a challenge.

Leaving temporal and cognitive factors aside, there are some important physical factors that contribute to the color of a surface: its *relative* spectral power distribution, chromaticity, its contrast to the surroundings and the illumination. Under normal and moderate changes of the illumination, these factors, and a physiological adaptation process that neutralizes the physical changes, ensure approximate color constancy. The details of this process are still unknown, but as we saw above, color induction seems to contribute to it.

Even if a white surface looks white in daylight and in artificial light, we should be aware that food (meat, vegetables, etc.), textiles, ceramic tiles, etc. can change their color in artificial light, although the color distortions may be less than anticipated. Particular spectral distributions of the light sources that illuminate the display stands in many shops may, for instance, boost the redness of meat, or of tomatoes, and enhance the attractiveness of food and vegetables. Such lighting with poor color rendering is frequently used in supermarkets. Poor, but less purposeful color rendering is also typical for some energy-saving illumination.

The von Kries hypothesis

Color induction and adaptation were for a long time explained at the retinal level by a relative reduction of the sensitivity of the cones, but more recent work points also to lateral interactions at several stages of the visual pathway. According to the von Kries hypothesis (1905), when a white paper looks white regardless of the color of the illumination, this is a consequence of the excitation of the three cone types being adjusted (normalized) to the prevailing illumination. The hypothesis states that sensitivity and excitation is reduced by a factor proportional to the receptor's initial excitation by the illuminant. For each receptor type, the factor is the same for all wavelengths, and therefore the relative spectral sensitivities of the receptors do not change. Compared with adaptation to equal-energy white light, adaptation to long-wavelength red light reduces the sensitivity and excitation of L-cones more than those of M- and S-cones. In real viewing situations, with close to achromatic illumination, the visual system also adjusts to some average luminance and chrominance across the visual field. We may call the latter induction effects *lateral adaptation*. Lateral adaptation and adaptation to the illuminant may in fact be treated as two different processes. In theory, adaptation by reduced sensitivities accounts for the neutralization of the chromatic component of the illuminant, at least for not too chromatic lights, and for not too strong lateral color-induction effects. If, for a certain surface and a neutral illumination, the excitations of the cones were L, M and S, in the new slightly different adaptation state, they would be changed to:

$$L' = \alpha L$$
$$M' = \gamma M$$
$$S' = \beta S$$

where $\alpha = 1/L_{\mathrm{III}}$, $\gamma = 1/M_{\mathrm{III}}$ and $\beta = 1/S_{\mathrm{III}}$. The cone excitations L_{III}, M_{III} and S_{III} due to the illuminant refer to those for an ideal, diffuse reflecting white surface under the new illuminant. In this way the cones adapt to every new illumination in such a manner that a white surface keeps equal cone coordinates $L'_{\mathrm{W}} = M'_{\mathrm{W}} = S'_{\mathrm{W}} = 1.0$ in that illumination, whereas the coordinates for all other colors will change. This transformation, which continue to give the illumination and a white paper the chromaticity coordinates $(\frac{1}{3}, \frac{1}{3})$, is often called a centering transformation.

Figure 5.41 demonstrates how one imagines that the von Kries hypothesis works. The color stimulus $\mathbf{F}$ has the cone excitations $(L, M, S) = (1, 1, 1)$. In the (non-realizable) case where only the M-cones are adapted so that their sensitivity is halved, the vector $\mathbf{F}$ will change to $\mathbf{F}'$ in a direction of 50 percent less M excitation, i.e. towards purple. A neutral gray surface affected only by lateral adaptation would thus be expected to appear purple if only the effectiveness of the M-cones was reduced and no other processes (e.g. a greenish illumination) compensated for this change. This is, of course, a hypothetical situation since it is difficult to adapt only one receptor type without affecting the others, but it illustrates the idea. If the same adaptation occurred only in L-cones, the vector would move towards blue–green, and if both L- and M-cone sensitivities were halved, the change would be in the direction of blue. In the theoretical case, where the coefficients α, β and γ are determined by the illumination alone and thus are inversely proportional to the excitation of the cones in a given illumination, the result is perfect color constancy. This is so because an increase in the intensity of the illuminant by a factor of 2 would lead to the cones being excited twice

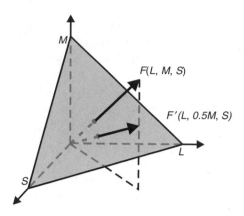

Figure 5.41 Chromatic adaptation can, according to the von Kries hypothesis, be described as a sensitivity reduction of the cones. The adaptation to a colored light (e.g. a chromatic illumination) will affect the receptors differently, depending on how strongly they are excited by the light. The figure illustrates what might happen in a simple (but unrealistic) case where green light reduces the sensitivity of the M-cone by a factor of 2 and leaves the sensitivity of the other receptors unchanged. The anticipated effect would be that the vector **F** changes to vector **F'** due to less excitation of the M cone. An embedded region that originally looked white, and which is not illuminated by the green light, would now appear purple. If, on the other hand, the embedded white surface were also illuminated by the greenish light, it would remain white.

as much, and the ensuing adaptation would then reduce their sensitivity to 50 percent of its original value. Taken together, excitation, being the product of power and sensitivity, would be unchanged, and thus the vector representing a white stimulus would also be unchanged. In other words, the von Kries hypothesis states that, if sensitivity is reduced by the same factor as the excitation is increased, the chromaticity of the illumination would be exactly compensated for.

Such total adaptation to the color of the illuminant is likely to occur only when its chromaticity deviates little from white. Even the transition from daylight to incandescent light is too large a color shift for this hypothesis to hold true; computing color shifts for this situation according to the simple von Kries scheme above gives results that do not agree with our experience. Therefore, the von Kries hypothesis can, at most, be a first linear approximation of a complex, probably nonlinear, process. Despite this, the von Kries hypothesis has played a prominent role within lighting engineering, and it is still used in the absence of a better method.

Color rendering

The CIE characterizes the color rendering of light sources by a single number, the color rendering index, CRI, computed as the average index of eight standard color test samples. A value of 100 is regarded as being excellent and a value below 70 is rather poor color rendering. However, the basis for computing the CRI is the long obsolete von Kries adaptation hypothesis, and the computations do not compare well with subjective judgements (Valberg *et al.*, 1979). Despite this rather unsatisfactory situation, the von Kries hypothesis is used by the lighting industry to calculate a CRI of light sources. This index is based on a calculation of the color shifts that occur when the illumination is changed from a standard source to the illumination under test. A standard source can be either the daylight or a blackbody radiator of the nearest color temperature (the temperature that gives a black-body color most similar to that of the test light). Manufacturers of light sources give the CRI in their catalogues. An index between 90 and 100 signifies a very good color rendering.

The reason why such an index is needed is that color constancy is only approximate, not absolute. Therefore, the pressing practical problem is to predict the deviations from constancy of a given illumination. Over the years there has been an abundance of experiments quantifying the color shifts for the most common light sources, in both simple and complex viewing situations, but the experimental results still await a good theoretical model. The centering transformation mentioned above for cone excitations can in many cases give a good first approximation to these data, and the approximation works better in broad-band, weakly chromatic illumination than for strongly chromatic light sources.

Apart from these problems, using a single value to characterize color rendering, such as the CRI, is not adequate for many modern light sources. For instance, many fluorescent lamps have a rather complex spectral distribution with many narrow spectral bands that may lead to unexpected shifts of surface colors and to problems with metamerism. It would therefore be an advantage to know, for each new light source, the color shifts of individual samples on a color circle, for instance for 24 hues of high saturation around the color circle, and for another circle with medium saturation. Such additional information would meet the requirement of being easily understood, and it would help the designers in their efforts to create esthetic environments with the help of light and color.

The centering transformation

Below we give an example of a centering transformation as it is used, for instance in TV broadcasting, to compensate for the chromaticity of (weak chromatic) illumination (although strictly it should be applied only for lateral adaptation), let us take a set of tristimulus values R, G and B that are normalized so that $R_W = G_W = B_W = 1.0$ for an ideal, diffusely reflecting white surface (made for instance from magnesium oxide or barium sulphate). R, G and B may be thought of as general tristimulus values; they can be either the tristimulus values of an additive color mixture, or the corresponding cone excitations L, M and S. Let Y signify luminance.

When changing to a new illumination, C, the white surface will have the tristimulus values R_C, G_C and B_C. If we want the white surface to keep its original tristimulus values

(1,1,1) in the new illumination, this can be achieved by dividing all tristimulus values by the tristimulus values for the new white surface. The new 'centered' tristimulus values are $R' = R/R_C$, $G' = G/G_C$ and $B' = B/B_C$. This is a simple color correction. Let us look at an example.

Table 5.1

Vector	R	G	B	Y	$(R - G)/Y$	$(G - B)/Y$	Color
W	1	1	1	2	0	0	White
F$_1$	1	1	0	2	0	0.5	Yellow
F$_2$	1.5	1.5	0.1	2	0.5	0.2	Red
C	2.5	1.5	0.1	4	0.25	0.35	Orange

The original color coordinates of Table 5.1 are shown in an opponent diagram in Figure 5.42(a), and the transformed coordinates, centered to the new *Ill. C*, are given in Table 5.2 and Figure 5.42(b). In Figure 5.42(b), the new white point, C', is at the origin. The old white chromaticity point, W (illuminated by the original illuminant, not *Ill. C*) has moved towards blue (W') due to lateral adaptation. This is as expected since the illumination C is perceived as orange when adapted to W and induction effects from extended surrounds are strong. A physical stimulus F_2 changes from being reddish in Figure 5.42(a) to become reddish-blue in Figure 5.42(b). The color stimulus F_1 does not change much.

Table 5.2

Vector	R'	G'	B'	Y	$(R' - G')/Y$	$(G' - B')/Y$	Color
C'	1	1	1		0	0	White
W'	0.4	0.67	10	2	−0.14	−4.67	Blue
F'$_1$	0.4	0.67	0	2	−0.14	0.34	Green-yellow
F'$_2$	0.6	0.33	1	2	0.14	−0.34	Red-blue

Edwin Land's retinex hypothesis

In the case where the tristimulus values R, G, and B above represent the cone excitations L, M and S, the centering transformation of Table 5.2 would correspond to the von Kries adaptation hypothesis. The sensitivities of the receptor types would have changed by factors equal to $1/R_C = \alpha$, $1/G_C = \beta$ and $1/B_C = \gamma$.

A similar description of color adaptation has been used by McCann *et al.* (1976) and Land (1983) in their *retinex hypothesis*. In the original version, all relative cone excitations were normalized to their values for a white surface, as in Tables 5.1 and 5.2. The assumption was that the surface with the highest luminance or reflection

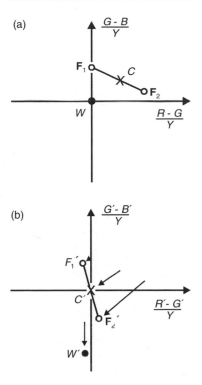

Figure 5.42 An example of how a centering of tristimulus values may give an approximately quantitative account of chromatic adaptation. Centering means to place the chromaticity of the prevailing illumination at the origin of the coordinate system. Changing the chromaticity of the illuminant from *W* to another, and slightly different illuminant, *C* (e.g. from daylight to incandescent light), will change the color coordinates of all objects illuminated. Centering shifts the 'neutral point' of the diagram, so that it coincides with the new illuminant. This manipulation of color space might, at a first glance, seem to explain why a white surface stays white regardless of small color shifts in the illumination. The centering, bringing illuminant *C* into the origin of the diagram, leads to close to parallel shifts of the chromaticity of all the other surfaces as well (arrows in the figure). However, this simple description of color shifts, as a result of the centering of tristimulus values (or even of cone excitations) to the new illuminant, is not quite in accordance with experimental data. However, in want of a better hypothesis, it can serve as a simple rule of thumb. Edwin Land's original retinex hypothesis was one such centering transformation.

factor contributes the most to adaptation. This led to a result that was identical to a centering transformation and to the von Kries hypothesis. Thus, the retinex and the von Kries hypotheses were essentially the same and suffered from the same weaknesses (Judd, 1960).

Later, Land changed the computational procedure. In the centering operation, the tristimulus values for the white surface were replaced by the geometrical mean of the cone excitations for all colored areas within the field of view. For a many-colored, complex scene this average came close to that for a gray stimulus, and this was now taken as the adaptation reference. All cone excitations were centered to (divided by) these averaged values, and such ratios changed relatively little when the illumination changed color. The increased redness of a banana in the evening sun is neutralized by the increased redness of the surrounding environment. In his impressive demonstrations of color constancy, Land used a composite image with many different sizes and colors, a so-called 'Mondrian' (Figure 5.43), named after the Dutch painter Piet Mondrian.

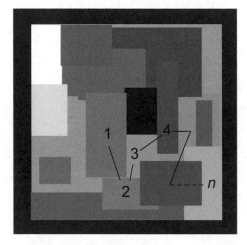

Figure 5.43   An example of a complex Mondrian of the kind used by Edwin Land to demonstrate his retinex hypothesis. (See also color plate section.)

However, many psychophysical experiments have shown that the retinex computations do not lead to the right solution. Constancy is not absolute, and there are significant differences between the experimentally determined color shifts of a surface and the color coordinates that follow from the retinex theory. The computational method leads to a result that resembles what happens under simultaneous contrast: an environment that has become reddish due to the illumination induces blue–green into darker, less reflecting areas, and the induced blue–green neutralizes some of the increased redness from the illumination. If these two opposite processes were linear and of equal magnitudes, the result would have been absolute color constancy. However, there is neither a complete neutralization of color strength, nor of hue. The deviations that occur are larger the more saturated the illuminant.

What would happen if we replaced the complex Mondrian outside the green area no. 1 in Figure 5.43 with a homogeneous gray area? The retinex hypothesis would predict that the green color of the no. 1 area would be maintained, provided that the surrounding gray gives rise to the same three cone excitations as the geometrical mean of all Mondrian patches. This is often called the 'gray world hypothesis'. Experiments have shown that it is indeed possible, by trial and error, to find a gray that, when replacing a particular Mondrian, does not change the color of area no. 1. Yet this surrounding gray is not that which is predicted by the geometrical mean, as required by the retinex hypothesis (Valberg and Lange Malecki, 1990). This is probably not too surprising, because we know for instance that proximity plays a major role in color induction. Areas nearby have a stronger inducing effect than those further away. This fact was not taken into account in the retinex hypothesis. All Mondrian patches were weighted equally in the computation of the average excitation values regardless their distance.

We can summarize Land's proposal for the normalization of the cone excitations in a Mondrian as follows: first take the ratio of the cone excitations L, M and S between areas that are adjacent (it was actually suggested to follow a path like the one shown in Figure 5.43, but this is not necessary). If we disregard threshold, for the L-cone the retinex calculates:

$$L'_1 = (L_1/L_2)(L_2/L_3)\ldots(L_{n-1}/L_n) = L_1/L_n$$

or

$$\log L'_1 = \log L_1 - \log L_n$$

Then summate the logarithms of all these ratios, L'_i, over all n fields in the Mondrian, and compute the geometrical mean:

$$\log L'_1 = (1/n)[n \log L_1 - \sum \log L_i]$$
$$= \log L_1 - (1/n) \sum \log L_i$$

The same procedure is repeated for the other cone excitations M and S. The normalized cone excitations L', M' and S' will be the new color coordinates for field no. 1 in the same way as R', G' and B' in Table 5.2. The values

$$(1/n)\sum \log L_i \qquad (1/n)\sum \log M_i \qquad (1/n)\sum \log S_i$$

are the logarithms of the geometric means of the cone excitations of all fields surrounding no. 1 in Figure 5.43.

An evenly distributed illumination, let us say a reddish one, leads to exactly the same changes in cone excitations for area no. 1 as for every field surrounding it. In the equations above, this will be factored out. Physiologically this means that the change in one field is neutralized by its surroundings. The 'redness' of the surround is subtracted from the 'redness' of the field in the logarithmic expression, and the net result is no additional effect by the new illumination on the relative cone excitations. Consequently, said Land, since this triplet of relative cone excitations of a patch remains unchanged, color constancy is the result.

Land published his first demonstrations and color vision theory in 1959 in a series of articles in *Proceedings of the National Academy of Sciences* (Land, 1959). In what turned out to be a consequence of simultaneous color contrast and induction, he described how to produce all hues by only mixing two colored lights in addition to white. This he considered a proof that the Young–Helmholtz trichromatic theory was wrong. Land's alternative theory was more or less refuted by Judd and other critics. For many years, there was little debate about Land's theory until his ideas experienced a renaissance in the 1980s. Land entered the scene at several international conferences with his Mondrian demonstrations and conducted a spectacular 'color show' with great bravura. This revival of interest initiated a series of neurophysiological measurements of the response to color stimuli of single cells in the retina and in the brain, using macaque monkeys as the experimental animal. However, in this wave of enthusiasm, the attitude towards Land's ideas was often somewhat uncritical (see the next chapter, Crick, 1994; and Zeki, 1993). All of Land's papers are published in *Science, Education and Industry*, Vols I–III, IS&T, 1993.

The problem that remains to be explained may be formulated as follows: how can one best separate the physical change in the direct excitation of cones from compensatory, indirect lateral processes like induction? When illumination changes, both object and surround change color coordinates. Even though adaptation and lateral induction contribute significantly to neutralizing the effect of this physical change, these processes do not suffice to cancel it completely. Most experiments designed to test the von Kries hypothesis have indicated that one needs to search for alternative ways to describe chromatic adaptation and color constancy. The quality and strength of the induction effects, represented by the curves for the induced colors in Figure 5.39, point to involvement of neural processes that are nonlinearly related to the stimulus changes (since a linear change in cone excitation would have been represented by straight lines). In natural viewing situations, simultaneous contrast (induction) and adaptation can be considered two sides of the same coin, and the nonlinearity displayed in Figure 5.39 shows that both processes arise beyond the linear excitation process that is so fundamental in color metrics.

On the background of Valeton and van Norren's measurements of receptor potentials under several conditions of light adaptation [Figure 4.15(b)], an alternative model would associate adaptation with changes in the half-saturation constants, σ, in the formula for cone responses (see p. 164). It would have been of great help to know more about the dependence of σ on adaptation level. Exploratory electrophysiological measurements on cones and on cone-opponent cells, have shown that, in 'M–L' and 'L–M' opponent cells, chromatic adaptation can be modeled by adjusting the half-saturation constants σ_L and σ_M of the cones that have inputs to the cell.

A possible further step towards a better model of chromatic adaptation might be to formulate excitation and inhibition in double-opponent cortical cells (such cells will be described on p. 393). Double-opponent cells are thought to also contribute to color constancy, because in such cells center and surround together are able to neutralize the chromatic component of the illumination. Thus, color constancy seems to be a composite phenomenon that incorporates adaptation and simultaneous contrast, with contributions from several processing stages in the visual pathway.

6 Color vision

Color between phenomenon and theory

As elements of our visual perception, color qualities are fundamentally different from the physical processes and object properties that bring them forth, such as electromagnetic radiation and the spectrally selective reflectivity of surfaces. While physical processes generate color stimuli that are imaged on the retina, there is no 'physical color' in the proper meaning of the term 'color'. A color's redness, for example, is not a physical entity. It is a subjective, perceived quality, sometimes called *qualia* when it is necessary to emphasize its difference from the physical and physiological processes generating it.

Color and color contrasts are important elements of an image. Everything we see has a color, or as Aristotle (384–322 BC) said 'what is seen in light is always color', and he continued 'without the help of light, color remains invisible' (Wade, 1998). The famous physicist J.C. Maxwell made a similar statement: 'All vision is color vision'. In keeping with this thinking, we will here use the word 'color' in a wide sense to include black, all shades of gray, and white. Consequently, with Maxwell (1872), we can say,

> 'for it is only by observing differences of color that we distinguish the forms of objects. I include differences of brightness or shade among differences of color.'

Color theory has occupied philosophers and scientists from a variety of disciplines through the ages. Today, the study of color vision has become central in the effort to understand the behavior of neural networks of the brain. In this context, a rather passionate debate has developed regarding the role of colors in our understanding of nature and of ourselves. The polemics of this debate have been no less fierce than the intellectual battles that took place between the adherents of Isaac Newton's (1642–1727) and Johann Wolfang Goethe's (1749–1832) theories of color. We shall

Light Vision Color. Arne Valberg
© 2005 John Wiley & Sons Ltd

return to this renewed interest in colors after a short review of the history of some color vision theories.

Although it was already pointed out by Newton (1979) that 'the rays are not colored', scientists have not always taken care to distinguish between the description of light rays and the description of qualitative experiences of colors. It also took a long time before it became clear which aspects of color phenomena needed a scientific explanation, and which were outside the scope of traditional scientific methods and language. The physical concepts of electromagnetic radiation, for instance, had to be developed before a scientific theory of color and color vision could be established. Then the whole complex needed to be sorted out and proper answers be sought to questions like 'Which parts of the color phenomena belong to the external, physical world?', 'Which parts are due to sensory processes?' and 'What roles do neural networks and cognitive processes play?' Even today this is a thorny issue (Chalmers, 1995a, b). Isaac Newton believed that every natural body reflects 'its own color' more conspicuously than the others, and many of us have been taught that color is caused by the wavelengths of light reflected off an object, i.e. that red is caused by long-wavelength light and blue by short. This is a simplistic view that is no longer maintained, except for within the darkness of the laboratory, as clearly demonstrated by Land (see Chapter 5). Since the time of Newton, many aspects of color phenomena have been omitted from a natural scientist's description of light, and Goethe's polemics against the Newtonian view can, at least in part, be interpreted as a fundamental disagreement on which aspects of experience should be dealt with by natural science.

Newton was more concerned with the properties of light rays than with colors as such, and in this field of optics his achievements were outstanding. The correlates he found between colors and the refrangibility of light rays were derived in a dark room illuminated only by a sunbeam. These correlates have a rather limited value for predicting the perception of colors in an everyday environment, but they have been intractably linked to color theory ever since. Goethe, on the other hand, was more interested in the phenomenology of color. He carried out many beautiful experiments and developed ideas and concepts that have had a great impact in art and culture. Particularly impressive are the demonstrations of boundary colors that appear when the slit used in Newton's *experimentum crucis* is wide. The colors are yellow and red on one side of the border, and blue and violet on the other. Goethe also devoted a lot of attention to the spectrum of colors that appears when the narrow slit in Newton's experimentum crucis is replaced by a narrow opaque strip. This Goethian inverted spectrum is complementary to the Newtonian spectrum and consists of a transition from blue via non-spectral magenta (purple) to a broad band of yellow. Goethe also experimented with *colored shadows*, an impressive manifestation of contrast colors that had no place in Newton's theory, and emphasized the important role of the experiment in serving as a mediator between object and subject.

One may call Newton's investigations analytic or theory-oriented. The predictive power of this approach has proved very successful in the natural sciences. It has given

Figure 1.7

Figure 1.11(b)

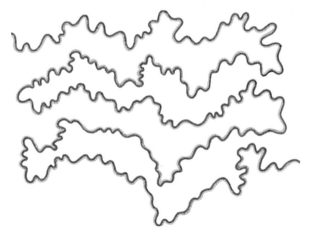

Figure 1.14

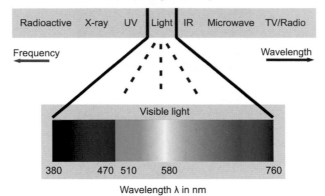

The electromagnetic spectrum

| Radioactive | X-ray | UV | Light | IR | Microwave | TV/Radio |

Frequency

Wavelength

Visible light

380   470 510   580   760

Wavelength λ in nm

Figure 2.1

(a)

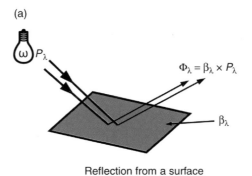

$\Phi_\lambda = \beta_\lambda \times P_\lambda$

β_λ

Reflection from a surface

(b)

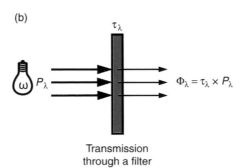

τ_λ

$\Phi_\lambda = \tau_\lambda \times P_\lambda$

Transmission
through a filter

Figure 2.26

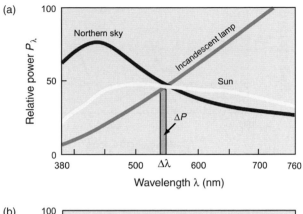

(a)

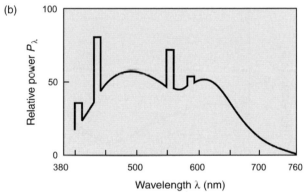

(b)

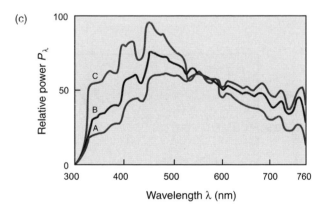

(c)

Figure 2.25

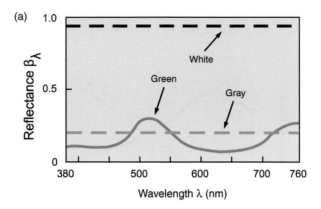

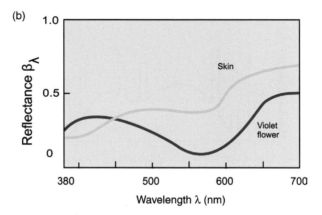

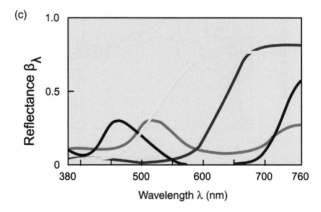

Figure 2.27

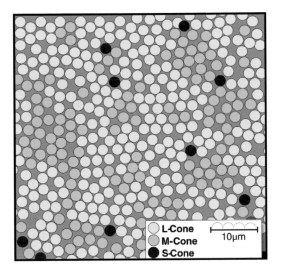

Figure 3.15

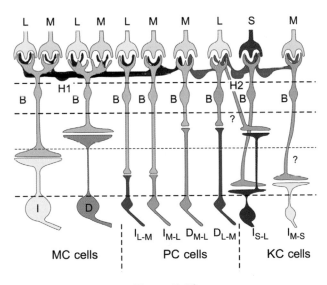

Figure 3.19

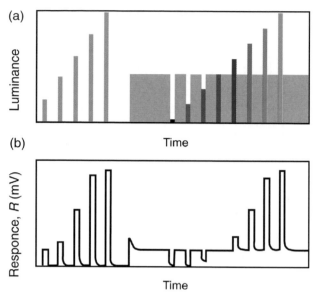

(a)

Luminance

Time

(b)

Responce, R (mV)

Time

Figure 4.14

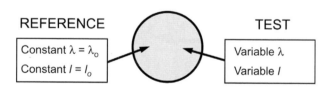

REFERENCE

Constant $\lambda = \lambda_o$
Constant $I = I_o$

TEST

Variable λ
Variable I

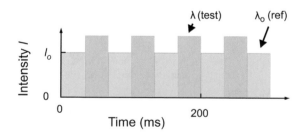

Intensity I

I_o

0

0   200

Time (ms)

λ (test) λ_o (ref)

Figure 4.17

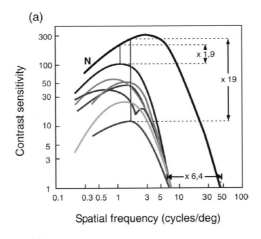

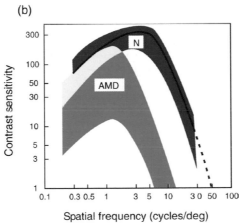

Figure 4.33

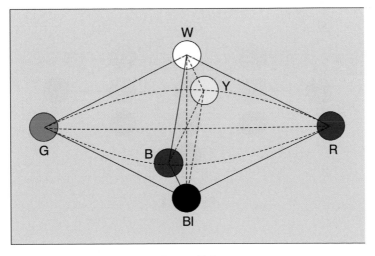

Figure 5.1

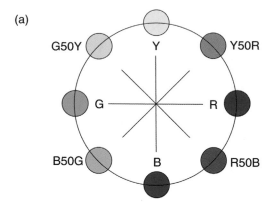

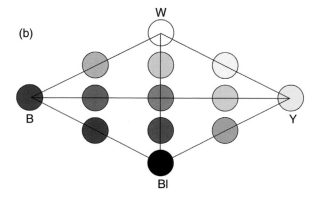

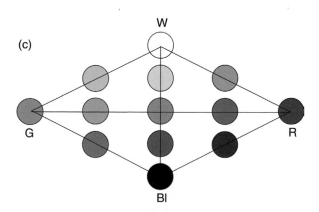

Figure 5.2

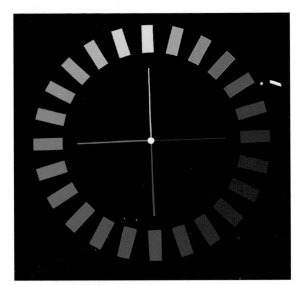

Figure 5.3

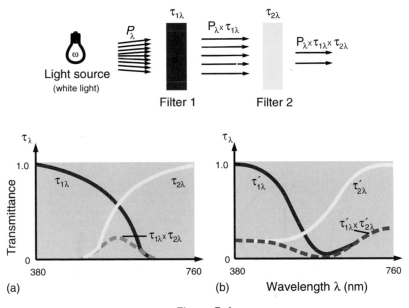

Figure 5.6

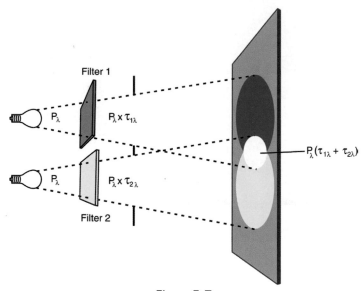

Figure 5.7

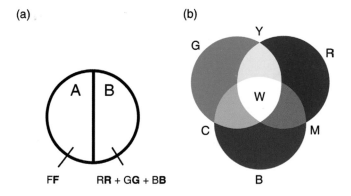

Figure 5.8

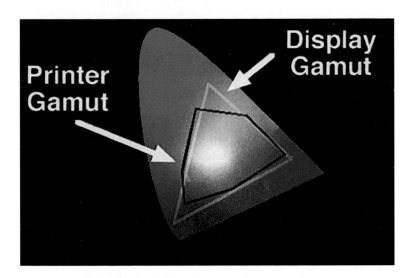

Figure 5.14(a)

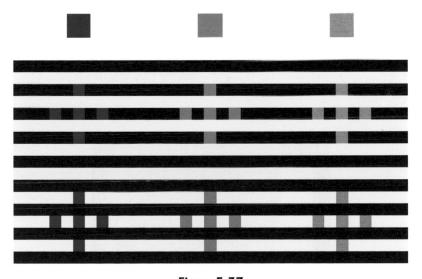

Figure 5.37

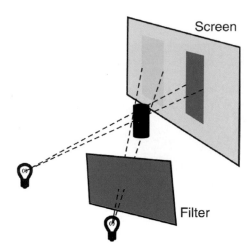

Figure 5.38

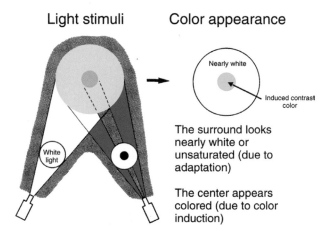

Figure 5.40

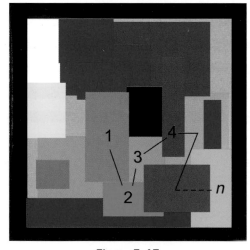

Figure 5.43

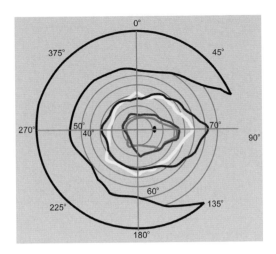

Figure 6.5

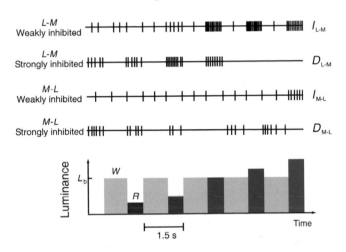

Figure 6.7

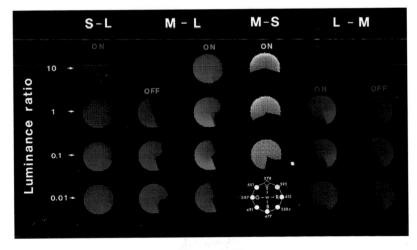

Figure 6.17

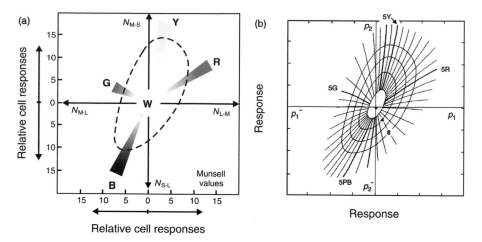

Figure 6.19

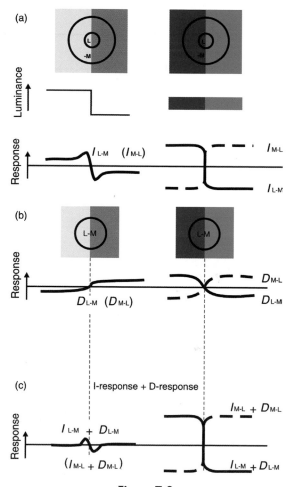

Figure 7.2

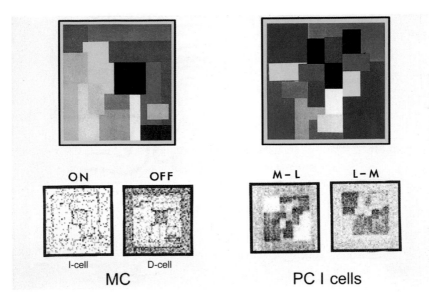

Figure 7.3

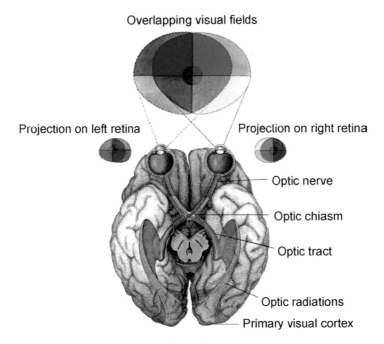

Figure 8.1

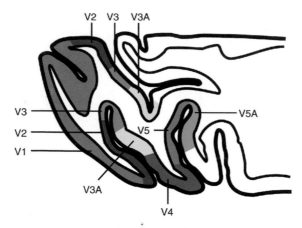

Figure 8.2(b)

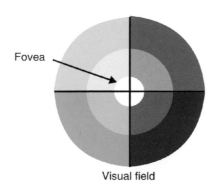

Visual field

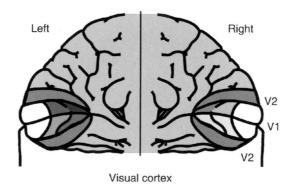

Visual cortex

Figure 8.3

us a tool with which to manipulate nature to such an extent that we today experience an increasing conflict between technology and science and culture. Goethe's experimentation was more intuitive and exploratory. This latter method has had prominent representatives in natural science, and Michael Faraday's investigation of electromagnetism is probably the most striking example (Ribe and Steinle, 2002). Exploratory experimentation typically comes to the fore in situations where a useful conceptual framework is not yet available and where the creative investigator must find ways and concepts along with experimentation (Fynn, 1979).

Thomas Young or George Palmer?

Some time before Goethe published his *Theory of Colors* (Goethe, 1810/1963), the physiological basis of color vision had caught the interest of the English physicist, physician and philologist Thomas Young (1773–1829). In his famous 1801 Bakerian Lecture at the Royal Society of London, he presented an idea of how color vision worked. There were three types of receptors in the retina, he said, each excited by light from a different wavelength band in the Newtonian spectrum, and each associated with one of the three primary colors. Differential excitation of these receptor types would lead to mixtures of the primary colors.

> Light mixtures of red, green and blue we can see every day on color TV. If you look more closely at the screen, for instance through a strong magnifying glass, you will find that the picture is constructed of many small red, green and blue dots or squares of varying intensity. All the colors you see on the screen some distance away are a result of the melting together (called an additive mixture) of these three 'primary' colors, modified by spatial contrast effects (simultaneous contrast).

It seems likely that this trichromatic hypothesis was developed from ideas that had become familiar to Young when, in 1795–1796, he studied for a doctor's degree in medicine at the University of Göttingen in Germany. The original idea probably came from an English glass manufacturer and carpet merchant, George Palmer, alias Giros von Gentilly (Walls, 1956; Mollon, 1993, 1995). This man was colorful in more than one sense of the word; he is supposed to have had some dubious business in France that made it necessary for him to use a pseudonym. Already in 1776 Palmer had published a pamphlet with the title *Theory of Colors and Vision*, in which he explained color vision by means of three receptors in the retina. This pamphlet was translated into French and published under the name of Giros von Gentilly.

Palmer's ideas must have been well known in the academic circles of Göttingen. His hypothesis for color vision, and the related thought that color blindness was caused by a missing or a defect receptor type, was discussed in 1781 in *Lichtenberg's*

Magazin, and color blindness was treated again in a new pamphlet in 1785 (Lee, 1991a). *Lichtenberg's Magazin* was devoted to popular science and it was edited by the brother of a famous professor in physics in Göttingen, Georg Christoph Lichtenberg (1742–1799). Lichtenberg was a well-known personality in German science in the second half of the eighteenth century and well acquainted with Thomas Young in Göttingen. The hypothesis about color vision that was later credited Young does not seem to have been entirely his own, but rather a development of ideas that he must have heard about in Göttingen.

Young–Helmholtz's three-receptor theory

The three-receptor theory lived a quiet life until, 60 years after Young's lecture in 1801, it was taken up by von Helmholtz and developed further into what today is called the Young–Helmholtz three-receptor theory of color vision. At about the same time, another well-known physicists, James Clerk Maxwell, developed an interest in color and color vision. He made experiments that prepared the ground for color photography. The following citation is from his writings (Maxwell, 1872): 'if the sensation which we call color has any laws, it must be something in our own nature which determines the form of these laws'.

In the years after von Helmholtz's (1866/1911) revival of Young's hypothesis, color perception was explained by light absorption in three types of cone photoreceptors in the human retina. The spectral sensitivities of these cone types are shown in Figure 4.12(A). The belief (that we now know to be wrong) was that three different classes of cone activated three basic color processes or sensations: red (L-cones), green (M-cones) and violet or blue (S-cones). Consequently, these receptor types were called the R-, G- and B-cones, respectively. All colors were said to be due to the excitation of these primary sensations in different proportions. The perception of yellow would, for instance, result from equal stimulation of the 'green' and the 'red process', whereas white resulted from equal stimulation of all three.

Referring to Young, von Helmholtz (1911, p. 119) writes:

> 'Es gibt im Auge drei Arten von Nervenfasern. Reizung der ersten erregt die Empfindung des Rot, Reizung der zweiten die des Grün, Reizung der dritten die Empfindung des Violett' ('In the eye there are three types of nerve fibers. Stimulation of the first one excites the sensation of red, stimulation of the second the sensation of green, stimulation of the third the sensation of violet'),

and on p. 120 he continues

> '(das Wesentliche in der Hypothese von Young ist), dass die Farbempfindungen vorgestellt werden als zusammengesetzt aus drei voneinander vollständig unabhängigen Vorgängen in der Nervensubstanz'. (The essence of Young's hypothesis is) that the sensations of color are imagined as composed of three mutually completely independent processes in the neural substrate'.

It is quite clear that Helmholtz here focuses on an explanation of the qualitative aspect of colors, and this is exactly the point where today's view departs from the Young–Helmholtz' theory.

Hering's opponent colors theory

Hering vehemently opposed the Young–Helmholtz theory, and he claimed that color vision was based not on three primary sensations, but on four chromatic and two achromatic elementary, or unique, color perceptions (*Urfarben*) and their corresponding physiological processes. This idea had support in Leonardo da Vinci's (1452–1519) observation in *A Treatise on Painting* (1906) that six particularly simple colors are found in nature. They are the four unique, or elementary chromatic hues, yellow, red, blue and green, together with the two achromatic colors, white and black. These colors serve as six qualitative references in subjective or phenomenological color space (Hering, 1964; Hurvich, 1981). The physical stimuli associated with these percepts depend on the viewing situation, and they vary from person to person. In the spectrum, unique blue is found around 470 nm, green at about 500 nm, yellow at 575 nm, and a saturated unique red is usually a mixture of the red and blue ends of the spectrum, with a complementary wavelength of about 495 nm. Unique yellow can be determined with an extraordinary precision. Although one person's selected wavelength for unique yellow may be found anywhere between 565 and 590 nm, the precision can be at the order of a few nanometers (Richter, 1969; Mollon and Jordan, 1997). These identifications seem to have little to do with culture and language (see peer commentaries to Saunders and van Brakel, 1997).

In Hering's opponent theory, the two pairs of chromatic unique colors, together with the achromatic pair, black and white, were associated with three pairs of hypothetical, antagonistic physiological processes. In agreement with Hillebrand (1888, p. 70), Hering came to associate unique red with the breaking down, or wearing out ('dissimilation') of a particular 'visual substance' – and unique green with the building up, or restoration ('Assimilation'), of the same substance. Similar antagonistic processes in two other substances gave rise to unique yellow and blue, and to white and black. The opposite nature of these paired hue qualities, as displayed in the hue circle of Figure 5.3, were thus associated with processes which mutually excluded one another. The 'visual substances' were not the photopigments of the receptors; they were unidentified physiological substrates at an unspecified level in the visual pathway (Trendelenburg, 1943, p. 81).

For quite some time, the three- and four colors theories were strongly opposed, even after the Austrian physicist Erwin Schödinger (1887–1961) showed, in 1925, how they could be reconciled into one. Schrödinger's idea was that the three-color theory could be valid at the level of the receptors in the retina, and that the receptors' reaction to light stimulation could be transformed to four color-coding signals later in the visual system.

In this context, it is of interest to note that G.E. Müller, in his zone theory of color vision, developed a more differentiated and modern view of the physiological processes underlying unique hues. He reserved the primary antagonistic 'neuronal processes' for the explanation of color contrast (such as simultaneous and successive contrast, induction, adaptation, etc.). With regard to unique colors, he envisaged four 'psychophysical excitations' [obviously referring to Fechner's (1966) 'inner psycho-physics'] at a central level that received their inputs from the lower-level antagonistic neural processes (Müller, 1930).

Dorothea Jameson and Leo Hurvich's extensive studies of the psychophysics of color opponency in the 1950s were based on the above concepts of physiological substrates, and the revival of Hering's ideas through 'hue cancellation' experiments (see Hurvich, 1981). In these experiments, additive mixtures of colored lights were used to determine a unique hue, and the theoretical framework implied that a neural mechanism was in an 'equilibrium state' whenever an opponent mechanism had been balanced. In this theory, unique yellow, for instance, was viewed as such an equilibrium state between a 'red process' and a 'green process'. However, the residual sensation (yellow in this case) did not need to be relevant, since the judgements were based only on the absence of redness and greenness, independent of other color attributes of the stimulus (it could also be white or blue).

When the Swedish physiologist Gunnar Svaetichin (1956) published the first recordings of spectrally dependent positive and negative potentials in the retina of fish, he believed he had proven Hering to be right. Later, when Russell De Valois (1965) had already found spectral activation and inhibition of cone-opponent cells in the lateral geniculate nucleus (LGN) of the macaque monkey, these findings were regarded as further confirmation of Hering's opponent theory in the primate. It became common among neurophysiologists to use color terms when referring to opponent cells, as in the notations 'red-ON cells', 'green-OFF cells', '+R–G' and '+G–R cells' (Wiesel and Hubel, 1966; De Valois and Jones, 1961; De Valois, 1965; De Valois et al., 1966). In the debate that followed, with traditional colorimetrists still clinging to the old trichromatic theory, some psychophysicists were happy to see what they believed to be Hering-opponency confirmed at an objective, physiological level. Consequently, little hesitation was shown in relating the unique and polar color pairs directly to the opponency found in these cells. In the so-called 'R–G' opponent cells, the activating and inhibiting inputs were mainly from L- and M-cones, and together with the cone-opponency 'B–(R+G)' these directions in cone excitation space were later called 'cardinal directions'. Despite evidence to the contrary (Valberg, 1971; Burns et al., 1984), textbooks have to this day repeated the misconception of relating unique hue perception directly to the responses of such opponent cells and the cardinal axes in color space. The apparent analogy with Hering's hypothesis was carried even further to imply that, for instance, red results from activation of 'R–G' cells and green from inhibition of the same cells [blue would be caused by activation and yellow from inhibition of the 'B–(R+G)' cells]. This use of color names on the early opponent responses was an understandable mistake in view of Hering's postulate, and will be discussed later.

The retinex theory

Parallel to the new neurophysiological exploration of color vision, scientists' interest in color vision had also developed in another direction. In 1958, 100 years after von Helmholtz and Maxwell, Land (1909–1991), the inventor of the polaroid filters and sunglasses, and of 'instant photography' (the Polaroid camera), gave some spectacular color demonstrations at the Annual Meeting of the Optical Society of America. The classical theory said that we would need three primary colors, for instance three slide projectors, one with red light, the other with green and the third with blue, in order to produce all other colors in an additive mixture on the screen. Yet Land surprised the world by projecting a black and white image of a bowl of fruit with only two primaries but nevertheless producing all colors. He had left out the red filter in one of the projectors, so that the projector projected only white light (the mixture on the screen being between green, blue and white). However, the image on the screen nevertheless contained red, although not so vivid a red as with the red filter in place. If he put the red filter back again and removed the green filter, green was still seen in the projected image, and the same happened with blue. This demonstration was not consistent with the belief that there was a close correspondence between the spectral distribution of a color stimulus and its appearance, i.e. that a patch of red color would require radiation from the long-wavelength (red) end of the spectrum, blue from the short-wavelength region, and so on. Land demonstrated that it was not necessarily so, and this came as a great surprise to many people (Optical Society of America, 1994). He claimed to have demonstrated that the Young–Helmholtz three-receptor hypothesis was wrong (Land, 1959).

In reality, Land had given another, and particularly striking, example of the strange phenomenon that had already been observed by Leonardo da Vinci, that adjacent areas of different colors within the field of view influence and change each other's appearance. This phenomenon was in no way new; it was a particular case of *simultaneous color contrast* that had also been observed by Goethe (1963) and made famous by him. Colors of surfaces are in reality, although this is not always easily observed, a complex product of all the light imaged on the retina, including those in the periphery. In 1839 the French chemist M.E. Chevreul (1969) devoted a whole book to this phenomenon, which has fascinated color researchers ever since.

> Chevreul worked in the famous Gobelins tapestry factory in France, and had to deal with customers' complaints that the colors of the final product came out differently from what had been agreed upon (for instance when a pattern or texture of the same material changed its appearance due to the background).

Thus, Land's demonstrations were a new twist to an old phenomenon. He nevertheless proposed his own theory, the *retinex* theory (*retin*a + cort*ex*), to explain it. When these effects were again brought to attention in 1983 after a long period of silence, Land (1983) had come to regard the retinex theory as an explanation of *color* *constancy.* Color constancy refers to the fact that, when illumination changes, say from bluish daylight to a yellowish-red evening sunset, the color appearance of the objects reflecting these different spectral distributions does not change in the expected way. A banana is yellow in either case. The bananas and the other objects resist the color change expected from the conventional three-receptor theory (which would be in the direction of the color of the illuminant). This ability of the visual system to adapt to and to neutralize the prevailing color of the illuminant is clearly not what one would expect from the Young–Helmholtz hypothesis. However, the retinex hypothesis was not able to predict the changes that actually occur any better than other hypotheses that had failed in describing color constancy and color adaptation.

According to Land, the Young–Helmholtz hypothesis was now obsolete. However, color scientists did not show much interest. Land himself admitted in private (personal communication) that, had he not been the founder of the Polaroid Cooperation, with great influence and considerable resources, the criticism would have defeated him. This fascinating story is worth closer study by someone interested in the sociology of science.

With his publications in 1983, a new chapter in the history of color science was about to be written, inspired by Land's stubbornness, but also by the need to develop algorithms for color constancy that could be used in artificial vision and digital photography. By this time, neurobiologists had developed methods that allowed them to record from single cells in the retina and in the brain. Many of them were fascinated by Land's demonstrations of color constancy in complex Mondrian-like images. Well-known neurobiologists became Land's protagonists, among them influential people such as the Nobel Laureate David Hubel in the USA, and Semir Zeki in the UK. Land cooperated with them both, and he therefore became quite influential in their thinking about color and color vision (Creutzfeldt *et al.*, 1979; Land *et al.*, 1983; Zeki, 1993; Crick, 1994).

Color in current neuroscience and neurophilosophy

It is tempting to see a link between the renewed interest in color vision and the extensive exploration of the separate visual pathways from the retina to the visual cortex that flourished at the end of the twentieth century. We have already mentioned the particularities of the parvocellular, koniocellular and the magnocellular pathways (see p. 121), and here we only want to emphasize once more the role played by the opponent cells in color vision. These cells combine cone inputs in an opponent manner, much like what was postulated by Schrödinger in 1925. The neurophysiological interest in color vision has not yet brought us much closer to an understanding of the physiological mechanisms behind color constancy, Land's important problem.

Until now neuroscientists have been concerned with the neural correlates of color perception, i.e. how cell activities could represent the intrinsic, abstract structure of color space. For instance, if a cell increases its firing rate to a patch of red spectral light of increasing intensity, and decreased its activity to mid-spectral green lights, one might conjecture that, either:

1. the cell is, via inputs from preceding neurons, excited by electromagnetic radiation from the long-wavelength end of the spectrum, i.e. to the physical property of the light; or

2. the cell responds to the red color; i.e. there is a correlation between the perceived quality and the cell's response, independent of the spectral composition of the stimulus.

Quantitative measurements in the retina and in the LGN point to the first alternative, except for a few special cases that have not been adequately confirmed. It has been reported, for instance, that cells in area V4 of the visual cortex respond to the color and not to the wavelengths of a stimulus (Zeki, 1980, 1983a, b). Whatever the case may be, the processes behind color qualities, say the redness of red, are still enigmatic.

Now someone is likely to ask if it is at all possible to come closer to the actual phenomenon of color qualia. The activity associated with perception can be measured at different levels in the nervous system, and we can localize specific activities to brain areas by means of positron emission tomography (PET) and functional magnetic resonance imaging (fMRI), but do we get any closer to the conscious experience of qualities by such means?

This discourse brings us to a debate that still lingers, about the role of colors and other qualia in understanding neural activity and conscious experience. On one side of the argument, we find Francis Crick, the discoverer of the structure of DNA and Nobel Laureate, together with the 'neurophilosophers' Patricia Churchland (1994) and Paul M. Churchland (Churchland, 1986; Churchland and Sejnowski, 1992). For them the mind was interesting and mysterious, just like matter, and their approach to the problem of consciousness was materialistic. With the rapid advances in neuroscience, the philosophical reflection about neural processes and mental activity could be related to a variety of experimental data. Crick and his colleges at the Salk institute in California mapped the functioning of the visual brain, and the working hypothesis was mechanistic. It can be described as follows (Crick, 1994, p. 3):

> 'your joys and your sorrows, your memories and your ambitions, your sense of personal identity and free will, are in fact no more than the behavior of a vast assembly of nerve cells and the associated molecules in the brain'.

Crick, and many with him, would have been satisfied if they were able to develop a neuron theory that answered the following questions: 'What are the co-variations (correlates) between the different color experiences and neural activity?' and 'Which neural representations of color perception exist?'

When you ponder over the link between your conscious experience of colors, e.g. the redness of an apple, and physical stimulation giving rise to it, maybe you will find that it is more appropriate to inquire into the relationship between your color experience and the activation of nerve cells in the visual system. Is there a chain of causal relationships between light, a particular physical property of the apple's surface, the intervening evoked neural activity in the retina and the visual pathways, and perception of red? Are the same nerve cells activated when, in my dream, I 'see' a beautiful bush of red roses in the evening sunset? Or should we, because of the 'explanatory gap', i.e. the problems of dealing with color qualities (qualia) and neural processes or physical stimulus properties in the same scientific language – talk not about cause and effect, but rather about correlation, co-variation and structural correspondence?

A surface with a certain spectral reflectance can take on virtually any color, less dependent on the illumination and the reflected spectral distribution than on the surround conditions and the adaptation of the observer's eye. These effects were well known in the last century, and they were again brought to our attention by Land's (1959, 1983) vivid demonstrations. The same object surface may look quite different in a normal visual environment than when illuminated in isolation in an otherwise dark laboratory. Generally, a particular color can neither be correlated with nor 'caused' by the spectral composition of the isolated patch, or even by the responses of the three cone types excited by it. Only in well-controlled laboratory experiments it is possible to establish a correspondence such as that postulated by von Helmholtz (1911, p. 119; see also p. 278).

At a first glance, the observation that color perception is a conscious, subjective and qualitative experience is less problematic than the attempts to relate it to some physical property of objects, or to neural activity. For those who regard consciousness as a subject for scientific investigation, now is the time to attack the problems of qualia by experimental methods. In Francis Crick's own words (Crick, 1994, p. 9):

> the problem of qualia – for example, how to explain the redness of red... is a very thorny issue... The problem springs from the fact that the redness of red that I perceive so vividly cannot be precisely communicated to another human being... This does not mean that, in the fullness of time, it may not be possible to explain to you the neural correlate of your seeing red. In other words, we may be able to say that you perceive red if and only if certain neurons (and/or molecules) in your head behave in a certain way.

If we substitute 'you perceive red' in the last sentence by 'you perceive a quality which you have learned to call red', we see that such an explanation of color qualia is indirect. It relies on hypothetical neural behavior and on representations – on linking hypotheses – in a symbolic language. Crick's proposal is not concerned with qualia in itself ('the redness of red'), but with the idea of a neural state leading to *the same subjective percept* in different situations.

A large part of primate cerebral cortex is devoted to processing information received from the retina, and much progress has been made in elucidating the organization and function of the visual cortex (Van Essen *et al.*, 1991). However, the various aspects of subjective qualities are less well understood (Gazzaniga *et al.*, 1998). Color perception is well suited to bring forth the pertinent aspects of qualities and conscious experience (see for instance Chalmers, 1995b; Crick and Koch, 1995). Today, the study of color perception is closely tied to experimental psychophysics and neuroscience, and even philosophical reflections about this issue are submitted to rather strong experimental constraints (Hardin, 1988; Thompson, 1995; see also the discussions of Saunders and van Brakel, 1997; Palmer, 1999). A neuroscientific program searches for collections of neurons in the retina, the LGN, and higher brain areas whose activities, in one way or another, are correlated with the dimensions of color perception. One records the activity of these neurons when color stimuli change along the psychophysical dimensions *luminance*, *dominant wavelength* and *purity*, in the hope of finding links to the related perceptive properties *lightness/brightness, hue* and *color strength* (*chroma*). Even if it should be demonstrated that a simple neural–perceptual parallelism or isomorphism is too naive a hypothesis, such attempts give us useful overviews of the neural representations of different color stimuli. This project would be completed when unequivocal correlates between the perceptual properties and accompanying neural activities were found.

However the neural activity correlated with my experience of a color consists of a stream of short electrical impulses within a network of neurons, i.e. a biophysical process. These correlates do not describe the nature of my subjective experience of, for instance, the redness of an apple and how it is brought about. My experience of the red quality depends on my living brain, but it cannot be explained or deduced in a causal manner from a biophysical process in and between its neurons. How brain activity gives rise to conscious experience remains an enigma.

Such questions, which were also investigated by the Austrian philosopher Ludwig Wittgenstein (1979) in his book *Remarks on Color*, has occupied the Australian David J. Chalmers (1995a, b; Tucson, 1996). Chalmers has provoked neuroscientists and neurophilosophers by statements such as:

> 'the one who is only interested in the functioning of the brain, he deals with the easy and resolvable problem. However, behind this lies the real problem, namely the one about conscious experience'.

Chalmers' point is that conscious experience is different from and something more than nerve activity; it is an enigma, and it will probably continue to be so for a long time to come.

Between the two extreme positions of Crick and Chalmers it is possible to claim that, when a color 'takes place', this is something more than a physiological reaction to a physical stimulus resulting in a percept. The color embraces both positions, it is a mediator between processes in outer nature and the subjective inner nature. In addition, it is a phenomenon in its own right governed by natural laws (like in color constancy, additive color mixtures, etc.).

It is not fair to reduce the genuine experience of a subjective reality to something else, to entirely substitute it for another experience, as in the following: 'sound is just a train of compressive waves traveling through the air'; 'pitch is identical with frequency'; 'light is just electromagnetic waves'; 'color of an object is identical with a triplet of reflectance efficiencies'; 'warmth and coolness of a body is just the energy of motion of molecules that make it up'. You may have heard these or similar statements from your high school teacher or read them in a text book. Phrases like these are forgivable when used in everyday speech, but in science, and especially in introductory texts in neuroscience and neurophilosophy, they can only cause confusion and serve as obstacles to understanding. Not only do they ignore the essential transformations that take place in sensory organs and in the neural system, but in equating physical concepts with subjective impressions they also deny the existence of qualia. In interdisciplinary research, one must carefully choose one's language, be it the terminology of physics for electromagnetic radiation, that of neurophysiology for describing neural activity in the brain, or a psychophysical or psychological account of our perceptions. In every field it is important to do justice to the relevant discipline – and to the phenomenon itself.

Colors continue to fascinate and to challenge our imagination. The many different color theories and models that have been developed to this day for explaining how colors 'take place' call for caution and skepticism towards such theories. They demonstrate the need to ascribe to subjectively experienced qualities a reality of their own, independent of culture and the level of knowledge which at any given time sets the background for our theoretical descriptions. In this way, the history of color theories illustrates the role that we have given ourselves, the roles we have played – and continue to play – in attempts to understand the physical world.

What needs an explanation?

How can we explain the following facts about color vision?

1. Metameric stimuli with different spectral distributions look alike (they have the same color).

2. Monochromatic lights of different wavelengths have different colors, and this qualitative difference cannot be eliminated by adjustments of intensity.

3. Three colors are sufficient to match all other colors in an additive color mixture.

4. Some people cannot distinguish red from green, and others confuse yellow and blue.

Below we shall see how these experiences can be explained by receptor physiology. In addition there are some other color phenomena that cannot entirely be explained at the receptor level, such as (a) the Bezold–Brücke phenomenon and (b) the Abney effect.

These two phenomena will find a sound description within a cone-opponent model of color processing. Other phenomena and facts of color vision that still have not found a physiological explanation in terms of neural correlates, are (c) color adaptation, (d) simultaneous color contrast, (e) color constancy and the deviations from it, (f) the unique hues, (g) the perception of white and black, and (h) border and area contrast (Mach- and Hering-type contrast; von Bekesy, 1968)

Defective and normal color vision

First, let us ask what has happened when a person has defective color vision. In popular language such people are often called 'color blind', although they are quite capable of seeing differences in color. A 'red–green color blind' person, for instance, can discriminate between yellow and blue as well as between white and black. In a classroom with 30 students there are often a few with defective color vision, usually males. In the whole population, there are about 18 times more males than females who have a color vision defect (8 vs 0.45 percent; Waaler, 1969; Wyszecki and Stiles, 1982). As we shall see, change in the cone pigments is the main cause of such problems. The reason for the sex difference will be explained later.

Monochromats

A rod monochromat is the rare case of a person having only one type of receptor, the rods. About 0.003 percent of the population are rod monochromats, and they have rhodopsin as their only visual pigment. A rod monochromat can distinguish bright objects from dark ones, but he has no chromatic color vision. Two arbitrarily chosen wavelengths cannot be distinguished by their colors, and it is possible to make them match merely by adjusting their relative intensity. Rod monochromats and persons with normal color vision use the same receptors at night, and therefore it is often assumed that their visual impressions at night are comparable to those of normal trichromats (although it is conceivable that subtle changes later in their neural system may give rise to some differences).

> Knut Nordby, a thoroughly studied monochromat, has described his visual experiences and his early childhood in the book *Night Vision* (Hess *et al.*, 1990). In his fascinating book *The Island of the Colorblind and Cycad Island*, Oliver Sacks (1997) reports a journey he made with Knut Nordby to a small island in the Pacific where a large number of the inhabitants were rod monochromats. This book gives a vivid account of the problems encountered by monochromats in daylight. In one respect, however, Knut Nordby had an advantage: in the dark he had a greater-than-normal contrast sensitivity for low spatial frequencies.

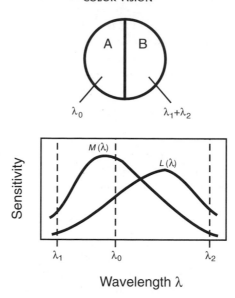

Figure 6.1 A dichromat is a person with defective color vision whose retina has only two different cone types. Therefore, only two wavelengths, λ_1 and λ_2, are needed for an additive color mixture in field B to match another wavelength, λ_0, in field A. The example in the figure is for a person with tritanopia, in whom the lack of S-cones leads to a yellow–blue color vision defect.

When a monochromat makes a color match, as in the fields A and B of Figure 6.1, the receptors that are excited by fields A and B are all of the same type. When they absorb equal amounts of light, fields A and B will therefore look the same, even if their wavelengths are different.

Another form of monochromacy is that of cone monochromats who have only one type of cone receptor in addition to rods. Theoretically, it is possible for such persons to have a rudimentary color vision. This would mean that they could see a qualitative difference between wavelengths, and that this difference could not be compensated for by changes of relative intensity. A requirement for this is that the rod and cone pathways are separate in the early stages of processing, allowing their relative inputs to be compared in the brain.

While having two independent receptor types with different spectral sensitivities is a necessary condition for color vision, it is not a sufficient one. If the signals from the two receptor types were to converge or to be added somewhere early in the visual pathway, in a manner that made differences irretrievable at a later stage, the possibility for color vision would be lost. Information about the difference, or the ratio, of the excitation of two independent receptor types must be preserved if later levels of processing are to take advantage of these inputs. This is just one example of how the principle of parallel pathways is exploited in vision.

For a cone monochromat, the excitation of rods and cones can be represented in an orthogonal, two-dimensional vector space, like that of Figure 6.2 for L- and M-cones

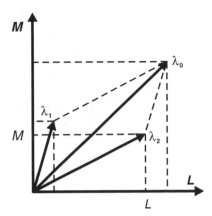

Figure 6.2 A vector representation of the cone excitations L and M for a dichromat and the additive color mixture of Figure 6.1. The spectral color λ_0 is matched by adding light of wavelengths λ_1 and λ_2.

(just replace one of them by rod excitation). When two different wavelengths (or two different spectral distributions) activate the two types of receptors to a different degree, this can be represented by two separate vectors. Their individual lengths depend on the relative magnitude of excitation of the two receptor types. It has been maintained that this vector difference somehow gives rise to a qualitative color difference for the cone monochromat (Stabell and Stabell, 1982a). For dichromats, with two cone classes, a two-dimensional scheme illustrates what happens, with the brain interpreting the vector difference as a difference in color. However, if rods were to contribute to color vision, this would complicate the simple two-dimensional scheme.

Dichromats

The most common color vision distortion is a reduced ability to discriminate between colors in the red and yellowish green regions of the spectrum;, in the most severe cases known as 'red–green blindness'. The latter problem affects people who lack either the L- or the M-cone types in the retina, giving rise to two equally large groups with slightly different red–green defects. For either type, this reduces color discrimination to two dimensions instead of three, and it is therefore called *dichromacy*. The lack of L-cones, or of the L-cone pigment, leads to a red–green color vision defect called *protanopia*, whereas a missing M-cone pigment leads to another, slightly different, purple–green defect, called *deuteranopia*. In both these cases the rods and the S-cones are thought to be normal. The rare case of missing S-cones, occuring only in about 0.002 percent of the population, is called *tritanopia*.

In the absence of the L-cone type, the photopic spectral luminous efficiency function, $V(\lambda)$, is determined by the M-cones. For deuteranopes it is determined by the L-cones. In both cases this leads to a spectral narrowing of the $V(\lambda)$ curve, with a sensitivity reduction in the long-wavelength region for protanopes, and at the short-wavelength end for deuteranopes.

One can learn a great deal about how photoreceptors work by considering what happens in color matching experiments. Let us go back to our matching experiment with two half-fields A and B (Figures 4.1 and 6.1). For a dichromat, say one that is missing the S-cones, the requirement for fields A and B having the same color is that the excitations L and M are the same in both halves. Then for each cone type, the following relations between the excitations caused by field A and B must hold:

$$L_A = L_B$$

and

$$M_A = M_B$$

In the photopic luminance range where we can disregard the contribution of rods, we obtain a simple relationship for the light absorption in each receptor type. Let us assume a simple case where the color of a wavelength, λ_0, in field A (Figure 6.1) is matched by the light mixture of the two wavelengths λ_1 and λ_2 in field B. Then the excitations L and M due to fields A and B obey the following relationship:

$$L(\lambda_0)\Phi(\lambda_0) = L(\lambda_1)\Phi(\lambda_1) + L(\lambda_2)\Phi(\lambda_2)$$
$$M(\lambda_0)\Phi(\lambda_0) = M(\lambda_1)\Phi(\lambda_1) + M(\lambda_2)\Phi(\lambda_2)$$

Here, $L(\lambda)$ and $M(\lambda)$ (with the subscripts 0, 1, and 2) are the spectral sensitivities for receptors L and M at the respective wavelengths λ_0, λ_1 and λ_2, and $\Phi(\lambda_0)$, $\Phi(\lambda_1)$ and $\Phi(\lambda_2)$ are the light intensities (either in W/m^2 or in relative units) of the three wavelengths λ_0, λ_1 and λ_2 when the colors in fields A and B do match.

A dichromat can thus achieve a color match of wavelength λ_0 by adding two other wavelengths λ_1 and λ_2, provided the intensities $\Phi(\lambda_1)$ and $\Phi(\lambda_2)$ are chosen correctly. Since a mixture of only two stimuli is sufficient to match a third, the color space of a dichromat is two-dimensional. This is illustrated in the vector diagram of Figure 6.2. The lengths of the projections of vectors λ_1 and λ_2 on the orthogonal L- and M-axes are proportional to the excitations of the L- and M-cones, respectively. The angles that each vector λ_1 or λ_2 forms with the L- or M-axis is then a function of the relative excitation of the two cone types.

So-called 'Rayleigh matches' on a Nagel anomaloscope are sometimes used in clinical testing of color vision. They serve to diagnose anomalous or missing L- and M-cone pigments. The subject is asked to match a mixture of red ($\lambda_1 = 679$ nm) and green ($\lambda_2 = 544$ nm) to a monochromatic yellow of 589 nm ($\lambda_0 = 589$ nm). In anomalous trichromats, the red–green ratio of the match falls outside the normal range, or, for a dichromat, the range of accepted matching ratios is drastically extended relative to normal subjects.

Can we say something about which hues deuteranopes and protanopes see? Normally, it is assumed that it is some kind of yellow and blue (Neitz *et al.*, 2001; Vienot *et al.*, 1995). We shall discuss some aspects of this intriguing question later, in the discussions of the limits of the trichromatic theory of color perception.

Trichromats

To make a color match in our photometric field of Figure 6.1, a *trichromat* would need three different color stimuli, and can therefore be said to have three-variant color vision. He or she has three different cone types each with a sensitivity that spans a large part of the visual spectrum [see Figure 4.12(a)]. The wavelength of maximum sensitivity of each cone type may vary somewhat from one person to the next, but they are grouped around 565 nm for L-cones, 530 nm for M-cones, and 420 nm for S-cones.

The above argument for the dichromat can be extended to apply also to the third receptor type, the S-cones. Two fields, A and B, match in color ($\mathbf{F_A} = \mathbf{F_B}$) when the light in field A affects all three cone types equal to the light in field B. This means that the signal from each of the three receptor types when activated by field A must equal the signal from the same receptor types when excited by field B. For an additive mixture of three monochromatic lights, of wavelengths λ_1, λ_2 and λ_3 that match λ_0, the excitations L, M and S of the three receptors are expressed by:

$$L = L(\lambda_0)\Phi(\lambda_0)$$
$$= L(\lambda_1)\Phi(\lambda_1) + L(\lambda_2)\Phi(\lambda_2) + L(\lambda_3)\Phi(\lambda_3)$$

and correspondingly for M and S.

Since a trichromat needs three color stimuli (here represented by the wavelenghts λ_1, λ_2 and λ_3), to match a given color stimulus (λ_0), the color space of a thrichromat has three independent variables and is, therefore, three-dimensional. The color space of a trichromat can also be represented by a vector space, now with the three axes **L**, **M** and **S**. This is shown in Figures 5.18 and 5.19.

This three-dimensional vector space of excitations can be used to illustrate why and how the two-dimensional space of a dichromat can be viewed as a reduction of the vector space of a trichromat. For instance, a deuteranope who lacks the M-cones will accept the color matches of a trichromat. For him, all colors that are distinguishable for a trichromat only by differences in the excitations of M-cones will look the same. The M-cone excitation does not matter to him. Such colors lie on a vertical line through the tip of the vector **F** down to the *(S,L)*-plane (marked with crosses in Figure 6.3). The figure illustrates how a set of stimuli for which only the M-cone excitation is different, leaving the *L*- and *S*-magnitudes unchanged, have vectors

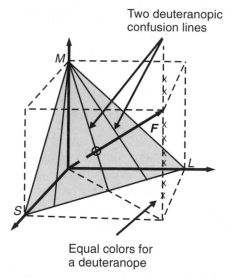

Figure 6.3 The figure illustrates, in LMS excitation space, the position of the chromaticities that are indistinguishable for a dichromat, here a deuteranope. The deuteranope does not have M-cones, and all the chromaticities he sees as equal to the color **F** are situated along the vertical line marked with crosses. This corresponds to a line in the chromaticity triangle with origin in the corner *M*. All such lines are referred to as 'confusion lines' since, seen from the point of view of a trichromat, he is prone to 'confuse' colors along these lines. Similar figures can be constructed for other pure dichromates, the protanopes and the tritanopes.

that, when projected onto the shaded triangle of the chromaticity diagram, form a line radiating from the upper corner. All these lines in the chromaticity diagram are called 'dichromatic confusion lines'.

A protanope who lacks the L-cones will also accept the matches of a normal trichromat. However, his confusion lines radiate out from the L-corner of the chromaticity diagram. A tritanope lacking S-cones has confusion lines radiating out from the S-corner of the diagram. It follows then, that when a dichromat makes a color match, it does not matter how much the color mixture would excite the missing cone class. A dichromat's color match for vector **F** in Figure 6.3, for example, might be found to project anywhere on the appropriate confusion line. Consequently, a trichromat will generally not accept the match of a dichromat. Figure 6.4(a–c) shows dichromatic confusion lines in the CIE (x,y)-diagram for a protanope (a), a deuteranope (b) and a tritanope (c).

In most cases, the lights in our colorimetric experiment with half fields is not monochromatic, but consists of composite spectral distributions of radiance $\Phi_A(\lambda)$ and $\Phi_B(\lambda)$ containing a whole range of wavelengths. In this case, the absorption in the

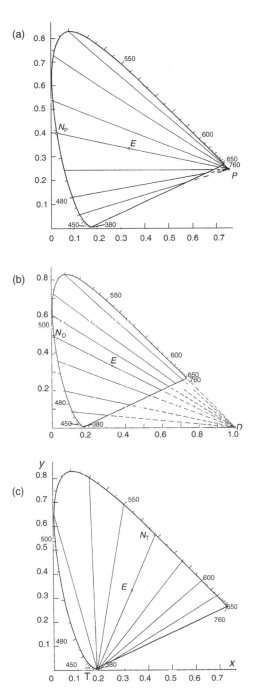

Figure 6.4 Confusion lines in the 1931 CIE (x, y)-diagram for protanopes (a), deuteranopes (b) and tritanopes (c), lacking the L-, M- and S-cones, respectively.

cones is an integral ($\int$, or a sum Σ) of excitations over all wavelengths of the light that enters the eye:

$$L = k_L \int \Phi(\lambda)L(\lambda)\, d\lambda$$

$$M = k_M \int \Phi(\lambda)M(\lambda)\, d\lambda$$

$$S = k_S \int \Phi(\lambda)L(\lambda)\, d\lambda$$

k_L, k_M, and k_S are normalizing constants, and the integrals are normally taken as between 380 and 760 nm. For a trichromatic color match, all three expressions must be equal for each of the fields A and B in Figure 6.1: $L_A = L_B$, and similarly for the cone excitations M and S. Many different spectral distributions, $\Phi(\lambda)$, in the integrals above will lead to the same absorption magnitudes L, M and S. Our visual system is not able to distinguish these physically different (metameric) stimuli from each other, and they will therefore have the same color appearance; the eye is not a good spectral analyzer and cannot distinguish between spectral distributions that have the same color appearance. Only when colors do not match can we be sure that the spectral distributions are different (provided equal viewing conditions). When colors match, their spectral distributions can either be equal or different. For two equal surface colors one can easily decide if spectral distributions are the same or not by comparing the surfaces under a different illumination (e.g. artificial light vs daylight). Except in special cases, metameric stimuli that have the same color in one illumination are usually unequal in another illumination.

Genes and color vision deficiencies

Authoritative sources hold that about 8 percent of the male population has some sort of congenital red–green deficiency, while this applies only to between 0.4 and 0.5 percent of the female population (Waaler, 1969; Birch, 1993). The two main forms of red–green color vision deficiency are associated with defects in the opsin gene on the X chromosome. In recent years, evidence has accumulated that what was once thought to be one single pigment group within each cone class actually consists of a number of pigments with slightly different absorption spectra. In the human population, normal color vision therefore includes slightly deviant forms of trichromacy.

Humans have 23 pairs of chromosomes, 22 of which, the autosomes, have pairwise analogous sets of genes. In the 23rd pair, the sex chromosomes, females have two analogous X-chromosomes while males have one X-chromosome paired with a Y-chromosome. The gene for the S-cone pigment is located on the chromosomes of pair no. 7, so both sexes have a double complement of the S-cone pigment gene. In contrast, L- and M-genes are found on the X-chromosome, with different consequences for males and females. While males have only one set of L- and M-genes, those passed on from their mother, females have two sets, one from each parent.

The two corresponding sets of parental genes may be the same or they may contain different versions of the pigment genes. Alternative forms of a gene coding for a similar functional unit or trait and found at the same location on a pair of chromosomes, are referred to as *alleles*. A pair of analogous genes is heterozygous when the two alleles are different and homozygous when they are the same. The relative effect of the two alleles on the phenotype of an organism in a heterozygous state is described in terms of a dominance relationship, for example dominant/recessive. A recessive allele is expressed in the phenotype only in the absence of a dominant allele. The allele for red hair, for example, is masked in the presence of an allele for dark hair (heterozygous state). In order to be expressed in an individual's phenotype, the allele for red hair must be inherited from both parents (homozygous state). The allele for red hair is therefore considered to be of a recessive character.

Since females have two sets of M- and L-genes, the effect of a defective gene inherited in a heterozygous state may be partially, or fully, compensated for by the presence of a normal copy of the gene. Therefore, these genes are sometimes described as being recessive, and heterozygous females are often considered to be unaffected carriers of a sex-linked red–green deficiency. This is not always an accurate description since her color vision may be subtly affected by the expression of both genes in the heterozygous pair. In males, the effect of the defective gene will always be fully manifested.

The four possible combinations of X and Y chromosomes in a child of a color-deficient male ($X'Y$) and a normal female (X_1X_2) are:

$$(X'X_1), (X'X_2), (X_1Y) \text{ and } (X_2Y)$$

X' being here the chromosome carrying the gene defect. All girls will be hetero-zygous carriers and all boys will have normal color vision.

In another case of a man with normal color vision (X_1Y) and a heterozygous woman ($X'X_2$) the result is:

$$(X_1X'), (X_1X_2), (X'Y) \text{ and } (X_2Y)$$

There is a 50 percent chance that the sons will be normal, and also a 50 percent chance that any daughter will be a carrier of a red–green defect. It follows from similar schemes that women are fully affected red–green color deficients only if they inherit a defective gene from both their parents.

There are several possible gene defects that lead to color deficiency. One common defect may be due to misalignment of two X chromosomes during meiosis that leads to unequal cross-over between misplaced L- and M-pigment genes. When maternal and paternal X chromosomes in a female are aligned during meiosis to form the X chromosome of an egg, the DNA strands are broken, exchanged and rejoined.

Sometimes, when the chromosomes are misaligned, the exchange is unequal. The result may be deletion of the M-pigment gene from one chromosome (leaving one L-pigment gene) and the addition of the M-pigment gene to the other chromosome (which will then have one L- and two M-pigment genes). A male with the former X' chromosome will be a deuteranope, and a male with the latter X chromosome will be a 'normal' trichromat. Between one and six M-pigment genes have been reported to reside on a single X chromosome.

In other cases, misalignment of the X chromosomes may lead to a faulty exchange of parts of a gene and not a whole gene. A hybrid gene may result, consisting of a part of the L-pigment gene and a part of the M-pigment gene. Such hybrid genes are either inactive in that no pigment is formed, or they may produce a functional cone pigment with a spectral absorption that is different from the normal L- and M- cone pigments (see also Sharpe *et al.*, 1999).

It is an interesting question if a red–green defective dichromat has lost one population of cones or whether they have normal numbers of cones filled with only one of the L- or M-pigments. Since dichromats have normal acuity, the latter replacement theory has been assumed, but a recent study using adaptive optics has given the surprising result that some subjects may have lost an entire class of cone photoreceptors and that this loss seems not to impair any aspect of vision other than color (Carroll *et al.*, 2004).

Limitations of the three-receptor theory of color vision

We have seen how it is possible to reduce the many color stimuli with different spectral distributions into stimuli of the same color. The existence of such 'metameric stimuli' is explained by light absorption in the three different cone types. Theoretically, there is no end to the number of different spectral distributions that yield the same triplet of cone excitations and are therefore represented by the same point in three-dimensional color space (implying that they have the same color). This three-dimensional vector space can either have real color stimuli as fundamental axes, the most common having spectral distributions concentrated in the long (red), middle (green) and short (blue) wavelength regions of the spectrum, or the axes can be defined by the cone excitations themselves. The two alternatives are equivalent since additive color mixtures are a consequence of linear absorption processes in the cones.

Today, we regard Young–Helmholtz three-receptor theory as a first step in explaining color vision. The theory's validity is restricted to the first level of visual processing, the cone excitations in the retina, and it deals only with the phenomena that are related to additive color mixtures, based on the principle of equality, i.e. on color matches. It is not a complete color vision theory in the sense of explaining essential qualitative color features of different stimuli. It is not possible to deduce the color appearance of a stimulus from the cone excitations alone. The triplet of relative cone excitations tells us little about the color, whether it is white, red or green. The same absorption magnitudes in the three cone types can, in fact, give rise to very different colors, depending on illumination, on the surroundings, and on temporal effects. The effect of chromatic adaptation on the color appearance of a stimulus has been dealt with earlier. Later in this chapter, when we describe the Abney effect and the Bezold–Brücke phenomenon, we shall see how nonlinearities in the cone responses correlate with changes in the appearance of color stimuli.

In popular explanations of color vision deficiencies, the color perception of afflicted individuals is nonetheless often inferred directly from the Helmoltzian idea of a close connection between color perception, available cone pigments and their excitation. An example of the inadequacy of this conjecture in explaining color vision, is one of the postulates of Young–Helmholtz theory stating that yellow arises from the stimulation of a 'red-sensitive' and a 'green-sensitive' receptor type (the L- and M-cones in our terminology). This is related to the fact that, in an additive mixture of lights, yellow is produced as a mixture of red and green colors. However, associating color perception with receptor excitations is problematic. Although L-cones have a maximum sensitivity around 565 nm, at a wavelength that is greenish yellow, they have often been associated with the perception of red. M-cones have a maximum sensitivity around 540 nm, also in the yellow–green range of the spectrum. For both receptors, then, most of the absorption is on the greenish side of pure yellow, which is found at about 575 nm. How can pure yellow arise from mixing greenish yellow with yellowish green? This paradox arises from trying to attach color perception to the activity of receptors.

The fact that yellow colors can be perceived in positions in the visual field where one sees neither red nor green is likewise a problem for the original Young–Helmholtz theory. When you move a colored light from the fovea towards the periphery of your visual field, red and green perception is lost before yellow and blue disappear. Figure 6.5 shows the borders for the perception of different colors as a function of eccentricity on the retina. The pair-wise limit for red and green, more central than that for yellow and blue, leads our thoughts in the direction of opponent processes (see also Stabell and Stabell, 1982a).

In general, it is problematic to speculate on the color perception of people with color-defective vision. Over the years, however, a few reports have been collected from people with a red–green defect on one eye and normal vision on the other eye. Using their normal eye as a reference, they were able to report on the colors perceived by the color-defective eye. The reports suggest that a dichromat with a red–green defect, where either the L- or the M-cone pigment is missing, perceives yellow and blue (Vienot et al., 1995; Neitz et al., 1999). If this is, indeed, the case, it is further evidence that receptor excitations alone have little predictive value for which colors are actually perceived. However, since the neural 'wiring' associated with the normal eye might not be altogether normal, one should exercise caution in interpreting such reports.

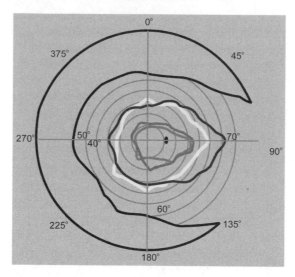

Figure 6.5 The maximum eccentricities that will allow identification of the color of a small spot of light projected onto the retina of the right eye. The blue and yellow curves show the maximum eccentricity for perceiving blue and yellow. These limits are about equally far out in the visual field, and they are more eccentric than the limits for the perception of red and green (adapted from Hurvich, 1981). (See also color plate section.)

It is not at all necessary that protanopes see the same yellow or the same blue as deuteranopes. How are their yellow and blue hues related to the unique hues of a normal trichromat? If a dichromat is able to see slightly reddish or/and greenish yellow, one would expect these hues to change when luminance is changed in what we may call a dichromatic Bezold–Brücke phenomenon. Some reports have indicated that this is, indeed, the case (Scheibner and Boynton, 1968; Wachtler *et al.*, 2004).

In the nineteenth century, Hering's opponent colors theory was a strong competitor to Young–Helmholtz tri-receptor theory. Today, much in the way Schrödinger anticipated, the fundamental ideas of both theories are accepted, but neither theory is considered valid in its original form.

Opponency and an opponent 'color code'

In Chapter 5, we presented an extended quantitative version of Hurvich and Jameson's simple, linear formulation of Hering's opponent colors theory (Jameson and Hurvich, 1955; see also Hurvich, 1981). Earlier quantitative forms of opponent theories, based on CIE tristimulus values, were published by Judd (1949, 1951b).

These included the suggestions of Hering, von Kries, Schrödinger, Ladd Franklin, Adams and Müller. The theories of Müller and Adams, in particular, had allowed for a nonlinear dependence of the responses on stimulus parameters. Following the work of Jameson and Hurvich, similar models became popular (Boynton, 1960; Hassenstein, 1968; Guth, 1980), much to the dismay of classical colorimetrists, who still believed the three-receptor ideas to be adequate. However, as science advanced, and neurophysiology delivered convincing evidence that cone-opponency was, indeed, a physiological fact in the retina, the LGN and striate cortex of primates (De Valois, 1965; Wiesel and Hubel, 1966; Gouras, 1968, 1974), the linear opponent models gave way to more realistic, nonlinear models (Seim and Valberg, 1986; Guth, 1991; De Valois and De Valois, 1993). Below, we shall take a closer look at a model emerging from relative recent neurophysiological recordings from single cells in the retina and the LGN of the macaque monkey *Macaca fascicularis*. This model deviates from others in that it accounts for highly nonlinear cell responses over a much larger range of stimulus intensities and chromaticities. The inclusion of such nonlinear behavior in models of color vision is essential for the understanding of, for instance, color scaling, the Abney effect, and the Bezold–Brücke phenomenon.

Some sensory qualities have a particular polar, or opponent, character, such as the mutually exclusive chromatic perceptions of red and green, blue and yellow. This applies to other sensory modalities as well. Temperature, for example is judged on a continuous scale between cold and warm, two extremes with very different, mutually exclusive, perceptual qualities. However, there is no intrinsic polarity in the corresponding physical magnitude. Physically, the temperature scale reflects the level of thermal energy on a scale from zero and upwards, there is no neutral or balancing point differentiating cold from hot. Nor is there any intrinsic polarity in the frequency spectrum of radiation that can account for the perception of unique yellow and the exclusive quality of blue. Mechanisms of our sensory organs somehow transform an one-dimensional physical magnitude into several different neural responses that end up as opponent sensory qualities. Opponency seems to be a consequence of the organization of activities in at least some sensory neurons.

The cone signals that are important for color vision, are mediated via the cone-opponent ganglion cells. In most cells in the retina and in the LGN, cone-opponency is in the simplest case opposing inputs from only two cone types. It is then a signal proportional to the difference of magnitudes proportional to cone polarizations, $V_1 - V_2$, of different cone types, one of which activates the cell by increasing its firing rate (1), while the other one (2) inhibits it. For example, a PC-cell with a 'L–M' combination of cone inputs responds with increased firing rate when its receptive field is illuminated by long-wavelength, red lights, and with reduced firing rate to medium- and short-wavelength lights. 'M–L' cells respond in an opposite fashion, while the response of KC cells is determined by opponent input from S-cones and some constellation of M- and L-cones.

A few words need to be said about the relation between inhibition and opponency. In weakly inhibited I-cells the activating cone mechanism dominates and in strongly inhibited D-cells the inhibitory one is the stronger. It is not clear what we should understand by a strongly opponent cell, but one can, perhaps, imagine a cell where the opposing cone mechanisms have equal weights. Such cells are rare (Lankheet *et al.*, 1998; Valberg *et al.*, 1985b, 1987). Relative cone weights vary substantially within the I- and D-types of cells, but they are seldom equal. All opponent cells show significant responses to increments or decrements of white, and to a luminance change in achromatic and chromatic lights. An opponent cell's neutral point in the spectrum (its zero crossing) is best defined as the wavelength for which the cell has the same response as an achromatic stimulus of the same luminance. This wavelength does not change with stimulus luminance (although the cell's firing rate does).

At the retinal level, cone opponency is an efficient code for transmitting information from the retina to the brain. Subtracting the signals of one cone type from the signals of another overlapping one, removes much of the information common to both cone types, thus reducing redundancy (Buchsbaum and Gottschalk, 1983).

Not all possible combinations of cone types are realized in the primate retina and the LGN. Functionally, there appears to be six main opponent cell types: four with L- and M-cone combinations and two types with major S-cone inputs, the 'Blue ON/ OFF cells' with activating S-cones and the 'Yellow ON' cells with inhibition from S-cones. The majority of opponent cells, however, differentiate between L- and M-cone signals, sometimes with minor S-cone input. There are four main classes of such parvocellular cells. Denoted by their dominating two-cone opponency, there are two types of 'L–M' cells and two types of the opposite 'M–L' cells, the ON- and the OFF-cells (Wiesel and Hubel, 1966; see Figure 3.22), which we have renamed Increment (I-type) and Decrement (D-type) cells, respectively. The main difference between I-type and D-type cells lies in their maintained firing rate in the dark and in the polarity of their response to increments and decrements of achromatic light, the latter being a consequence of the relative weights for activation and inhibition by the cone mechanisms. The I-cells are weakly inhibited, whereas D-cells have a higher maintained firing rate in the dark and are totally inhibited and silenced for high luminance ratios. The responses of I-cells asymptote for high intensities where all wavelengths elicit the same positive response, and the effect of opponency is lost. Another difference is the higher contrast sensitivity of D-cells, being responsive for darker colors than I-cells (Tryti, 1985; Lee *et al.*, 1987). Because of difference in strength of inhibition, D-cells usually have a smaller maximum firing rate, and the position of their maximum response occurs at a lower luminance ratio than for I-cells. Typical responses of these cells to monochromatic lights are shown in Figures 6.7–6.11 and will be discussed later.

We do not feel that the terminology of 'ON-center' and 'OFF-center' cells is appropriate for understanding the response model to be presented here. Information

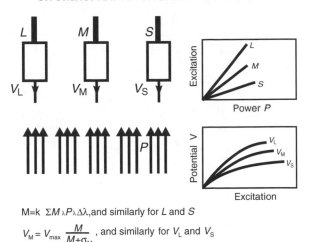

M=k $\Sigma M_\lambda P_\lambda \Delta\lambda$, and similarly for L and S

$V_M = V_{max} \dfrac{M}{M+\sigma_M}$, and similarly for V_L and V_S

Figure 6.6 There is a linear relation between excitations (absorptions) L, M and S of the cone receptors and the power, P, of the light. The relationship between cone excitations and cone signals V_L, V_M and V_S, is, however, nonlinear.

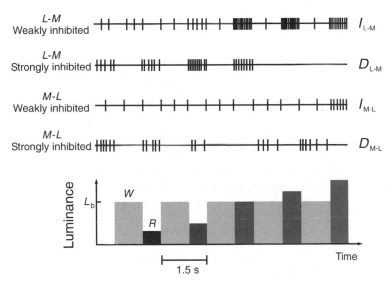

Figure 6.7 The spike trains of four PC cell types that receive inputs from the L- and M-cones in response to a stimulus covering the center and surround of the receptive field. Long-wavelength, red light (R) of various luminance (L) is interchanged with a coextensive white adaptation background of constant luminance $L_b(W)$. The luminance ratio, $Y = L/L_b$, increases towards the right in the figure. Weakly and strongly inhibited 'L–M' cells have the greater response to red light, but for different ranges of luminance ratio. Note that the letter 'L' is here used in two different meanings; as luminance in the lower drawing and as L-cone excitation in the upper part of the figure. Green light would have given about the same response in 'M–L' cells as red light in 'L–M' cells. See also Color Plate Section.

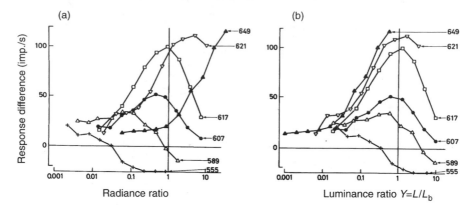

Figure 6.8 The responses of a strongly inhibited, 'L–M' decrement cell when the whole receptive field is alternately illuminated by a white reference light and high-purity chromatic lights of different wavelengths, as shown in Figure 6.7. The two alternative plots of the same data demonstrate the difference between using the energy based unit radiance as the intensity variable (a) instead of luminance (b). Using luminance gives the intensity–response curves an orderly appearance with thresholds that are grouped together and maximum responses at roughly the same luminance ratio, $Y = L/L_b$ (the firing rate of this cell was unusually high).

about positive and negative contrasts of luminance as well as color is transmitted within the opponent channels, always relative to an adaptation level determined by the background (Du Buf, 1994; Chichilnisky and Wandell, 1996).

Despite what has been suggested by many opponent models of color vision and is often repeated in textbooks, cells with such cone opponencies are not directly responsible for the perception of the elementary or unique colors. Unique hues do not correspond to the activity of the 'cardinal', opponent cells found in the retina and LGN. The perception of unique red, for instance, is not linked to the response of 'L–M' cells alone. Nor does the perception of unique green correlate with the inhibition of 'L–M' cells or with the excitation of 'M–L' cells. Later, we shall present data that prove that cone-opponency in the retina and LGN is not a direct correlate to the elementary hues.

However attractive this idea of color specific ganglion cells (Red-ON, Green-OFF, etc., or 'labeled lines') might have been in earlier days, it can be considered a parallel to associating color qualities with cone excitations (R-cone, G-cone, etc.). Color perception is believed to correlate better with signals in the visual cortex than with the excitations of LGN cells, although we still do not know enough details to describe how (De Valois and De Valois, 1993).

Cone responses

Correlates between stimulus magnitudes, physiological responses, and color perception are found at many levels of processing, first at the cone level and later in the

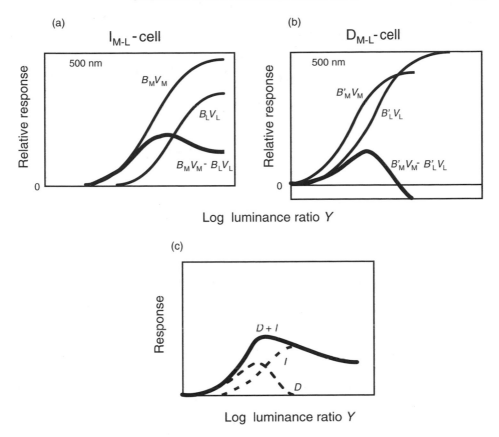

Figure 6.9 A model of how signals proportional to the receptor potentials, V, are combined in opponent PC ganglion cells. The examples are for a weakly inhibited, 'M–L' Increment cell (a), and a strongly inhibited, 'M–L' Decrement cell (b). For both cell types, the final I–R curve is a difference between the weighted response inputs from the M- and the L-cones, $B_M V_M - B_L V_L$, where the inhibitory weighting factor, B_L, is largest for the Decrement cell in (b). The 'L–M' cells behave symmetrically. This model corresponds nicely to the experimental data shown in Figure 6.10. (c) We postulate further that the responses of I- and D-cells both contribute to color vision, and that the responses of I- and D-cell types with the same cone input, e.g. 'L–M' are summated (and similarly for 'M–L' cells of the I- and D-types). This particular example is for the 'M–L' cells of (a) and (b).

responses of opponent cells. In order to establish such correlates, one must chart the responses of opponent ganglion cells in the retina and in the LGN to the several dimensions of color stimuli, such as wavelength, purity and luminance. Two phenomena may serve to illustrate such relationships: the *Abney effect* (also called the Kohlrausch effect) and the *Bezold–Brücke phenomenon*. The Abney effect describes how hue changes when white light is added to a stimulus of high purity, for example to a monochromatic red light. As purity starts to decrease, the hue changes towards bluish-red, and when white is added to monochromatic blue of a short wavelength, the hue becomes more reddish-blue. The Bezold–Brücke phenomenon describes the hue change we experience when a light undergoes a change in

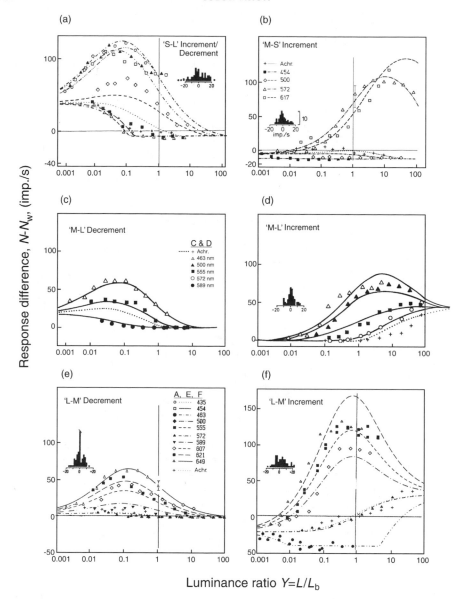

Luminance ratio $Y = L/L_b$

Figure 6.10 Examples of the I–R curves of two KC cells and four PC geniculate cells to chromatic and achromatic lights. The symbols represent the measured firing rate in impulses/s for a given wavelength and luminance ratio. The curves are derived from a mathematical simulation of these responses, based on a linear combination of receptor potentials as in Figure 6.9 using Estevez (1979) fundamentals. More precisely, the responses of 'M–L' cells can be expressed as $N = B_M V_M - B_L V_L + N_o$, where B_M and B_L are weighting factors that represent the strengths of activation and inhibition by the two opposing cone types. N_o is the empirically determined maintained activity for the achromatic adapting reference stimulus. All responses are plotted relative to N_o, the response to the white adaptation light of 100 cd/m^2. The small insets within each panel show the distribution (number of data points) on the y-axis of a given deviations between the data points and the mathematical simulation in impulses/s. The rms deviation was normally between 5 and 10% of the maximum response of each cell.

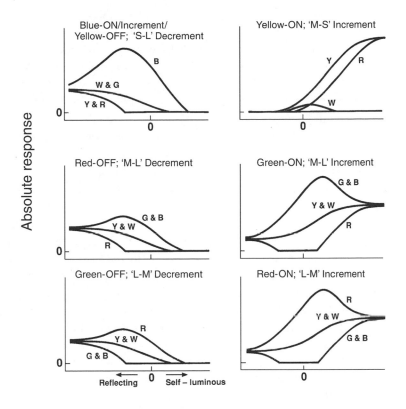

Log luminance ratio $Y = L/L_b$

Figure 6.11 Schematic drawings of the luminance–response curves for the same six types of opponent cells as shown in Figure 6.10. The curves represent responses to the exchange of an adapting white reference stimulus of luminance L_b with a test light of luminance L, as in Figure 6.7. The letter associated with each curve is short hand for the spectral region of the test light (B = blue; G = green; Y = yellow; R = red; all of which are near-monochromatic, while W = white). Log $Y = 0$ corresponds to a test light with a luminance equal to the adaptation luminance, L_b. The I- and D-types of cells share the luminance scale between them. Decrement cells have excitatory responses for the darker colors (for example object colors), while Increment cells respond preferentially to higher luminance ratios (for example for self-luminous colors and light sources). Since Decrement cells are unresponsive to (silenced by) bright lights covering the whole receptive field, they cannot alone cover the whole intensity range of color vision.

luminance. For instance, a wavelength between 580 and 650 nm, which looks orange for a relatively low photopic luminance, becomes more and more yellow as luminance increases. These two phenomena, which are clear examples of a nonlinear relationship between cone excitations and hue, saturation and lightness (and brightness), have eluded a satisfactory, quantitative explanation.

Later we shall try to provide plausible physiological explanations for these phenomena, based on what we now know about low-level processing. This model is fairly general in that it also gives a background for understanding the scaling of color differences, another area where linear color metrics fails. However, first let us, on a general basis, discuss why the nonlinear responses of cones are more important than their excitations in accounting for hue shifts when luminance changes.

We have already established that cone excitations L, M and S are linear functions of the intensity of the incident light (as long as pigment bleaching is small). If the radiance is doubled, the excitation of a cone is also doubled. This does not hold true for its receptor potential, V. Nor does it apply to the responses of other cells in the retina, such as ganglion cells that receive their direct inputs from bipolar cells, which, in turn, have inputs from cones. From Figure 6.6 we see that a change in the magnitude of the power, P, of a light leads to a proportional increase in the excitations L, M and S. As a result, there is no change the ratios $L: M: S$, between these receptor absorptions. This means that, in linear vector space, there is a change in the length of the color vector, but not in its direction. Therefore, the chromaticity coordinates (x, y) are constant as intensity changes. However, the ratio of the receptor potentials, V_L: V_M: V_S, i.e. of the signals that the cones transmit to the next cells in the retina, changes with intensity because of the nonlinear relationship between absorption of light quanta in each cone and the magnitude of the receptor potential (see the equation in Figure 6.6). As intensity increases, one of these potentials may change more than the others, altering the ratios of outputs from the three cone types. For instance, if increasing the luminance of a short-wavelength light leads to a doubling of the S-cone signal, V_S, while there is little change in the other two signals, V_M and V_L (Figure 4.13), this is likely to modify the color appearance of the stimulus. Since the ratio between the receptor potentials changes with a pure intensity change (although the ratio of cone excitations does not), one should not be surprised if the hue of the stimulus also changes (as it does in the Bezold–Brücke phenomenon). As we shall see, the same arguments can be used to explain the Abney effect of hue changes as white light is added to a monochromatic stimulus. This does not mean that changes of hue can be understood at the receptor level, only that the early nonlinearity will be conveyed to the next level in the visual pathway where it determines activation and inhibition of opponent cells, and that this has consequences for hue perception.

Responses of retinal and LGN opponent cells

In order to better understand the role of retinal and geniculate opponent cells in color processing, let us devote some space to a description of experiments that have mapped out quantitative relationships between stimulus magnitudes and cell responses. The physical stimulus sequence used in these experiments is illustrated at the bottom of Figure 6.7. It is similar to that described in Figure 4.14 for measuring receptor potentials. Cell responses were evoked by 4°, 300 ms duration stimuli of different dominant wavelengths and luminances, alternating in a temporal sequence with a fixed-luminance, white adaptation light of the same size and 1.2 s duration.

Figure 6.7 illustrates schematically the responses of 'L–M' and 'M–L' cells when their respective receptive fields were illuminated by long-wavelength red light. Each of these categories is subdivided into ON and OFF variants, which, for reasons mentioned earlier, we shall call Increment cells (I-cells) and Decrement cells (D-cells). A short-hand notation for these cell types would be I_{L-M}, I_{M-L}, D_{L-M} and D_{M-L} cells. The top of Figure 6.7 shows that the firing rate of an I_{L-M} cell rises as the luminance ratio between the long-wavelength stimulus and the white adaptation background increases. For the highest luminance ratio, however, the response decreases slightly because the M-cone inhibition becomes more effective. Because L-cones are more responsive than M-cones to light at this wavelength, the nonlinear (saturation) effect of high intensity manifests itself sooner in the L-cone response than in the M-cone response. The next trace shows the firing of a D_{L-M} cell. This cell will respond in the dark and, due to a stronger inhibitory input, it reaches maximum response at a rather low luminance ratio. For each of the I- and D-cell types, the working range covers only part of the luminance ratio range. Together, however, they span more than a 4 log unit range of luminance, from black to bright colors.

The next two traces show the response of the opposite 'M–L' cells to the same long-wavelength light. Because L-cones are more sensitive then M-cones in this part of the spectrum, the I_{M-L} cell is inhibited by the red light at most luminance ratios, but gives a weak response at the highest luminance ratio, where the response of L-cones has leveled out and the M-cone input becomes relatively stronger. The response of the D_{M-L} cell is dominated by the maintained response in the dark and the strong inhibitory L-cone input (analogous to the behavior of the D_{L-M} cell to 555 nm light in Figure 6.8).

Examples of intensity–response curves of a D_{L-M} cell are shown in Figure 6.8. Despite the large excitatory responses to long-wavelength light, this cell is characterized as a D-cell because it has a relative high maintained activity in the dark and responds to decrements of white light (which for this cell, lacking S-cone input, is equivalent to about 571 nm). Also, it has no positive response asymptote for high intensities as Increment cells do.

Figure 6.8 also demonstrates the difference between using the photometric magnitude luminance (cd/m^2) as the intensity variable instead of the radiometric magnitude radiance (W/m^2). Since we are interested in the response to the test stimulus relative to that of the white adaptation stimulus, the response magnitude plotted on the ordinate ($N - N_W$) is the difference in firing rate for the test stimulus and for the white adaptation background (which was approximately 20 impulses/s). The field diameter was 4°, large enough to cover the receptive field of the cell.

Figure 6.8(a) gives the relative response as a function of the radiance ratio between stimulus and adaptation background, while Figure 6.8(b) shows the response as a function of the luminance ratio of the same stimuli. Using luminance ratio as the independent variable leads to better grouping of the intensity–response curves along the x-axis, and gives a surprisingly simple picture of the family of the cell's spectral intensity–response curves. For cells combining the outputs from L- and M-cones, this is a consequence of luminance sensitivity being proportional to the sum of L- and M-cone sensitivity. Even for cells with S-cone input, using the luminance ratio gives an orderly appearance of the response curves [Figure 6.10(a) and (b)].

For long-wavelength light of 649 nm, the response of this 'L-M' cell increases steadily with increasing intensity. For this wavelength, the L-cones dominate the response for all available intensities, and the effect of M-cone inhibition is hardly noticeable. For shorter wavelengths, the subtractive M-cone mechanism brings the response down to absolute zero (e.g. 555 nm) at a relative low luminance ratio. For wavelengths in between, there is a changing balance between activation and inhibition, with activation dominating for wavelengths longer than 571 nm (the spectral neutral point) at low intensities and inhibition taking over at higher intensities. An illustration of how the difference, between cone potentials can lead to responses like those in Figure 6.8 is given in Figure 6.9.

Figure 6.9 demonstrates that, for a weakly inhibited I_{M-L}-cell, the effectiveness of the inhibitory receptor inputs is less than for activating receptors ($B_L < B_M$). For high stimulus intensities, the difference of the cone potentials, $B_M V_M - B_L V_L$, reaches a constant, positive asymptotic value. For the more strongly inhibited D-cell, the potential difference, $B'_M V_M - B'_L V_L$, is negative at high intensities, resulting in total inhibition of the cell's activity. The cell's maintained activity will, of course, influence its absolute firing rate.

Figure 6.9(c) illustrates the response of a hypothetical cell that sums the outputs of an I- and a D-cell of the same opponency. Since I- and D-cells divide the lightness/brightness dimension between them, such a sum is required to account for continuous color scaling at all lightness levels, as in the model of color coding presented below.

Modeling intensity–response functions of opponent cells

PC cells

In Figure 6.10 are shown recorded data at different luminance ratios together with intensity–response curves derived from a mathematical model described by the equation below. Panels (a) and (b) are for KC cells having S-cone inputs, and (c) and (d) for two 'M–L'-cells, an I_{M-L} and a D_{M-L} cell. Similar results for two 'L–M' cells are shown in Figure 6.10(e) and (f). The different symbols are for different wavelengths of the test stimulus. The physical stimulation conditions were the same as in Figure 6.7, with the 4° stimulus covering the receptive fields of these cells.

The responses of the PC cells in Figure 6.10 can be described by a relatively simple response equation based on a linear combination of receptor potentials. The firing rate of PC cells in the retina and in LGN is essentially proportional to the differences of cone potentials. This means that all synaptic processes and interactions in the retina, between photoreceptors, bipolar and ganglion cells, can be lumped together and incorporated in the constants of the equation.

In most cases, the responses, N, of the four main classes of PC cells can be modeled by combining the membrane potentials, V, of two or more cone types in the following way

$$N_{\text{L-M}} = A_{\text{L}}V_{\text{L}} - A_{\text{M}}V_{\text{M}}(-A_{\text{S}}V_{\text{S}}) + N_{\text{o}}$$

and

$$N_{\text{M-L}} = B_{\text{M}}V_{\text{M}} - B_{\text{L}}V_{\text{L}}(+B_{\text{S}}V_{\text{S}}) + N_{\text{o}},$$

where N_{o} represents the maintained firing in the dark. Provided the cone excitations have been normalized to be equal for white light ($L_{\text{W}} = M_{\text{W}} = S_{\text{W}}$), and the receptor potentials, V, are normalized to have the same maximum, the constants A and B account for the relative weights of the cone inputs. For I-cells $A_{\text{L}} > A_{\text{M}}$ and $B_{\text{M}} > B_{\text{L}}$, and for the more strongly inhibited D-cells $A'_{\text{L}} < A'_{\text{M}}$ and $B'_{\text{M}} < B'_{\text{L}}$ (Lee *et al.*, 1987; Valberg *et al.*, 1987). The cone potentials, V, are described by the common nonlinear, hyperbolic Naka–Rushton function of the cone excitations L, M and S, being $V_{\text{M}} = M^n/(M^n + \sigma_{\text{M}}^n)$ for the M-cone. σ_{M} is the half-saturation constant determined by the state of light adaptation, and the exponent $n = 0.7$. In some instances, an additional inhibitory S-cone input to 'L–M' Increment cells and a synergistic M-cone and S-cone activation of 'M–L' Increment cells would improve computation (Valberg *et al.*, 1985b).

If we assume that these M/L cells have inputs from all L- and M-cones within their receptive fields, the weighting factors may mirror the relative numbers of L- and M-cones within the receptive field. This has been called the random connection hypothesis (Diller *et al.*, 2004). The selective connection hypothesis, on the other hand, would regard the weighting factors as reflecting spatially averaged relative strength of activating and inhibitory synaptic contacts in the center and surround of the receptive field.

In terms of the model in Figure 3.22, the excitatory input to the $I_{\text{M-L}}$ cell of Figure 6.10(c) comes from the receptive field center, and the inhibitory input mainly from the receptive field surround (see also Figure 6.12). As mentioned before, there are two important and typical differences in the responses of the I-type and the D-type cells. First, the maintained activity in the dark is much higher for the D-cell than for the I-cell, allowing for a greater range of inhibition of the D-cell before the firing rate drops to zero. An I-cell has little or no maintained activity. This is one of the few aspects in which retinal ganglion cells and LGN cells differ. The maintained activity, or the spontaneous firing rate, of retinal cells is generally higher than for the cells in LGN. Second, the effect of inhibition is stronger in D-cells than in I-cells. As a result, the maximum response of an excitatory wavelength (e.g. 463 nm in the figure) occurs at a lower luminance ratio for D-cells than for I-cells, and the response of the former is completely suppressed at high luminance ratios. Despite these differences, there

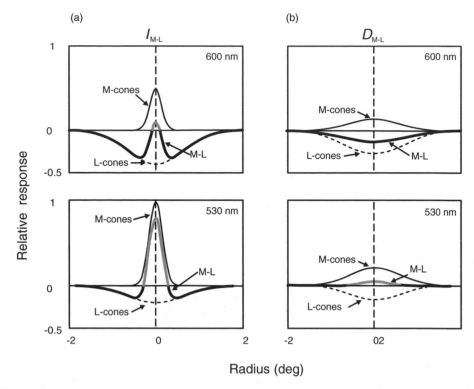

Figure 6.12 The relative size of the center and surround of a receptive field depends on stimulus wavelength. The response to a small spot of either 600 or 530 nm traversing the receptive field is mapped in (a) and (b) (thick solid line). The broken line and the thin solid line indicate a model for the underlying excitatory and inhibitory mechanisms.

appears to be an overlapping distribution of sensitivity and strength of inhibition of I- and D-cells, suggesting that they may possibly be the extremes of a continuum (Lee *et al.*, 1987; Rodieck, 1991).

Let us compare the responses of I- and D-cells in Figure 6.10 for a luminance ratio $Y = L/L_b = 1.0$, i.e. for a luminance that is equal to that of the white adaptation field. We see that, for all excitatory wavelengths, and for white light, the I-cell response increases for luminance increments in this region, whereas the D-cell response increases for luminance decrements. The different behavior of these two cell types to increments and decrements, relative to the adaptation luminance, points to opposite or complementary roles in the coding of positive and negative contrasts.

Figure 6.10 demonstrates the combined wavelength and luminance dependency of the firing rates of opponent cell types. We see that, for a particular wavelength, the initial luminance ratio determines whether a luminance increment results in a response increment or a response decrement. The wavelengths that are strongly excitatory give rise to non-monotonic response curves. For the D-cell in Figure 6.10(c), increasing the luminance of an excitatory wavelength (e.g. 463 nm) from black first results in an increased firing rate. However, if the luminance increase has a starting point of a luminance ratio close to 1, it results in a reduced response. This cell, often called a 'Red OFF-cell' is a pure Decrement cell for white light and for inhibitory wavelengths longer than about 570 nm. Starting with black up to a certain luminance ratio, for excitatory wavelengths, the response will increase for luminance increments. Further luminance increments result in less activation.

The responses of these Increment and Decrement cells contain information about spectral region and about relative luminance (as well as about stimulus size). Such entangled or multiplexed information can, in principle, be disentangled by comparing it with other cells' responses to the same stimulus.

KC cells

Figures 6.10(a) and (b) give response curves for two opposite types of cells with S-cone inputs recorded under the same conditions as for the PC cells described above. The activity of these koniocellular cell types can be modeled by the following combination of cone potentials:

$$N_{\text{M-S}} = C_{\text{M}}V_{\text{M}} - C_{\text{S}}V_{\text{S}}(-C_{\text{L}}V_{\text{L}}) + N_{\text{o}}$$

and

$$N_{\text{S-L}} = D_{\text{S}}V_{\text{S}} - D_{\text{L}}V_{\text{L}}(+D_{\text{M}}V_{\text{M}}) + N_{\text{o}}.$$

The ratio of the constants $C_{\text{M}}/C_{\text{S}}$ and $D_{\text{S}}/D_{\text{L}}$ determines the relative strengths of excitation and inhibition. The particular 'Yellow ON' ('M–S' Increment) cell of Figure 6.10(b) could be modeled by an 'M–S' cone combination, but others would need an additional L-cone inhibition (Valberg et al., 1985b, 1986b).

The cell with excitatory S-cones input in Figure 6.10(a) is normally called a 'Blue ON' cell. It is progressively more activated by increments of short-wavelength light of moderate intensity and increasingly inhibited by increments of long-wavelength light. The modeling of the cell responses, based on their high maintained firing rate in the dark, their positive response to white-light decrements and total inhibition of all wavelengths at high luminance ratio, has suggested renaming these cells 'D-cells', instead of 'Blue-ON' cells. Their opponent cone inputs are best simulated as 'S–L'. Recent anatomical studies have shown that these cells have dendritic fields in both the ON and the OFF strata in the retina, hence the name 'bistratified' (Dacey and Lee, 1994). We have therefore called these cells Increment/Decrement cells.

The rarely encountered 'Yellow ON' cells, for which an example of response data is given in Figure 6.10(b), are peculiar in that they respond only to middle-wavelength light, and not to white. These cells discriminate effectively between yellow and white. We have called them I-cells because of little or no maintained activity in the dark, thereby excluding decrement responses or disinhibition at low relative luminance. However, they do not show an excitatory asymptotic response at high luminance ratios, as do I-cells of L/M opponency. For a long time they have been elusive to anatomists, but recent findings indicate that they correspond to a distinct class of large, sparse, monostratified retinal ganglion cells (Dacey and Packer, 2003).

The quantitative experiments leading to these cone combinations of PC and KC cells are described in Valberg *et al.* (1983, 1985b, 1986a,b, 1987), Derrington *et al.* (1984), Tryti (1985), Lee *et al.* (1987) and Lankheet *et al.* (1998). The model curves in Figure 6.10 fit the data points rather well. The curves have not been individually optimized; they all result from a given spectral sensitivity for each cone type and a given weighting of the cone inputs for each cell. Typical deviation between data and simulation is a root mean square deviation between 5 and 10 percent of the maximum response (Tryti, 1985; Creutzfeldt *et al.*, 1986). From Figures 6.9 and 6.10 we conclude that

- D-cells have a stronger inhibitory cone-input than I-cells;
- response and sensitivity to low relative luminance (dark colors) is generally greater for D-cells than for I-cells;
- the response at high relative luminance (bright colors) is greater for I-cells than for D-cells.

As we have seen, a relative wide range of spectral stimuli will evoke a positive response in opponent cells. At the same time, these cells respond to the lightness contrast of chromatic as well as achromatic stimuli (black, gray and white). The size of the stimulus is also important for their response specificity and magnitude. A PC Increment cell, for instance, responds positively for increments in all wavelengths if the stimulus is too small to activate significantly the inhibitory receptive field surround. Therefore, opponent cells of the retina and LGN cannot be associated with a single attribute of the stimulus, such as color. Opponent cells are not color-specific at this level. Only a comparison between responses of different types of cells at a higher level can determine which stimulus evoked the neural firing, whether it was a luminance contrast, a change in color, a particular stimulus size, or something else. At this level of processing, multiplexing is the rule, meaning that more than one property is signaled along the same channel (Martinez-Oriegas, 1994). For much the same reason, it is not appropriate to use color names to characterize these retinal and geniculate cells, like 'Red Increment cell', 'Red ON-cell', 'Red-sensitive cell', etc., even though it may make reading easier to use such abbreviations. One needs to watch out for oversimplified concepts that imply that cells in the retina or LGN directly code for color qualities.

Figure 6.11 summarizes schematically how responses to spectral stimuli depend on luminance ratio for the major six types of opponent cells in the macaque retina and LGN. The curves refer to the experiments described above, where the stimuli covered the whole receptive field, thus resulting in a greater degree of opponency than with a smaller stimulus (see below).

Electrophysiological recordings from single cells in the macaque LGN (Zrenner, 1983; Valberg et al., 1985b) has given the following distribution of cell types: D_{S-L}, 10 percent; I_{M-L}, 15 percent; D_{M-L}, 15 percent; I_{M-S}, 3–4 percent; I_{L-M}, 20–25 percent; D_{L-M}, 15 percent. For the rest of the encountered cells the cone inputs could not be sufficiently identified, or they fell between these classes.

Correlates of related and unrelated colors

We have seen that for each type of M/L opponency, 'L–M' or 'M–L', there are two types of PC cells; a D-cell and an I-cell. Color coding in one of these two opposite directions of color space is divided between I- and D-cells depending on relative luminance. The strongly inhibited D-cells can only transmit information about colors of low relative luminance, for example of surfaces of objects with reflection factors lower than about one. Such reflecting surfaces are the most frequent object surfaces in a natural environment. The color appearance of such surfaces is strongly dependent on the adapting surround, and opponent cells have been demonstrated to adjust their responses and sensitivity to such surrounds (Valberg et al., 1985a).

I-cells give spectrally selective responses for higher luminance ratios, far above a luminance value that completely suppresses the responses of D-cells, albeit with less opponency at the higher luminance ratios. This qualifies I-cells for responding to bright, unrelated colors of self luminous areas, or isolated reflecting areas in the dark, also called *void colors*. Because, for a given adaptation, these different physical domains of lightness and brightness together span more than four decades of relative intensity, the need for different coding units seems natural. This is analogous to the sensory temperature scale being divided between 'warm coding units' and 'cold coding units' (Hensel and Kenshalo, 1969). Combining the responses of I- and D-cells with the same opponency (as in Figure 6.9(c)) allows us to cover the whole range of relative luminance values, from black surfaces to bright light sources.

It has been suggested that only cells with relative larger and coextensive excitatory and inhibitory receptive fields can be involved in color vision processing (Rodieck, 1991), thus excluding the retinal midget ON-ganglion cells and geniculate PC Increment cells with their small receptive field centers and highly spatial selective responses. However, the consequence of limiting the M/L dimension of color vision to processing by the Decrement cells would be to restrict color discrimination to related colors of rather low luminance ratios (see Figures 6.10 and 6.11). In the account of the Bezold–Brücke phenomenon to be presented later, we shall see that the Increment cells are indeed needed in order to explain changes of color strength and hue over a several log unit range of luminance ratios.

Based on psychophysical evidence, we shall further assume that constant hue of a chromatic stimulus is determined by the relative response, or the response ratio of two neighboring cells with different opponency, e.g. the response ratio of 'L–M' cells and 'M–S' cells. A constant orange hue, perceived to be midway between unique red and unique yellow, will be related to the response ratio between $I_{M\text{-}S}$ cells and the summated response of the hypothetical '$I_{L\text{-}M} + D_{L\text{-}M}$' unit. Later, we shall see how these conjectures allow us to account for several perceptive color phenomena and the scaling of color differences.

Antagonistic receptive fields of opponent cells

The concept of receptive fields was introduced in Figure 3.22. Here we shall discuss possible receptive fields structures of opponent cells. Figure 6.12 shows a model of probable spatial sensitivity of the receptive fields of PC-cells. Figure 6.12(a) is for a weakly inhibited I_{M-L} cell (type I in the terminology of Wiesel and Hubel, 1966), and Figure 6.12(b) is for a strongly inhibited D_{M-L} cell (type II with coextensive center and surround fields). Possible spatial sensitivities of the center and surround mechanisms are shown by the thin solid and dashed curves. In a test situation, where the spatial sensitivity is probed with a small spot of light that traverses the receptive field along its diameter, these two sensitivities summate to a resultant sensitivity shown by the solid green curve (see also the Gaussian model in Figures 3.22 and 3.23). In Figure 6.12, the excitatory input from M-cones is restricted to the center while the inhibitory mechanism of the L-cones extends throughout the receptive field, both center and surround. This need not always be the case. In the central fovea, for example, the excitatory center mechanism is likely to consist of only a single cone. To determine the spatial profile of the excitatory center mechanism, the M-cone response must be isolated. This can be achieved by adapting the cell to light of a suitably chosen wavelength that strongly excites and saturates (adapts) the L-cones, but not the M-cones. With the surround mechanism rendered insensitive, the center mechanism can be probed selectively. A corresponding isolation of L-cone responses allows the surround mechanism to be mapped.

In Figure 6.12(a) and (b), the spatial boundaries of the two cone mechanisms are only slightly dependent on stimulus wavelength. However, in the spatial profile of the M–L response, the boundaries of the excitatory center and the inhibitory surround are very much a function of the probing wavelength, as shown by the examples of response profiles for 600 and 530 nm. The greater the relative sensitivity of the center M-cone mechanism to the chosen wavelength, the larger the excitatory center of the corresponding M–L receptive field. In the top panel of Figure 6.12(a), for example, the center is small and the excitation is weak at 600 nm and much larger at 530 nm.

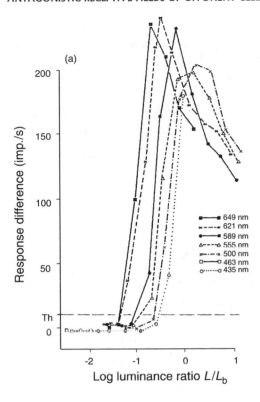

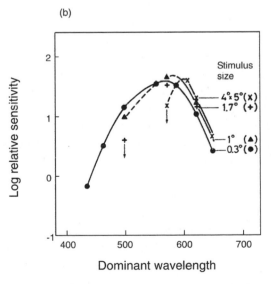

Figure 6.13 (a) I–R curves of an 'L–M', Increment cell stimulated in the receptive field center by a 0.3° spot of different wavelengths. (b) The spectral sensitivities derived from the I–R curves in (a) are shown by solid circles. This sensitivity corresponds to that of the L-cone alone. Increasing the stimulus size leads to a narrower spectral sensitivity, indicating opponency due to increased activation of the M-cone surround.

There are several possible cone-combinations within a receptive field center of a size extending beyond a few cone diameters. The same applies to the larger antagonistic surround. Not all of these possibilities are utilized. Surrounds with a single cone type seems to be the rule (Lee, 1999). The evidence for single cone centers is not equally strong, except for the small foveal receptive field centers. Figure 6.13(a) shows the family of intensity–response curves for a single cone-center, geniculate 'L–M' I-cell, stimulated by a 0.3° spot in the center. The small field sensitivity of this cell is that of the L-cone with no sign of opponency. The opponency becomes prominent when field size is increased, and is strongest when the stimulus covers the whole receptive field, indicating different cone inputs to center and surround. Increasing the diameter of the stimulus of Figure 6.13 resulted in a shift of maximum sensitivity towards longer wavelengths and a reduction of spectral sensitivity at short wavelengths [Figure 6.13(b)], indicative of the activation of a subtractive M-cone mechanism. In this case, the spectral sensitivity curve did not imply a summed input of L- and M-cones in the center, in which case the spectral sensitivity would have been broader. Nor is it likely that the subtractive M-cone mechanism extended through the center, in which case the spectral sensitivity curve would have been narrower. Pure L- or M-cone centers may be a genetically programmed rule, but they can also result from drawing center cones from a cluster of neighboring cones of the same type, as would often be possible with a cone distribution like that of Figure 3.15. Pure cone surrounds, however, cannot be explained by random selection.

For some opponent D-cells, the excitatory and inhibitory fields seem to be of about the same extension, and these cells therefore lack a typical center-surround structure [they are type II cells, see Figure 6.12(b)]. The yellow sensitive I_{M-S} cells and the blue sensitive D_{S-L} cells often seem to have spatially coextensive excitatory and inhibitory understructures. For such type II cells, the difference between excitation and inhibition is less dependent on the position of a small probe within the receptive field, and these cells are therefore less sensitive to the size of the stimulus than type I cells. This has given rise to speculations that maybe only type II cells are the color coding cells, whereas type I cells, with their prominent center-surround structures, have a main task in coding spatial relationships (Rodieck, 1991). However, in our model both I- and D-cells contribute to color vision, each for a certain range of reflectance value and relative light intensity. Types I and II may not be distinct classes, but the extreme cases on a continuum of relative center/surround sizes. Since the same ganglion cells are members of several different neural networks, they also contribute to information about spatial and other visual dimensions (multiplexing).

Spectral sensitivity and response

Figure 6.14 shows typical spectral responses of the six most common opponent cells of the macaque retina and LGN of the macaque monkey for the same experimental

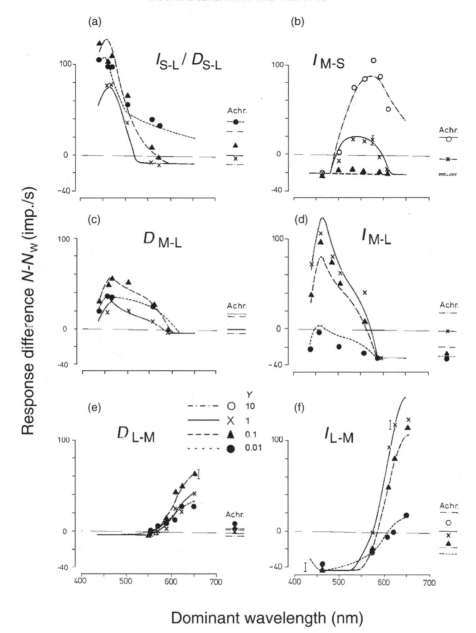

Response difference N-N_W (imp./s)

Dominant wavelength (nm)

Figure 6.14 Spectral response curves for the six types of opponent cells represented in Figure 6.10. Data points refer to responses to stimuli such as those presented in Figure 6.7, while the curves are derived from the opponent response model described in the text. The ordinate represents the response difference, i.e. the response to a spectral test light, N, minus the response, N_W, to the white adapting reference field. The horizontal line at 0 impulses/s therefore represents the response to the white adaptation field. The responses to achromatic stimuli of different luminance ratios, Y, are given to the right of each plot.

conditions as those used to obtain in Figure 6.10. The different symbols and curves represent different luminance ratios Y of stimulus/adapting background. The responses have been derived from plots like those of Figure 6.10, for different luminance ratios $Y = L/L_b$ (where $L_b = 100 \, \text{cd/m}^2$):

$$Y = 10 \; (\text{open circles})$$
$$Y = 1 \; (\text{crosses})$$
$$Y = 0.1 \; (\text{solid triangles})$$
$$Y = 0.01 \; (\text{solid circles})$$

The responses to white light can differ from cell to cell and, to emphasize the chromatic and luminance response differences relative to the achromatic adaptation stimulus, we have once again subtracted the response to the white adaptation light (for $Y = 1$) and plotted $N - N_W$ along the y-axis. The adaptation stimulus is represented by the horizontal line, for which $N - N_W = 0 \, \text{imp/s}$ (for $Y = 1$). The responses to achromatic light of the given luminance ratios are shown to the right in each graph. Note that for PC cells, without an S-cone input, the response to white corresponds to that of 571 nm of the same luminance; 571 nm is the PC cells' *spectral neutral point*. This means that these cells are tritanopic and cannot distinguish 571 nm from the white adapting light. The fully drawn curves in the graphs are the calculated responses from the simple response equation applied to the results of Figure 6.10.

The neutral wavelength 571 nm represents the transition point in the spectrum between activation and inhibition for cells without S-cone inputs. An 'L–M' cell is activated for longer wavelengths, and it is inhibited for shorter. For an 'M–L' cell it is the other way around. The blue-sensitive 'S–L' cells have a neutral point at about 515 nm, and the 'M–S' cells have two neutral points, one at about 600 nm and the other one for short wavelengths near 495 nm.

The same response equation that has been used to calculate the curves in Figure 6.10 and 6.14 has also been used to compute the responses of these cells to colorimetric purity (metric saturation between white and maximum chromaticity) for selected wavelengths in the spectrum. Examples of such responses are given in Figure 6.15, and examples of the combined responses to luminance ratio and purity for an 'L–M', I-cell cell are shown in three-dimensional plots in Figure 6.16. Again, the plot is in terms of response difference relative to the adapting background so the value for the white adapting background will therefore be zero.

The results of such computations for an isoluminant plane are shown in Figure 6.17. The color circle in the inset shows the approximate orthogonal directions of yellow (Y), red (R), blue (B) and green (G), with white (W) in the center. Computations for the same cell, and for different luminance ratios, are displayed as one vertical column, and the brightness of the disks illustrates response strength. For the 'S–L' cell to the left of the figure, we see that the brightest sector is at the bottom of the disk, centered around 477 nm. For the 'L–M' cells in the two right-most columns, maximum activity is in a sector between red and blue, and thus shifted towards purple relative to the

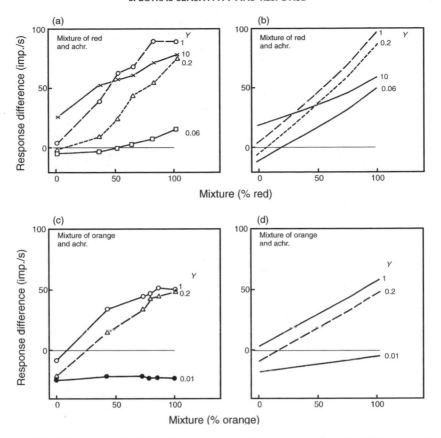

Figure 6.15 Left panels: the data points represent cell responses of an 'L–M' Increment cell to the exchange of a white adapting field with a coextensive red (a) or orange (c) test light. The response magnitude is a difference measure, where the response to the white adapting field is subtracted from the response to the test light. For each data set, the ratio, Y, between the test light luminance and the adaptation luminance is constant, while the colorimetric purity of the test varies. Right panels: the prediction of the model for the cell's response magnitude for the same stimuli.

unique red hue. The approximate hue of maximum response under our adaptive conditions is indicated in the figure by the color of the disks in the column for each cell type (Valberg and Lee, 1992).

'M–L' cells show the largest response for blue–green colors. In other words, neither for 'M–L' cells nor for 'L–M' cells does the response maximum correspond to a unique hue. For the 'S–L' cell, the maximum response is not far from unique blue, shifted only slightly to the violet side, and for the 'M–S' cells the maximum is close to unique yellow.

The I-cells shown here are activated, to a greater or lesser degree, by white light and by a surprisingly wide range of wavelengths. Because all cell types are activated by a broad spectrum of wavelengths, the discrimination of a color's hue most likely

'L-M' Increment cell

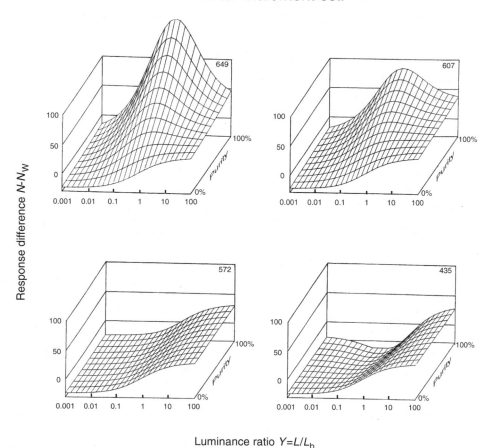

Figure 6.16 Model response magnitudes of an 'L–M' Increment cell for combinations of colorimetric purity and luminance ratios at four different wavelengths. Such three-dimensional plots summarize the results of Figures 6.10 and 6.15.

depends on the relative responses of at least two opponent cell types from neighboring groups in this figure. We postulate that, when this response ratio is constant and does not change with purity, hue will remain constant [see also Figure 5.32(c) and the Abney effect, p. 327].

Of all the opponent cells (about 1000) recorded in Barry B. Lee's laboratory at the Max–Planck Institute of Biophysical Chemistry in Göttingen over a 10 year period, between 75 and 80 percent belonged to one of the six groups described above. Among the remaining 20 percent, some cells had so little inhibition that an activation could be elicited by all wavelengths and all luminance ratios. Earlier reports of such cells (Padmos and van Norren, 1975) described how their 'hidden opponency' can usually be revealed by selective spectral adaptation. These cells may represent one extreme in

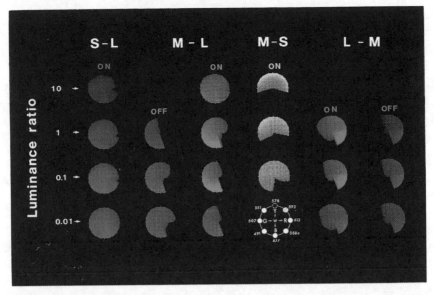

Figure 6.17 A representation of the responses of the six types of opponent cells of Figure 6.10 are given here as the density of bright spots in a polar diagram representing isoluminant color stimuli of different dominant wavelengths (indicated by radial angle) and spectral purity (indicated by distance form the center). Achromatic stimuli are represented at the center of each disk. Brighter areas indicate stronger responses than dark ones. The figure visualizes the neural responses shown in Figures 6.10 and 6.14 (except that the response to the white adapting reference has not been subtracted). I-cells (ON-center cells) fire more strongly to a bright color than D-cells (OFF-center cells) and all cells respond to a relatively broad range of wavelengths. The colors used for each disk indicate the spectral region/hue to which the cell responds best. Luminance ratio, $Y = L/L_b$, is between stimulus and adaptation background (adapted from Valberg and Lee, 1992). (See also color plate section.)

a continuum of different degrees of inhibition and might be called 'L–(M)' or 'M–(L)' Increment cells. At the other end of the scale, cells were so strongly inhibited that, for all wavelengths, only decrement responses were elicited. Consequently, they might be called '(L)–M' or '(M)–L' Decrement cells, depending on the dominant cone type. Other cells were hard to classify in terms of three cone inputs, either because their response characteristics were between those of the established groups, or because wavelength and luminance responses did not resemble anything that the model could simulate with a simple combination of cone potentials.

Quantitative simulations aimed at determining the cone inputs to opponent cells have demonstrated a great deal of variation (Derrington *et al.*, 1984; Valberg *et al.*, 1987). For instance, even some of the relatively simple 'L–M' and 'M–L' cells may have additional weak S-cone inputs (DeMonasterio, 1984; Valberg *et al.*, 1985b). De Valois (1969) used an S–L combination for his 'B–Y' cells, and L–S for the 'Y–B' cells, whereas Wiesel and Hubel (1966) concluded, as we have done, that the latter cell type received inputs mostly from M- and S-receptors. We have found a substantial number (between 3 and 4 percent; Valberg *et al.*, 1986b) of LGN cells with clear M–S inputs that were for a long time regarded as a missing link in opponent colour theories (Gouras and Zrenner, 1981). This cell type may now have been identified morphologically (Dacey and Packer, 2003).

After this introduction to a quantitative model of the responses of retinal and geniculate opponent cells across several stimulus dimensions, let us see how these ideas relate to psychophysical measures of unique hues, the Abney effect and the Bezold–Brücke phenomenon (Abney, 1910, 1913; von Bezold, 1873; von Brücke, 1878).

The opponent model and color perception

In the following we shall see how the functionally important nonlinearity in cone-opponent cells can be used to construct a physiological model of color vision. While linear models account nicely for near threshold data and for relative small changes in stimulus intensity and chromaticity, a nonlinear model is valid over a much larger stimulus range. Furthermore, a nonlinear model can successfully account for many color phenomena where linear models fail.

Munsell color scaling

Using the model of Figure 6.9, Figure 6.18 shows the combined response magnitudes of macaque geniculate 'L–M' cells to chromatic stimuli of Munsell hue 5R and different chroma and values. Solid circles connected with lines represent constant Munsell chroma. Figure 6.18(a) shows the chromatic responses of I-cells and Figure 6.18(b) that of D-cells to the same stimuli. The Increment cells in A are more responsive to the lighter colors of higher value, leading to tilted iso-chroma lines with positive slopes, while Decrement cells are the more responsive to darker colors of lower value, giving chroma lines with negative slopes. Because of these tendencies to prefer colors of different lightness, the averaged responses of these two types of cells, shown in Figure 6.18(c), yield close to optimal chroma lines, parallel to the achromatic axis with a few exceptions for the darkest colors.

Results similar to those of Figure 6.18 were also obtained for value–chroma planes of other hues, confirming that, along the reddish-greenish dimension of color space, the output of both I- and D-opponent cells are necessary to account for chroma scaling of colors of different lightness.

Unique colors

The colored wedges in the diagram of Figure 6.19(a) represent the model responses to the four unique hue stimuli, Y, R, B and G of increasing chroma (denoted 5Y, 5R,

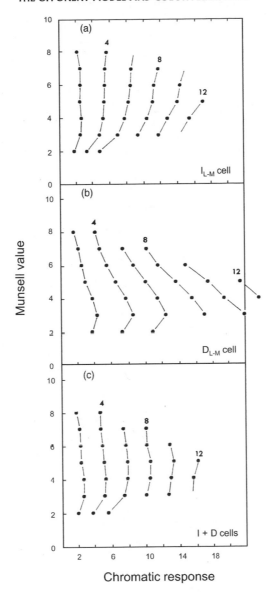

Figure 6.18 (a, b) Model responses of macaque I- and D-cells to the chromatic stimuli of Munsell hue 5R, varying in chroma and value. I-cells (a) have the higher firing rate to bright colors while D-cells (b) respond better to dark colors. (c) The linearly combined responses of (a) and (b) [as in Figure 6.9(c)] give close to equal spacing of chroma.

5PB and 5G in the Munsell hue system). For a given light adaptation, these stimuli correspond to the unique hues as determined by human subjects. The dashed ellipse in the same figure represents the loci of colors of constant chroma 8. The response magnitudes to the Munsell color stimuli have been computed for each opponent direction and plotted along the *x*- and *y*-axes. The response magnitude of the summed

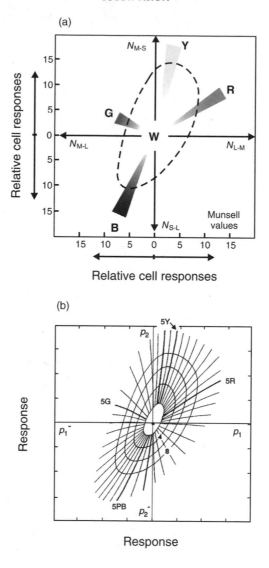

Figure 6.19 (a) The loci of the unique hues yellow, red, blue and green in a diagram combining the chromatic model responses, N, of the six opponent LGN cell types mentioned in the text. (b) The combined model response of the macaque opponent cells to colors in the Munsell color system, varying in hue and chroma (value 5). Loci of constant hue and varying chroma are approximately straight lines, while equal chroma is represented by quasi-elliptical loci. See also Color Plate Section.

I- and D-, 'L–M' cells is plotted along the positive x-axis and the response of the 'M–S' cells along the positive y-axis. 'M–L' cells are represented by the negative x-axis, and 'S–L' cells by the negative y-axis. This figure shows only the chromatic responses, i.e. the difference in firing rates between a chromatic stimulus and an achromatic stimulus of the same reflectance factor (same Munsell value).

In order to simplify the computation, the variability of the response within each class of opponent cells has been dealt with in Figure 6.19(a) by using data from cells that represent typical averages of each type (Valberg *et al.*, 1986b). This is equivalent to narrowing the spectral bandwidth of the 'population response' represented by each axis, and for this reason the unique hues have been represented by fan-like distributions in the figure.

Several properties of Figure 6.19(a) should be noted. The center of the coordinate axes represents the response to achromatic stimuli (that need not be zero) for all cell types. This magnitude has been subtracted from each response to obtain a 'pure chromatic response'. Activation responses are related to an increase in the magnitude of a certain percept, whereas inhibitory responses of the same cells imply less of the same, *not* the opposite quality.

It is immediately clear from Figure 6.19(a) that the cone opponent directions, represented by the four cardinal coordinate axes, cannot individually represent the unique hues. However, as is shown in Figure 6.19(b), a constant unique hue, and all other constant hues, approximate straight lines radiating from the white point, and can be approximated by a constant ratio of opponent cells responses. It may come as a surprise that unique red, for instance, falls nearly midway between the L–M and M–S axes. Chromatic stimuli that activate only one group of opponent cells, without causing any differential excitation in the other independent cell types, do not correspond to unique hues. Such stimuli, with responses along one axis and zero response in the orthogonal directions, have binary hues; we perceive their color as being composed of two unique hues. Increasing the chroma of one of these binary hues (e.g. a bluish red of Munsell hue 5RP along the L–M axis) would lead to differentiated responses in only the 'L–M' cells, whereas the orthogonal 'M–S' and 'S–L' cell types would react as if the stimulus were achromatic of the same value. For our adaptation condition, the same would be true for stimuli of hue 10G (activating only 'M–L' cells, stimuli of hue 2.5GY activating 'M–S' cells, and stimuli of hue 2.5P activating only 'S–L' cells). The neural correlate to white (and other achromatic colors) appears to be related somehow to normalized responses in all opponent cell types. Each hue – whether it is of elementary, binary or any other hue – depends on vector coding of separate opponent mechanisms. The hue lines radiating from the white point in Figure 6.19(b) are closer to straight lines in this cell representation than in the CIE chromaticity diagram (see Figures 5.28 and 5.29). This is a consequence of the nonlinear stimulus–response relationships of opponent cells. The fact that straight lines in the diagram of Figure 6.19(b) are reproduced as curved lines in the CIE (x,y) diagram gives an obvious explanation of the well-known Abney effect (see below).

The fact that unique hues are not represented by the four cardinal axes in the diagram may come as a disappointment to those who have grown up with the opponent theory of color vision. Nevertheless, and not surprisingly, the ratios of the 'chromatic responses' of different types of opponent cells are clearly important for hue perception. A straight line, radiating from the white point, represents a constant ratio of the responses of orthogonal cell types. At the same time, it represents a single hue where distance from the white point is proportional to chroma. These results are in accordance with earlier psychophysical data which also demonstrated non-correspondence between cardinal axes and unique hues (Valberg, 1971; Krauskopf *et al.*, 1982; Burns *et al.*, 1984; De Valois *et al.*, 1997).

Under other experimental conditions, with a different adaptation or with a different surround, a unique red color, for instance, would be associated with another wavelength or spectral composition. Despite this, there is a theoretical possibility that the same local and lateral feedback that contributes to color constancy would counteract adaptation changes in the receptors and single cells – and thus provide for the same response ratio between opponent cells whenever we see the same hue. We do not yet know to what extent these opponent cells' relative firing patterns are linked to the physical properties of the stimulus projected onto their receptive fields, or are being modified (adapted) by long-range interactions, as would be necessary for 'color constancy' to occur (but see Valberg *et al.*, 1985a). If the phenomenon of color constancy is, in part, governed by processes beyond the retinal and geniculate levels (according to the results of Land *et al.*, 1983), changing adaptation would lead to unique red being represented by different response vectors under altered states of adaptation. So far, unique hue directions have not been found to have a particular status in the cortex.

In conclusion, in a normalized viewing situation, there is a covariation of a specific chromatic response ratio of opponent cells and our perceiving *a constant hue* irrespective of chroma. Generally, however, we cannot know which hue is associated with a particular ratio, only that the hue stays the same as long as the cell response ratio does. The chroma scale of increasing color strength corresponds to equal increments along each hue vector [but not the same increment for all vectors; see Figure 6.19(b)]. Theoretically, a general correlate to color strength can be approximated by a linear combination of the cardinal axes p_1 and p_2 of Figure 6.19(b), transforming the elliptical shapes to circular shapes:

$$F_1 = 1.5p_1 - 0.5p_2$$
$$F_2 = 0.7p_2$$

The p_1 axis is negative in the M–L direction and the p_2 axis is negative in the S–L direction. As shown in Figure 5.32, such a transformation leads to steps of equal hue approximating equal angles, and equal color differences becoming nearly equal geometrical differences. However attractive such a mathematical/cortical solution might appear, nature nevertheless may have adopted an alternative strategy in primates. For instance, neurones tuned to several chromatic directions other than the cardinal axes have been found in primary visual cortex of primates (Lennie *et al.*, 1990) and at later stages (Kiper *et al.*, 1997; Wachtler *et al.*, 2003).

Under normal conditions, no color is at the same time perceived as red and green or as yellow and blue. The mutual exclusiveness of the unique red and green hues, and of unique yellow and blue, has no simple geometrical counterpart in the cell response spaces of Figures 5.32 and 6.19 (thus being different from complementary colors in stimulus space). Although the cardinal directions p_1 and p_2 (as well as F_1 and F_2 of Figure 5.32) seem to account for variations in chromatic scaling and color discrimination, unique hues are still without a neural representation, and their physiological origin remains enigmatic. The argument could be made, of course, that the reason for our bewilderment about colour qualities, and particularly about the unique hues, is that monkeys are different from humans, or because neurons in anesthetized animals respond differently from those in awake animals. The first objection is hard to maintain, considering the many demonstrations of psycho-physiological similarities of color vision in primates. Except at the peripheral levels of visual processing, the second argument may, however, be relevant.

The Abney effect

Anyone with experience of additive color mixtures will have noticed that, when white light is added to a colored light of high purity, the hue of the mixture will change (Abney, 1910, 1913). As noted earlier, provided luminance is kept constant, even a small amount of white added to unique red will give the mixture a bluish-red appearance. Increasing amounts of white will intensify the blue shift up to a point, after which the blue component diminishes. When white is added to a saturated orange, the hue becomes more reddish, and when added to yellow–green, it becomes more green. These hue shifts, also known as the Kohlrausch effect, are represented in the chromaticity diagrams of Figures 5.28 and 5.29.

The calculations for hues by the opponent model are shown in Figures 6.19(b) and 5.32(c). Perfectly straight lines in Figures 6.19(b) and 5.32(c), like those representing slightly yellowish-red or a yellowish-green, and purples and blue-greens, are reproduced as curved hue lines in the (x,y)-diagram of the CIE system of Figures 5.28 and 5.29. Although not all the hues are represented as perfectly straight lines by the model, straightening them would lead to a slightly stronger curvature than they already have in the chromaticity diagram, in the direction predicted by the Abney effect. We therefore conclude that the nonlinearity of the cell responses in the model describes the Abney effect adequately.

The Bezold–Brücke phenomenon

Another well-established phenomenon is that, when the luminance of foveal chromatic light increases from zero, there is first an achromatic interval between light detection and the identification of its hue, after which chromatic strength increases and the perceived hue steadily changes. Long-wavelength red light becomes more yellowish and short-wavelength violet light becomes more bluish. This latter phenomenon is known as the Bezold–Brücke phenomenon (von Bezold, 1873; von Brücke, 1878; Purdy, 1931a; Stabell and Stabell 1982b).

It is less well known that this hue shift is accompanied by significant changes in the perceived chromatic content of the stimulus. When the luminance of a spectral light increases, the chromatic strength first increases and then decreases in a wavelength-specific manner (Haupt, 1922; Purdy, 1931b; Vimal *et al.*, 1987). Such nonlinearities of color vision occur, for instance, in colored light sources and in self-luminous color displays of various kinds where luminance ratios relative to an adaptive surround or a background can vary over a much larger range than for (related) surface colors. Here we shall describe how the combined changes of hue and saturation affect the color appearance of near-monochromatic lights over a 4–5 log unit range of intensity above the chromatic threshold.

A common theoretical framework for explaining these phenomena has long been sought by the vision science community. Bezold–Brücke hue shifts have been explained by the saturation of the intensity–response curves of the cones themselves (Walraven, 1961; Cornsweet, 1978), or by the relative activities of 'chromatic and achromatic signals' (Judd, 1951b; Hurvich and Jameson, 1955). Explanations in terms of compressive nonlinearities of receptor excitations (e.g. von Helmholtz, 1962) are rather general, and such models have not been demonstrated to predict the experimental data for the combined hue and saturation appearances of individual chromatic lights.

In the discussion below, these hue shifts are linked to the Abney effect by a common physiological mechanism. The hue shifts are readily accounted for at a postreceptoral level by the combination of non-monotonic intensity–response curves of different types of cone-opponent cells. From the assumption that both the chromatic content of a stimulus and its hue are related to the relative activation of parvo- and koniocellular opponent cells (Valberg *et al.*, 1986a), it follows that the dependence of the perceived magnitudes on stimulus intensity must be understood in the same framework.

In order to demonstrate the validity of this hypothesis, let us first present psychophysical data on the combined color shifts that occur when the relative luminance changes. Figures 6.20 and 6.21 exemplify the estimates of hue and chromatic strength of several different stimuli, in which the luminance changes relative to a white adapting reference background. The experimental results show significant inter-observer variation in the details, but the general trend is in accordance with the physiological conjecture that chromatic changes can be accounted for by the relative activation of opponent cells.

The observers viewed 4° stimuli of increasing luminance flashed on a screen for 300 ms in alternation with a white (100 cd/m^2) adaptation field of 1.2 s duration. The white stimulus in this successive contrast paradigm served as a constant reference and set the adaptation level during the experiment. The apparatus and the procedure were the same as that used in our neurophysiological recordings described earlier (see

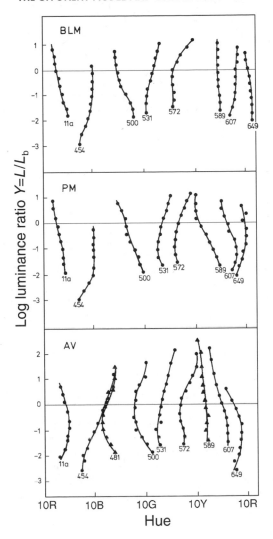

Figure 6.20 The hues of the spectral colors change as their luminance ratio relative to an adaptation light increases. In this experiment, the hue of a light with a given wavelength is assessed at different luminance ratios. A wavelength of 572 nm normally appears yellow–green when the luminance ratio is low. As the luminance ratio increases, it become less green and approaches the yellow elementary hue at a high luminance ratio. No wavelength maintains a constant hue for luminance changes over a range of 3–4 log units (although 589 nm comes close to doing this for subjects B.L.M. and A.V.). These data were obtained in a stimulus situation like that used in the neurophysiological experiments described in Figure 6.7.

Figures 4.14 and 6.7 and Lee *et al.*, 1987). The CIE 2° (*x,y*) chromaticity coordinates of the white light were (0.388, 0.406). The presentation was monocular and foveal. The luminance of the chromatic stimuli was changed with neutral density filters over a range of 5 log units in steps of 0.3 log unit from 0.1 to 10 000 cd/m². The surround was dark in the experiments reported here. No artificial pupil was used.

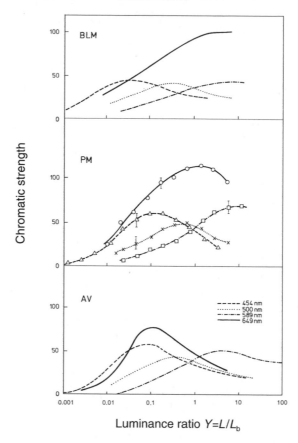

Figure 6.21 Chromatic strength also changes with luminance ratio. Starting at low values, color strength increases with luminance ratio, typically reaching a maximum which is followed by a slow decline. Maximum color strength is reached at different luminance ratios for different wavelengths: at low ratios for short-wavelength, blue lights (e.g. 454 nm) and at high ratios for mid-spectral yellow (589 nm).

The observer was asked to estimate two attributes of each stimulus, the hue and the chromatic strength, but only one attribute in each session. Hue was judged in terms of the contribution of two neighboring elementary (unique) hues (yellow and red, red and blue, blue and green, or green and yellow) in terms of two values that added up to 10. Elementary hues, 10Y, 10R, 10B, 10G in Figure 6.20, obeyed the usual neither–nor criterion (e.g. elementary yellow, 10Y, is neither reddish nor greenish). In this way a reddish orange light could be characterized, for example, as 3Y and 7R. Stimulus luminance were presented in a pseudo-random order.

In scaling the chromatic content of a stimulus, the observer was instructed not to pay attention to the intensity changes in the stimuli, but to estimate only 'the perceived chromatic difference relative to an equally bright achromatic stimulus'. This is essentially the definition of Munsell chroma (Wyszecki and Stiles, 1982), but in this context we call it 'chromatic strength'. We deliberately did not use the term 'saturation', since saturation is often used in the sense of a ratio between chromatic and achromatic components of light (either perceptual or colorimetric).

The magnitude of the perceived chromatic strength of the stimulus was rated by the observer on a scale from 0 to 100, where 100 was supposed to be the absolute maximum for the most saturated hue. In one case a number greater than 100 was necessary. Further details are found in Valberg et al. (1991b).

Following an established tradition, the hue ratings in Figure 6.20 are plotted on the abscissa while the ordinate refers to the ratio between the luminance of the stimulus (L) and that of the white (100 cd/m^2) adaptation field (L_b). With increasing luminance ratios above 1.0 (the ratio when the chromatic stimulus and the white adaptation field both had a luminance of 100 cd/m^2) the curves of the stimuli between 531 and 649 nm converge towards yellow. For luminance ratios below 1.0, orange and greenish or greenish-yellow hues dominate increasingly, although some inter-individual variation was seen.

Even though we measured color changes as a function of (a successive) luminance ratio, and not as a function of absolute luminance, the directions of hue shifts are in general agreement with earlier observations reported in the literature for comparable luminance ranges. However, several studies have concluded that the elementary hues are invariant with intensity, whereas others have provided evidence that they generally do vary (Purdy, 1931a; Savoie, 1973; Nagy, 1979). None of the stimuli of Figure 6.20 were hue-invariant for all observers. There is, of course, a slight possibility that invariance was missed due to the sampling of wavelengths, but it is also possible that hue-invariant wavelengths are only found for a restricted luminance range; we used a constant white adaptation stimulus and investigated a much larger luminance range than usual. Note also the different positions of the stimuli on the hue axes of Figure 6.20 for the three subjects. Both position and the relative hue shifts with intensity were subject to inter-observer variation.

In Figure 6.21 the ratings of chromatic strength are illustrated for a few selected wavelengths and for the same observers as in Figure 6.20. Reproducibility was good for all observers, and similar to that shown for subject PM. The curves show a rise and a fall of chromatic strength as luminance increases. Short-wavelength stimuli reached maximum chromatic strength at low luminance ratios (L/L_b) of about 0.1 or below. For the 500 nm stimulus, maximum chromatic strength was found at a luminance ratio close to 0.3 for all observers, and for the midspectral yellow stimulus of 589 nm, the maximum has moved further to the right to about 10 times the luminance for the white adapting reference. Such a phenomenon of hue shift with intensity can be observed in colored light bulbs, where the bright filament appears nearly white and the darker colored glass, at some distance away from the filament, has a much higher color strength.

The ratings of hue and chromatic strength for some of the wavelengths of Figures 6.20 and 6.21 are combined in polar plots in Figure 6.22. In this diagram,

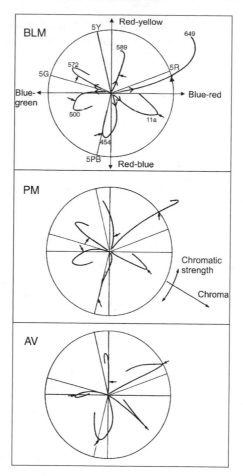

Figure 6.22 In this figure the results of Figures 6.20 and 6.21 are combined in a polar diagram. In these plots, each curve represents the hue and color strength of a given wavelength for a range of different luminance ratios. Subjective color strength is proportional to the distance from the center, and constant hue corresponds to a particular radial angle. The elementary hues, here denoted 5*Y*, 5*R*, 5*PB* and 5*G*, have the same position as on the hue circle of the Munsell system. The small arrows point to the coordinates for the subjective impression of color strength and hue for a luminance ratio $Y = L/L_b = 0.1$. Since color strength decreases for lower luminance ratios, all the curves converge towards the origin as Y drops from 0.1 to 0. For stimuli of constant chromaticity we perceive a change in color strength and hue as luminance ratio increases from zero.

the relative chromatic strengths are proportional to radial distance from the origin. These distances have been normalized so that for all observers they have equal length for the 649 nm stimulus at a luminance ratio, L/L_b, of 0.1, as indicated by the circle. Luminance increases from the origin as is indicated by the chevrons in the top figure.

From this figure, we see that, as luminance increases from zero, short-wavelength violet light becomes more bluish (and finally bluish-green for AV), whereas long-wavelength stimuli usually turn more yellowish at high relative luminances. All five observers found that the green 500 nm stimulus became more bluish as luminance increased, and all but one found that the purple stimulus 11a became reddish. Despite the obvious inter-individual differences in the color estimates, the similarities outweigh the variability.

The relative magnitudes of maximum chromatic strength deserves some comments. The magnitude was largest for the red 649 nm light. The yellow 589 nm light, which is normally the least saturated in a low-luminance spectrum, had a surprisingly high maximum chromatic strength, but it peaks at a higher luminance ratio than for the other stimuli (see Figure 6.21).

Below we shall explain these color shifts within the same framework as that used to describe color scaling. In Figure 5.32(c) it was demonstrated that color strength (chroma) and hue are closely related to a linear transformation of the response magnitudes of primate color-opponent cells. The predictions of this model are in agreement with experimental results above, without additional assumptions.

Theoretical implications

The essential assumptions of the model are:

1. Opponent cells adapt to an extended light surround (the condition of related colors).

2. Responses of I- and D-cells with the same cone opponency are added.

3. Orthogonality (independence) exists between 'L–M' and 'M–L' cells on the one hand and 'M–S' and 'S–L' cells on the other.

4. Perception of the same hue corresponds to a certain ratio of responses of independent opponent cells irrespective of their absolute response (this would mean that perceiving a certain orange hue of varying chromatic strength implies that the ratio of the responses of 'M–S' and 'L–M' cells is constant).

5. Chromatic strength corresponds to a vector sum of the responses of the opponent cells.

6. The response to achromatic white light can be subtracted in a way that allows us to define a pure 'chromatic response'.

7. The coordinate axes (p_1, p_2) of retinal and geniculate opponent cells combine linearly to new coordinates (F_1, F_2) so that equi-chroma ellipses are transformed into circles about the origin.

The example of Figure 6.23 explains how the model is applied. The figure traces the chromatic responses of two postulated opponent units, F_1 and F_2, to a stimulus of wavelength 649 nm (upper and lower left panel). Chromatic response is defined as the

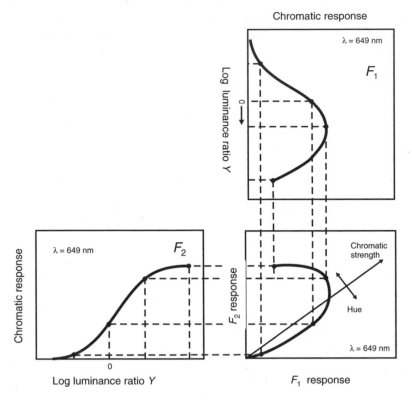

Figure 6.23 A nonlinear neural vector model of color vision that accounts well for results like those in Figure 6.22. This schematic figure explains how two postulated opponent cell types, F_1 and F_2, respond to a 649 nm stimulus at different luminance ratios. The response of the F_2 system is shown on the lower left, and the response of the F_1-system is shown at the top. When combined in the diagram to the lower right, these two responses lead to a curve resembling those of Figure 6.22 for the same wavelength. The model predicts that a constant ratio between the responses of the two hypothetical cell types, which gives a straight line in the response diagram to the lower right, would represent the same hue. This postulate is confirmed by Figures 6.19 and 5.32(c).

difference in firing rate of an opponent cell to a chromatic and an achromatic stimulus of the same luminance. In the lower right panel, the responses of these two units are added as vectors. With increasing luminance, the resulting response curve turns left towards the F_2-axis, a consequence of the different dependency on luminance of the two units' responses. For example, F_1-units reach saturation at much lower luminance than do the F_2-units. Since the response to white light is already subtracted, the figure gives a pure chromatic response (the responses to white light are shown in Figure 6.10, and schematically in Figure 6.11).

Since for every luminance level 'L–M' and 'M–L' cells have rather similar responses to white light, one could imagine that pure L/M chromatic responses can be found by taking two differences of responses of these cell types ($N_{L-M} - N_{M-L}$ and

$N_{M-L} - N_{L-M}$). Whenever these differences were both zero, some putative higher-order mechanism would signal a correlate for an achromatic color, with the relative responses of I- and D-cells determining the actual gray percept. A sum of the responses of the L/M cell types, in some combination with opponent cells with S-cone inputs, might correlate with brightness (since the response of a chromatic light would be higher than that for the achromatic one of the same luminance). Where in the visual pathway such a possible separation of 'white responses', 'chromatic responses', and 'brightness responses' would occur, if they occur at all, is an unsolved issue, but nothing like it has been observed in the retina and the LGN.

The response magnitudes of the model (composed of the six most typical cells, of which examples are shown in Figure 6.10) to stimuli of constant chromaticities but different luminance ratios are shown in the polar plot of Figure 6.24(a) and (b). While Figure 6.24(a) covers the same luminance range as Figure 6.22, Figure 6.24(b) extends the responses of Figure 6.24(a) to very high luminance ratios. The curves in the diagram are projections on to a plane, and radius vector in this plane is proportional to chromatic strength. The orientation of the vectors relative to the coordinate axes is related to hue. The figure shows that color strength initially grows as luminance increases from zero, reaches a maximum for a luminance ratio that is characteristic for each hue, and then decreases again. The model predicts [Figure 6.24(b)] that at very high intensity all chromatic response is lost, and the stimulus will appear achromatic.

The predictions of the model for the combined changes in chromatic strength and hue are very similar to the experimental results plotted in Figure 6.22. In accordance with the measured Bezold–Brücke hue shifts, the model predicts that a stimulus that appears reddish at low luminance will turn more orange and finally yellowish at very high relative intensity. At some intermediate intensity, chromatic strength reaches a maximum, and for the highest intensities, the light becomes whitish. The typical directions of the hue shifts in Figure 6.22 are well predicted for all hues, with the exception of the 500 nm stimulus. This overall agreement demonstrates that the Bezold–Brücke hue shift and the related change of chromatic strength have a common origin in the non-monotonic responses of retinal and geniculate opponent cells.

We have described the perceived changes of chromatic strength and hue which occur as stimuli of constant chromaticity increase in luminance. The successive contrast paradigm resembles normal viewing conditions, with eyes successively fixating objects of different color and relative luminance. Together with the color scaling, the Abney effect and the Bezold–Brücke phenomenon can be well accounted for by the nonlinear and non-monotonic responses of opponent cells in the retina and the LGN.

Judd, as well as Hurvich and Jameson, has referred the correlate of hue perception and the Bezold–Brucke phenomenon to the relative activity of two neighboring, red–green and yellow–blue 'primary processes' (Judd, 1951b; Hurvich, 1981). Here,

(a)

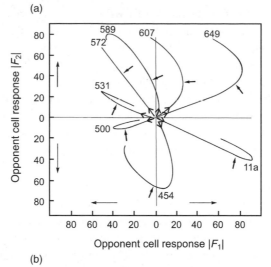

(b)

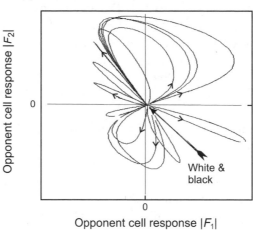

(c)

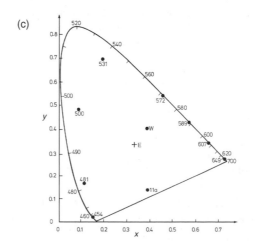

we have related the joint hue and chroma changes to a combination of cone-opponent system responses, which, individually, exhibit no simple relationship to the elementary hue qualities.

The decreasing chromatic strength of chromatic stimuli at high luminance has interesting theoretical implications. It has been suggested that it is caused by the relative activity of separate chromatic and achromatic processes, the achromatic signal being dominant at high intensities. Our model, however, points to an alternative unifying explanation for the Bezold–Brücke phenomenon and the related changes of chromatic strength. These phenomena, and the Abney effect, were here accounted for by combining non-monotonic outputs of low-level opponent cells, a process that may take place at the cortical level. The non-monotonic, composite responses of the model of Figure 6.24(a) fit the same functions of luminance as do psychophysical chromatic strength and hue in Figure 6.22. These responses are modeled by a sum of activating and inhibiting cone inputs to opponent cells. There is no need to invoke an early achromatic mechanism to explain these results; both achromatic and chromatic response components can be derived from combinations of outputs from the same opponent cells. However, one would like to find the physiological substrate that isolates and processes the 'achromatic component', but this appears currently to be as difficult as localizing the physiological correlates to unique hues. The ideas presented here support the notion that, once a chromatic threshold is overcome at low luminance, it is the response magnitude of the combination of chromatic responses of opponent cells that alone sets the correlate for chromatic strength and hue. In addition, the excitatory profile across these color-selective cells probably contains the code for lightness and brightness attributes.

We can thus view the joint changes in chromatic strength and hue as a result of low-level, opponent cell activities and their combination at a higher, cortical level. Both chromatic and achromatic colors and intensity attributes are implicit (multiplexed) in the activity of these early cells, and cortical mechanisms may derive several perceptual attributes of the retinal image by analyzing such multiple converging and diverging pathways. The 'cell opponent stage' of Figure 6.24(a) suggests that, after the outputs of I- and D-channels have converged on some cortical mechanism, they may again be related in new opponent pathways. Theoretically, this is one simple possibility for arriving at an equidistant color space [see Figure 5.31 and

----

Figure 6.24 (a) Predictions made by the neural color vision model for the changes in perceived hue and color strength of eight stimuli of constant chromaticity (shown in c) when luminance ratio is raised from 0 (origin) to 0.1 (solid arrows) and beyond. The distance from the center of the diagram is proportional to model color strength, and the radial angle represents hue. The units for cell responses are scaled in such a way that they correspond to Munsell chroma. The model gives a similar picture of variations of hue and chromatic strength as Figure 6.22. (b) Extrapolation beyond the experimental data shows that for extremly high luminance ratios monochromatic stimuli will appear achromatic.

5.32(c)]. A relationship between achromatic and chromatic 'primary processes', such as the one postulated by Judd and by Hurvich and Jameson, is not likely to be found at the early levels since retinal and LGN opponent cell responses cannot be directly linked to the perception of elementary hues (Valberg, 2001).

It may be of interest to ask what consequence the relative number of L and M cones in the retina may have for hue perception? Does the excitation and response amplitude of opponent cells, and hence chromatic strength, depend on the relative number of each cone type having input to the cell, or maybe to the number of synaptic contacts? In a retina where the L-cones dominate (or have more L-cone synapses), would the 'L–M', Increment and Decrement cells (the red ON-center cells and green OFF-center cells) be more excitable (except in the fovea where there is a one-to-one relationship), or would there be more such cells? Would this lead to a stronger input to higher-level 'L–M'cells, and to stronger inhibition of high-level 'M–L' cells. The relation between cone proportions and their synaptic contacts to ganglion cells and hue perception is by no means clear. For instance, there is no clear numerical relationship between relative cone density and elementary hue.

Summary

In this chapter, we have explained why a match of any colored light can be achieved using only three base colors, and that this is readily explained by the excitations of three types of cone. A match is obtained when, and only when, each of the three cone types is excited equally by the two lights being compared. Systems for color measurement, like that of CIE, rest upon this principle of equality. Such systems, however, say little about the colors actually perceived in a match.

In accounting for perceptive color phenomena, receptor excitations and linear models of neural processing are bound to fail except for near threshold. Neurophysiological research has brought major advances in our understanding of the underlying mechanisms of color vision, but much remains to be discovered. We still have no complete physiological model of color vision, but the model we have described here has many promising features. In addition to describing color differences and color scaling in an encouraging way, it accounts well for constant hue perception, the Abney effect, and the Bezold–Brücke phenomenon (considering the large individual differences that exist in the psychophysical data).

When a spectral light increases in luminance, the hue changes. Normally, long-wavelength light becomes increasingly yellow, and short-wavelength light turns blue or blue-green. Less attention has been paid to the change in relative chromatic content (saturation or chromatic strength) that accompanies these hue shifts. As luminance increases from zero, chromatic strength grows to reach a maximum at a luminance that is wavelength-dependent. Short-wavelength bluish light reaches this maximum at low relative luminance, whereas mid-spectral yellowish stimuli need several log units higher luminance. Red and green fall somewhere in between. For a luminance above this maximum, the chromatic strength usually diminishes, and most wavelengths become more whitish. Both phenomena can be accounted for by a physiological interpretation in terms of the nonlinear and non-monotonic responses of cone-opponent cells in the retina and lateral geniculate nucleus of the primate. Our model, that combines the outputs of six opponent cell types, accounts fairly well for observers' estimates of hue and chromatic strength.

However, further explanation is required. For instance, we would like a more complete model to explain:

- color adaptation;

- simultaneous color contrast;

- color constancy, or more specifically the deviations from color constancy (being of greater practical importance);

- border and area contrasts.

If one believes, like Ernst Mach did, that a unique color perception has a unique neurophysiological correlate, then one must also expect a future model to incorporate a structure of

- unique hues and elementary colors; and

- of white and black

The last two points refer to qualities of immediate color experiences, phenomena that relate to a different level of experience than do those mentioned earlier.

We have argued that, for the same adaptation, combined neural activities of opponent cells correlate well with the perception of *constancy* of hue and chroma, whereas the elementary hues qualities cannot be associated with these neural processes. This is a general statement, not restricted to the model presented here. The old idea that the elementary colors have a direct relation to activation and inhibition of early opponent cells, either in the retina or LGN, does not hold. As yet, no other physiologically plausible model has come up with good neural correlates. What conclusion should we draw from this lack of correspondence? Is the program that Crick (1994) announced to find functional parallels between neural activity and perception proving more difficult than one had anticipated at least for color vision? Perhaps neural correlates for the qualitative basis of color vision – such as the unique colors – do not belong to Chalmer's easy problems after all (Chalmers, 1995a, b; Valberg, 2001).

When the experience of colors is related to physical stimulation (even though qualities cannot logically and causally be derived from it, since physics and perception are on different sides of the 'explanatory gap'), it should be possible to establish neurophysical and neuropsychological relationships at different levels in the visual process. For instance, a number of phenomena and experiments point to underlying opponent mechanisms. This applies to adaptation to the prevailing illumination, color discrimination, the organization according to attributes (dimensions) in a three-dimensional color space, and movement. In all these cases, the neuro-scientific program still seems to be adequate.

Lightness, hue and color strength (chroma) are abstract dimensions (attributes) of colors, and they have a simpler connection to known opponent neural processes than perceived color qualities. These three dimensions can be given geometric representations, and color discriminations made along these dimensions can be derived form neural activities. In blind-sight it is possible for a person to discriminate between colors without having a conscious knowledge of seeing them. If asked, the person will firmly deny that he has seen any color at all. These rare cases suggest that stimulus discrimination may depend on rather peripheral processes, since some information can be available without consciousness perception. The processes in the primary visual center, V1, for instance, seem largely to escape consciousness (Zeki, 1993; Crick, 1994; see also Chapter 7).

We have offered a physiological explanation of color discrimination in terms of equal differences in responses of cone-opponent units, but the qualitative opponency and independence of elementary colors still remains a problem. Referring to what we have just said about blind-sight, this is perhaps not so strange, considering that elementary colors are qualities associated with conscious experience, whereas electrophysiological recordings are done on anesthetized monkeys. Unique colors may be regarded as an inner, subjective, but nevertheless common reference system for color perception for all people. However, we are not primarily seeking an explanation of these qualities, only for the particularities of some unique directions and structures in psychophysical color space. We had hoped – and maybe expected – that they would be revealed in the firing of visual neurons.

We are free to interpret colors as expressions of physical and chemical processes, and also as a product of the reaction of the brain and the neural system to stimulation of the eye by light. Colors reveal themselves as one of many attributes of our visual world, but they can also be treated quantitatively and be included in an objective, scientific description at the same level as form, movement, depth, texture, orientation and direction. When I experience colors, organize them, and explore their relation to physical stimuli and neural activity, this happens within logical structures and with reference to my culture's understanding and knowledge. While I would maintain that my theoretical analysis of color phenomena and my qualitative experience of them are two complementary activities, and that they require different concepts and verbal expressions, I recognize that this is rooted in my preliminary understanding of the phenomenon color. Neuroscience makes it possible to investigate these and other relationships experimentally, and, as culture and science develops, we expect our understanding to change, as has been the case from ancient times to today.

Color and art in a scientific perspective

When focusing on the evolutionary advantages of trichromatic color vision, one usually explains its development from previous achromatic vision by the advantage it provided for better discrimination of fruits from foliage, to discover enemies, etc. Beside this biological advantage, is there no other value to color vision? What about perception and esthetics? The auditory sense, for instance, together with our intellectual and emotional capacities and the faculty to develop culture, has made it possible for us to appreciate a rich world of sound and music. Color vision obviously provides humans with yet another esthetic dimension. It has enabled us to produce new visual experiences and to admire and appreciate colors and texture, on clothes, in art and images, on artifacts, and in nature. The access to the dimension of color has opened our minds to a wealth of new esthetic adventures, much in the same way as our ability to perceive sound has allowed us to enjoy music.

7 Neural correlates

Low- and high-level neural correlates

Throughout this book, there have been several references to neural coding of visual attributes and to the co-variation or correlation of perceptual properties and neural responses. Before we go on to present some more data on such correlations, we need to consider in more detail what correlation and co-variation implies. We shall discuss some of the principles and hypotheses that link neural activity to perception. There are several general issues that should be addressed. For instance, what are the relevant coding principles and strategies? How does the brain represent the visual world? Another, more specific question is: how does the brain represent objects and object properties, such as movement and color?

In an answer to such questions, Horace B. Barlow (1972) postulated that

> ... perceptions are caused by the activity of a rather small number of neurons selected from a very large population of predominantly silent cells. The activity of each single cell is thus an important perceptual event and it is thought to be related quite simply to our subjective experience .... A description of that activity of a single nerve cell which is transmitted to and influences other nerve cells, and of a nerve cell's response to such influences from other cells, is a complete enough description for functional understanding of the nervous system.

This *neuron doctrine* has led to great advances in the study of the function of single cells. In vision, the idea that the function of the nervous system can best be described at the level of single cells has gained support from a series of experiments on the detection of object properties using weak stimulation, close to visual threshold. Comparisons of psychophysical and neural sensitivity to several dimensions of light and color indicate that, for a particular task, it is the most sensitive cells that determine psychophysical threshold.

Light Vision Color. Arne Valberg
© 2005 John Wiley & Sons Ltd

The alternative, and more general hypothesis, that objects and their properties are represented by the activity of an ensemble, or a network of cells (that can be distributed over several areas of the brain), is attractive for more complex stimuli and for stimulus intensities above threshold which engage more cells. Generally, the most striking support for Barlow's hypothesis has been provided by comparison of psychophysical sensitivity with the threshold sensitivity of peripheral neurons, for example, in the retina. Cortical recordings have been more successful in finding correlates to higher level functions, where some form of convergence and distributed processing is likely.

Neural representations

As we shall see, different parts of the brain form specialized areas and handle different kinds of visual information. Recent studies using functional magnetic resonance imaging (fMRI) have shown that many of the locations that respond during active vision also are active when memorizing the same events (e.g. in mental imagery with closed eyes).

Neurons at and after the ganglion cell level in the retina, respond to varying stimulus magnitudes with a sequence of discrete and identical action potentials (spikes), and they are generally more selective and stimulus-specific than retinal cells. These neurons respond to optical stimulation related to certain stimulus properties, and their firing rate in impulses/s changes with stimulus specificity, intensity and contrast. Activation is balanced by inhibition, and adaptation and habituation appear to prevent the cells from being excessively stimulated. We have indicated the usefulness of applying detection and discrimination sensitivity as a means of comparing psychophysics and physiology. For a reliable physiological measure of sensitivity we must consider the cell's responsiveness to a particular stimulus and determine the change in firing rate in response to a stimulus increment or decrement. The noisiness of the cell and the level of the maintained discharge must also be considered when defining a threshold criterion.

One may ask if the number of nerve pulses per unit time is a complete and sufficient description of the neural code, or whether other forms of information transfer between nerve cells are possible? The traditional view is that a cortical neuron, for instance, is an integrate-and-fire device. However, the time interval between successive spikes from one cell and the temporal correlation of action potentials from different cells can also be considered possible sources of information. Synchrony between spikes or impulse trains from different cells (and cell assemblies) might be registered by neural elements serving as coincidence detectors. However, at the level of single cells in the retina and in the geniculate it looks like firing rate is a fairly optimal code. At the higher levels of more composite and organized neural networks, it seems that additional means of information transfer are possible.

A distinction is often made between local and parallel distributed processing – or between local representation and vector coding. In the field of artificial intelligence (AI) there has been a great deal of attention to the properties of so-called 'neural networks' and their achievements in pattern recognition. These theories have also influenced our thinking about brain processes. Neural networks consist of an assembly of interconnected units, and they can be used to simulate processes in simple biological systems and to test theories of how parts of the nervous system interact. Groups of related neurons in the brain can be thought of as equivalent to local processing units. The functional state and the output of a neural net depend on the *weights* allocated to the interconnections between the different elements. The weights determine the relative contribution of each of the individual elements to the total response. These weights remind us of the Hebbian synapse (Hebb, 1949) between nerve cells, i.e. of connections that are strengthened by particular stimulus or activation patterns and frequent use. Plasticity can be represented by changing weights within a network, resulting in another input vector to output vector transformation. Whether or not this theory applies to the brain, the concept of an abstract multidimensional vector representing an object, instead of single elements, is central to distributed representation. This idea has proven extremely successful in image analysis and pattern recognition. The use of several dimensions allows us to define different n-dimensional feature spaces (Figure 8.9), and the use of numerical values (i.e. a set of numbers signifying magnitudes) allow a simple definition of similarity within a certain feature domain and between feature domains. In biological networks, with different areas of the brain being active simultaneously by composite stimuli, this leads to the so-called 'binding problem' of synchrony of integrated information distributed over several feature spaces (see Chapter 8).

The representation of the excitation of three types of cone in a three-dimensional color space is a particularly simple example of a sensory vector space. At the level of retinal ganglion cells, cone outputs are linearly transformed into another, opponent color vector space with cardinal axes, which is retained in the LGN and is further transformed in area V1. In this particular case, the early cone transformations lead to individual ganglion cells representing the resultant vectors.

At a subcortical level, a wealth of data has confirmed the idea of local coding, i.e. that correlates of stimulus attributes can be found at the level of single cells, as Barlow (1972) suggested. Several examples will be given later. It is central in Barlow's dogma that 'there is nothing else looking at what the cells are doing, they are the ultimate correlates of perception' (Marr, 1982). Let us use color as an example: local coding would imply that there is a particular, specialized cell type with a narrow spectral bandwidth, which responds in conjunction with the perception of a certain color, e.g. an 'orange coding cell'. Another type of cell would respond to a similar, but slightly different orange color (e.g. carrot orange vs apricot orange). Such local hue coding would be analogous to orientation coding with highly orientation-selective cells such as one finds in area V1 (see Chapter 8), where different cells respond to different orientations with a narrow angular resolution (of about 10°). However, is this feasible in the case of color?

Imagine an alternative form of orientation coding similar to our postulated low-level hue coding using cardinal axes. It would utilize combinations of the relative responses of two populations of neurons sensitive to orthogonal orientations, e.g. one for the vertical and one for horizontal (these orientations being defined with respect to the retina and not to the outside world). Orientation selectivity might then be coded by the relative responses of orthogonal groups of cells, e.g. 45° would correspond to equal inputs from both groups.

The first principle of early local coding combined with convergence would imply the existence of a hierarchical system of distinct cell types for every relevant feature. In the case of face recognition one would, for example, assume the existence of a central locus (the so-called 'grandmother cell') that integrates the particular features of your grandmother (her eyes, ears, nose, mouth, etc.). This hypothetical cell would be activated only by the presence of the right shapes and the right combination of these features. The concept of signals from local processing units converging on a central processing site, giving rise to extreme stimulus selectivity in higher level single cells, arose after Hubel and Wiesel first used a hierarchical model to explain orientation selectivity. Such hierarchical organization seems, for instance, to underlie the strange cases of 'face blindness' (*prosopagnosia*; see the Chapter 8) where the afflicted person cannot even recognize the face of close family members. Although this hierarchical model has played a certain role in the history of neuroscience, most neuroscientists have found the concept of a specific cell for every combination of stimulus features to be unacceptable as a general principle. A strategy of a different cell responding to every possible combination of nuances of general attributes, such as color, shape, structure, position, distance, movement, etc., would require an unrealistically huge number of specific cells to ensure that no nuance would be missed. For color alone, one would need between 5 and 10 million neurons since this is the number of shades that can be discriminated under optimal conditions. An alternative explanation in terms of distributed processing would imply that proso-pagnosa somehow results from a disturbance of the established connections and weights between neural elements within a neural network.

In the context of color representation, vector coding could mean that the attributes of a color, e.g. orange (its hue, saturation and lightness), depend on the relative activity of different, cardinal cell types. It could, for instance, depend on the response ratio between 'L–M' and 'M–S' cells in the LGN, or on the ratio of the inputs of these cells in area V1, or later. In associating vector coding with parallel distributed processing and neural nets, we have not addressed the question of whether there are cells at a higher level that detect this relation. In the next chapter we shall see that, for color-coding in area V1, recent evidence indicates a coding strategy with hue-selective cells for many more directions than in the LGN (although not with a narrower wavelength tuning). This suggests that the vector coding at lower levels might have been transformed into local coding in V1, or to vector coding with a greater number of base vectors.

In an even wider context, distributed processing and vector coding could mean that every perceived object, or even an object attribute, is represented by the pattern of activity in the neural network, or by an abstract state vector, with a specific magnitude and direction in a multidimensional space. The active dimensions of this space could be flexible and depend on the stimulus' specificity and contrast, and on how many different cell types influenced the perception of the object in question at any given moment. The resulting state vector, whether represented by a separate neural locus (convergence) or not ('open' solution), would depend on how the activation was distributed among all component cells of the network (and on their relative weights). Continuing along this line of thought, it appears that one can also imagine consciousness as being distributed over many brain areas and cell types, each one with its 'partial consciousness'.

Single cell recordings from brain cells in humans are possible only under special circumstances, for instance when one needs to monitor cortical activity during brain surgery. Normally, single cell recordings from the cortex are carried out on anesthetized animals, such as fish, frogs, cats, rats, monkeys and others. Monkeys belong to the order of primates, the same branch of vertebrates as humans and, so far, psychophysical studies have shown that the visual systems of macaque monkeys and humans are similar and function in much the same way. Therefore, macaques are often used as a model system for human vision. In the following we shall describe experiments that demonstrate correspondence between a human subject's performance in psychophysical tasks and the activity of macaque neurons in response to similar test stimuli, but first let us take a look at some important methodological problems.

Class A and class B observations

In science it is useful – often necessary – to develop tentative theories, or a working hypothesis as a framework for designing experiments and interpreting their results. In neuroscience, an important set of working hypotheses are linking hypotheses, postulates about the relationship between sensory stimulation, the subsequent neural activity, and the perception it gives rise to. A linking hypothesis should, perhaps, be regarded as a theory of correspondence between stimulation, neural activity and observer's response, rather than a statement about cause and effect.

The simplest and most unproblematic hypothesis states that equal signals from a peripheral sensory organ to the brain give rise to equal sensations (matching stimuli) under equal conditions. In the words of Brindley (1960),

'whenever two stimuli cause physically indistinguishable signals to be sent from the sense organs to the brain, the sensations produced by these stimuli, as reported by the subject in words, symbols or actions, must also be indistinguishable'.

This is true, for instance, in the case when three primary colors in an additive mixture of lights are adjusted to match a fourth one in a color match. When a match is achieved, two unequal physical stimuli lead to equal neural activity, and consequently to the same percept. Observations of matches or equalities, or of differences, and even thresholds, are methodologically simple, and they are not subject to the aforementioned 'explanatory gap' (p. 284). Brindley called such experiments 'class A observations'. A conservative attitude towards psychophysics would require that one deals only with judgements such as these.

Observations that cannot be expressed simply as matches or thresholds are called 'class B observations'. In such experiments the subject is required to describe the quality of his experience, for instance the hue of a light or its intensity. It is more difficult still to extract a common sensory aspect from two or more unequal qualities. Examples of such class B experiments might be comparing the brightness of differently colored lights, selecting colors of the same hue when saturation and/or lightness differs, or determining which stimuli have the same saturation when their hues and lightness are different. However, the ability to experience perceptual qualities and make class B observations is often a precondition for designing class A experiments. For instance, without the ability to distinguish colors along all qualitative dimensions, we would not be making color matches. Qualitative experience comes before its scientific exploration. The inner order and structure of perceived qualities will often give us an indication of how to understand the underlying physiological mechanisms. Without such clues, neurophysiologists would be lost when confronted with seemingly chaotic neural activity. Therefore, visual science cannot exclusively rely on class A experiments. To do so would mean to steer clear of many interesting phenomena and to declare class B type observations as off-limits to scientific inquiry. Both classes of experiments have a legitimate place in visual science, but we need to be aware of the fundamental difference between them.

Ideally, and whenever possible, psychophysical and electrophysiological experiments should be designed as class A experiments. Throughout this book, we have repeatedly discussed the problems that arise with class B experiments. Below we shall enter even deeper waters and describe interesting co-variations or analogies between simple physical properties of objects and the associated neural activity and perception. Analogies and correlates are not proofs of causation or interdependence, only a demonstration of a common structure. The study of such co-variations gives a broader basis for understanding the functional aspects of visual processes than would the isolated study of psychological, psychophysical or physiological relations or processes.

The interpretation of psychophysical results in terms of neural processes is fraught with caveats. However, it does not seem unreasonable to assume that the evolution of neural networks favors solutions taking best possible advantage of the high sensitivity of peripheral sensory receptors. This would imply that the early stages of sensory processing set the limits for detection. That being the case, one would expect to find that the behavioral sensitivity in various situations be traced back to the functional limits of receptors and retinal cells. At later stages of visual processing, higher visual areas would add their particular flavors of perceptive features such as color and form – qualities that come fully into play for supra-threshold stimuli.

B- and D-types of cells

ON- and OFF-cells are first encountered in the retinal bipolar and ganglion cells. Earlier, when describing receptive field models in Figures 3.29 and 3.32, we discussed how ON-center cells, the Increment cells in our terminology, convey information about a stimulus becoming brighter, whereas OFF-center cells, or Decrement cells, signal the opposite, that a stimulus has become darker. In the context of ON-center/ OFF-surround and OFF-center/ON-surround systems in the cat, these two cell systems have been called B- and D-systems (Jung *et al.*, 1952), where B stands for brightness and D for darkness. Below we shall take a further look at phenomena that gave rise to these labels (Spillmann, 1971; Magnussen, 1987).

The *Hermann grid* is shown at the top of in Figure 7.1. At the intersections of the black stripes in the top-left image, one can see diffuse bright spots, and at the intersections of the white stripes in the top right image, one sees dark shadows. The clarity of the illusion depends on the width of the stripes and can be modified by changing viewing distance or by fixating to the side of an intersection.

This illusion is thought to result from the joint activity of ON-center/OFF-surround and OFF-center/ON-surround cells [Figure 7.1(b)]. Take for instance an ON-center cell with its receptive field center placed at the intersection of two white stripes (right image). This cell is inhibited more by the surround than are other cells positioned elsewhere along the stripe away from an intersection (since a larger area of the surround is illuminated). Therefore an ON-center unit with its receptive field centered on the intersection will respond less vigorously than those centered elsewhere on a stripe. If ON-center units contribute to a brightness-coding system, the B-system, the former unit would signal 'less bright' than the latter, resulting in faint shadows in the intersections. For a particular ON-cell, angular size of the stimulus would be significant for the strength of the effect. As pattern elements grow, the illusion should be strengthened in the peripheral visual field since receptive field center size increases with retinal eccentricity. A similar argument holds for the OFF-center cells, the hypothetical darkness- or D-system. Such cells would be activated more strongly in positions where the surround of the receptive field is illuminated more extensively, i.e. in the white intersections, and consequently the intersections would appear darker. B- and D-systems thus pull in the same direction, rendering the white intersections less bright, or darker than the adjoining stripes.

In the black line intersections (left) we see lighter spots, and, according to the same reasoning as above, B-cells are less inhibited at the black intersections than in a black stripe. Here too, the higher activity when the receptive field center is positioned on the black intersections signals brighter, and the smaller response of D-cells would

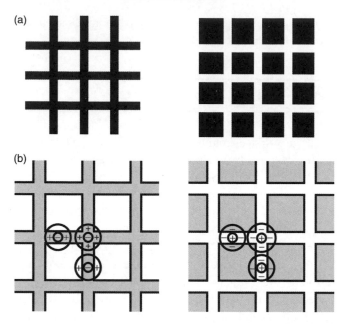

Figure 7.1 (a) The Hermann grid. In the black crossings one can see lighter spots, and in the white crossings darker spots. This phenomenon has been attributed to perceptive fields generated by ON- and OFF-center cells as they form brightness and darkness systems. Brightness neurons (b; right) are more inhibited in the white crossing than along the white stripe because a greater part of the inhibitory receptive field surround is exposed to light. When placed in a black crossing, these cells are less inhibited by the white surround than when they are placed on a black stripe. This behavior is thought to correspond to the darker shades in the white crossings and lighter spots in the black crossings. The same argument, when applied to a darkness system with a reversed organization of the receptive field (b; left), leads to a similar result. Both systems are pulling in the direction of darker for the bright crossings and brighter for the black crossings. The strength of the effect is dependent on the size of the grid stripes in terms of visual angle.

pull in the same direction. ON- and OFF-cells work in concert, but in principle the illusion can be explained in terms of either cell system alone, provided the receptive fields have spatially antagonistic centers and surrounds.

If this theory were correct, measuring the stripe width that gives the most conspicuous effect could indicate the size of the underlying 'perceptive center fields'. Such fields would be made up of several cells with overlapping receptive fields and need not correspond to the receptive fields of single cells (Spillmann, 1971).

Contrast and contour enhancement

The interplay of activation and inhibition within a receptive field can accentuate a luminance gradient by enhancing the difference between two adjacent stimulus fields.

Figure 3.32 illustrates the responses of ON- and OFF-center cells placed in different positions close to a luminance step. This example is for the cat. One of the classical examples of lateral inhibition leading to contour enhancement was found in the facet eye of the horse-shoe crab *Limulus*, as described by Hartline (1940), a Nobel Prize winner in 1967. Although the process is probably more complicated in the human visual system, lateral inhibition in the *Limulus* eye has long served as an example and a model of a biological contrast-enhancing mechanism (see p. 133).

Two juxtaposed areas of different colors, but with no luminance contrast, do not give rise to border enhancement in the same way that a pure luminance contrasts does. Chromatic Mach bands seem not to exist. Opponent cells with a center-surround antagonism in their receptive fields provide for a spatial response profile for achromatic borders that correspond to Mach bands. This is analogous to a band-pass filtering of the stimulus and is different from the mechanism described in Figures 3.32 and 3.33. There are other possible strategies for generating border effects similar to Mach bands, and we shall take a closer look at one interesting alternative that might apply to primates. A possible design for luminance border enhancement without concurrent enhancement of equiluminant chromatic borders is described in Figure 7.2.

In Figure 7.2(a) the responses of I_{L-M} and I_{M-L} cells to a moving border of achromatic luminance contrast and another of isoluminant red–green chrominance contrast are shown. Since both types of I-cells respond in the same way to the luminance contrast (upper left image), their summed response will be larger, roughly twice that of each cell system alone. In the case of the isoluminant red–green border (upper-right image), the responses of the two opponent I-cells are opposite; while I_{L-M} cells are excited by red and inhibited by green, the reverse is true for I_{M-L} cells. Taken alone, each of these two I-cells give a relatively smooth transition from one color to the other such as that shown in the upper-right drawing. Since activation is higher than inhibition, summation of the responses of the two I-units could provide information about brightness. On the other hand, a response difference of the two symmetric types of PC I-cells, where activation dominates over inhibition, would remove the luminance response in the achromatic case and extract two different chromatic dimensions (L–M and M–L; not shown in Figure 7.2).

Examples of opponent, PC D-cell responses to the same two border stimuli are given in Figure 7.2(b). Because of stronger inhibition in the D-cells, they have generally smaller responses than I-cells, and if they were mainly type II cells with coextensive center and surrounds, they would round off sharp contours and serve as low-pass filters for both luminance and chrominance contrasts.

In the physiological model of color processing that was presented in Figures 6.9 and 6.23, the activity of PC I- and PC D-cells of the same opponency were summated, a process that may occur in the blob clusters of area V1. Summation was considered necessary to preserve opponent responses over the whole range of luminance ratios for which we see color, from dark surfaces reflecting only a few percent of light to bright, self-luminous lights. A sum preserves and accentuates color differences

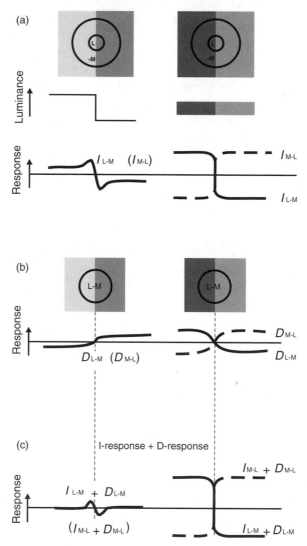

Figure 7.2   The responses of opponent PC cells with different receptive field organizations to achromatic luminance contrast (left column) and isoluminant red–green contrast (right column). The response is plotted as a function of the horizontal position of the receptive field center as it traverses the stimulus. (a, b) Increment and Decrement, 'L–M' and 'M–L', cells respond equally to a luminance step and have mutually inverted responses to the chromatic difference. Summing these I- and D-cell responses, as in (c), leads to contour enhancement for luminance contrast, but no such enhancement for an isoluminant chromatic difference, in agreement with observations. This demonstrates that the combination of cell responses suggested by the neural color vision model can signal spatial luminance contrast as well as differences in color (note that isoluminant red–green differences also give a pronounced transient response in MC cells; see Figure 7.22). (See also color plate section.)

(bottom right image) over a larger luminance range than is possible with only one cell type alone. The sum of the band-pass and low-pass filters (of the PC I- and the PC D-cells) to the left in the figure removes the DC-level and enhances luminance contours.

As demonstrated earlier, this model of sums and differences of opponent cell responses, and other linear transformations of cell responses, account rather well for several aspects of color vision, especially the scaling of colors. One feature of this model is that it predicts contour enhancement equivalent to Mach bands for luminance contrasts, but none for chrominance differences. This agrees nicely with psychophysics. The model of Figure 7.2 is a little more complicated than that of Figure 3.32, but it explains the lack of chromatic Mach bands without the need for auxiliary hypotheses. The response magnitude of MC I-cells to borders would be much like the combined I + D response of PC cells, as shown to the left of Figure 7.2(c), and MC D-cells would respond in the opposite phase, as shown in Figure 3.32. They are therefore both capable of adding to the achromatic border contrast. Thus, Mach bands to achromatic stimuli may well arise from activity of both PC and MC cells. Figure 7.3 shows the responses of monkey geniculate MC and PC cells to areas of different luminance and color, the 'Mondrians' (Land, 1983). Whereas MC cells are obviously more interested in the borders between colored areas, PC cells respond over the whole area and to several different colors. This difference is quite distinct in the response profiles below the Mondrians.

The receptive field of each cell scanned the Mondrian, and each spike of the response was imaged as a black dot on the monitor screen. This figure clearly illustrates the difference between these two cell systems; the transient response of MC-cells occurs only at the borders between the squares and rectangles, and it is up to the PC cells to 'fill in the rectangles' by responding inside the borders.

It should be noted that an enhancement of border contrast need not spread over a large area, but in some cases it does, as in the Cornsweet–Craik–O'Brien illusion (see for instance Jung, 1973). This, and similar phenomena have been described as due to some sort of 'filling in process'. Historically, border enhancement and area contrast have been treated separately, and they have even been given their own names: Mach contrast (Mach bands) for the former and Hering contrast (Von Bekesy, 1968; Nobel Laureate, 1961) for the latter.

As we see from these figures, the responses of opponent PC cells are ambiguous, being influenced by changes in luminance, in color, and in the spatial and temporal parameters of the stimulus. In signal theory this is called multiplexing (Martinez-Oriegas, 1994). Only by comparing these responses with those of other, more or less specific cells, can the visual system arrive at an unequivocal judgement about the stimulus.

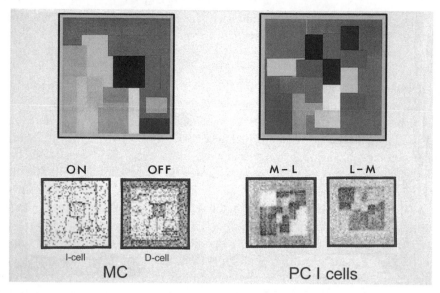

Figure 7.3 The processing of contours and extended areas. The two black and white panels to the left show the responses of two MC cells, an Increment and a Decrement cell, to the Mondrian pattern above. This pattern was moved across the receptive field of the cell, and each nerve impulse was registered as a black dot on the screen, in a position corresponding to that eliciting the response. The firing rate is thus visualized by the density of black dots and the degree of blackness on the screen. As one can see, the MC cells are not concerned with the colors of the squares and rectangles of the Mondrian, but respond in a transient fashion to the borders between different chromatic areas. Outside the responses to the Mondrian is a frame with maintained activity without stimulation. To the right are two panels mapping the responses of two Increment PC cells, an 'M–L' and an 'L–M', to another Mondrian pattern. Here, too, the degree of blackness is proportional to the firing rate in impulses/s. Dark areas reflect a high firing rate and lighter areas inhibition of the maintained firing. The I_{M-L} cell responds relatively homogeneously within areas that are green, blue and white, but its maintained firing is inhibited by red, and it is unresponsive to black. The I_{L-M} cell is activated by red, yellow and white areas, but its maintained activity is inhibited by green and blue. Again black causes no difference relative to maintained firing. (Reproduced from Nothdurft and Lee, 1982, *Exp. Brain Res.* **48**, 43–54 by permission of Springer.). (See also color plate section.)

Psychophysics and the parallel pathways

In this section we shall try to be more quantitative in our comparison of the activity in macaque magnocellular, parvocellular and koniocellular pathways with human psychophysical sensitivity. The examples have been chosen from investigations of the opponent PC, KC pathways and the non-opponent MC pathways in primate vision and their subdivisions into Increment and Decrement cells. We shall present some of the possible different functional tasks of opponent cells as compared with those of non-opponent magnocellular cells. Because these cell groups often have quite different reactions to weak stimuli, one assumes that it is possible in psychophysical tasks to probe the sensitivity of each cell system alone. Caution is needed when inferences are made for high-contrast stimuli (see Kaplan *et al.*, 1990; Valberg and Lee, 1992).

Generally, the fast responding MC cells are regarded as important in conveying information about low luminance contrasts, about depth vision, movement detection and analysis, about stimuli with a wide range of temporal and spatial frequencies, and also for contours. The opponent PC cells code for high luminance contrasts and are, perhaps, responsible for high spatial resolution. Both PC and KC pathways code for color at relatively low spatial frequencies. Other attributes, like texture, shape and area contrasts may be shared between all neural systems, depending on the contrast and subsequent signal strength.

In this context, the definition of threshold sensitivity is important. Earlier in this book we defined threshold sensitivity, $s - \Delta R/\Delta I$, using a fixed threshold value of ΔR and varying ΔI. The psychophysical threshold, ΔR, can, for instance, be based on an individual's subjective criterion for when the stimulus is seen or not, or it can be determined statistically as, for example, a detection rate of 75 percent of all presentations (poor chance would be 50 percent). The neural threshold is less well defined. The response of a cell to a repeated stimulus is not always the same, and with this variability it is not easy to define threshold. One way out of the problem is to present the same stimulus many times and to compute an average response. Threshold can be determined statistically as a certain signal-to-noise ratio. Alternatively, a measure termed 'contrast gain' can be used, which is the inverse of the slope of the cell's stimulus–response curve (with the slope having, for instance, the unit impulses/s/contrast). A fixed response criterion, ΔR, of 10 or 20 impulses/s is another alternative that is similar to using the slope. This criterion, when applied to the most sensitive cells, is often seen to correlate well with the psychophysical threshold and hence with sensitivity. Consequently, we must conclude that the initiation of firing in only a few cells may be sufficient for detection. In the following, both the latter methods will be used.

Increment stimuli and spectral responses

The behavior of six typical opponent cells in a stimulus situation such as that shown in Figure 6.7 was dealt with previously. This situation is comparable to that which occurs when the eyes are freely moving around, and when points on the retina are successively exposed to areas of lower and higher luminance than its surroundings (positive and negative contrasts).

Figure 7.4 shows responses of opponent cells in another situation where stimuli are close to monochromatic incremental lights of different intensities projected upon a steady white background. The light stimuli were 4° in diameter, added to an adapting background of the same size. The stimulus period was 1.5 s, with 300 ms stimulation and 1.2 s adaptation to a 110 cd/m^2 white stimulus. Although such additive mixtures

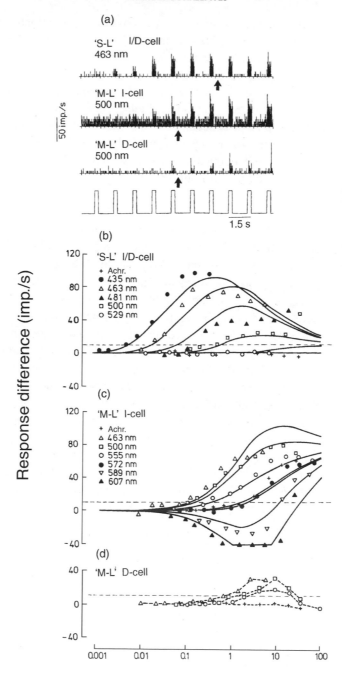

Response difference (imp./s)

Luminance contrast $\Delta L/L$

are rare in nature, this type of stimulation has a long tradition in neurophysiological and psychophysical research, and several important experimental studies have been performed under such conditions. The data points in Figure 7.4(b) give the increment responses of an 'S–L', I/D cell (it would usually be called a 'Blue ON', but it shows response features of both an ON- and an OFF-cell, see Figures 6.10 and 6.11). The two lower panels, (c) and (d), show the responses of an 'M–L', I-cell and an 'M–L', D-cell (a 'Green ON' and a 'Red OFF'). We see that the 'S–L' cell was already activated by very low luminance contrasts below 1 percent for the 435 nm light. The threshold of 10 impulses/s (dashed horizontal line) was already reached below a Weber ratio of 0.01, a factor of 10 higher in sensitivity than was typical for MC cells in the same situation (see Figure 7.5). The 'S–L' cell was inhibited by white light at all luminance ratios and for lights of wavelengths above 500 nm. For the most effective wavelengths, the luminance threshold sensitivity of the 'M–L', I-cell shown in Figure 7.4(c) is about 10 times lower than that for the 'S–L' cell. The response of the D-cell in Figure 7.4(d) is much diminished, due to inhibition by the adapting background, and its response is shifted towards even higher luminance contrasts.

For the same stimulus conditions as in Figure 7.4, Figure 7.5 shows the intensity–response curves of two MC cells, one I-cell and one D-cell, for different wavelengths. The stimulus strength is given as luminance contrast (Weber ratio), and with this scale the responses to all wavelengths follow largely the same curve. This demonstrates that MC cells do not distinguish between wavelengths when stimulus intensity is defined in terms of large-field luminance, demonstrating that MC cells have a $V(\lambda)$-like spectral sensitivity (here equal to the CIE-$V(\lambda)$ for 10° fields).

In these cases, a steady white background provided a constant adaptation of the retina. For opponent D-cells, such a background will lead to a strong inhibition (see Figure 7.4), reducing the cell's responses to excitatory wavelengths. We still do not have a good model of adaptation, and it is therefore not quite clear how to give a general account for a cell's response to increment stimuli on a steady, adapting background. Notwithstanding this limitation, one can use the simple model presented

Figure 7.4 (a) The incremental responses of three different opponent cells: an 'S–L' Increment/ Decrement cell; an 'M–L' Increment cell, and an 'M–L' Decrement cell. The stimulus was a spectral test light projected on top of a coextensive, 4° white background with a fixed luminance of 110 cd/m². The test light luminance increased in 0.3 log unit steps. The short arrows below the histograms are positioned at a luminance contrast of $\Delta L/L = 1.0$. The Decrement cell is strongly inhibited by the white background. (b–d) The responses of the cells in (a) for test lights of different wavelengths. Responses are plotted as a function of luminance contrast, after subtraction of the response to the adapting white background. Responsivity is strongly dependent on wavelength. The fully drawn curves represent the simulations made by the neural model, similar to that presented in Figure 6.10 in another stimulus situation. The 'M–L' Decrement cell has about 1/10 of the sensitivity of the corresponding I-cell. The horizontal dashed lines indicate a threshold response criterion of 10 impulses/s. This threshold is used to plot the relative spectral sensitivity of these cells in Figure 7.6 (from Valberg and Lee, 1989).

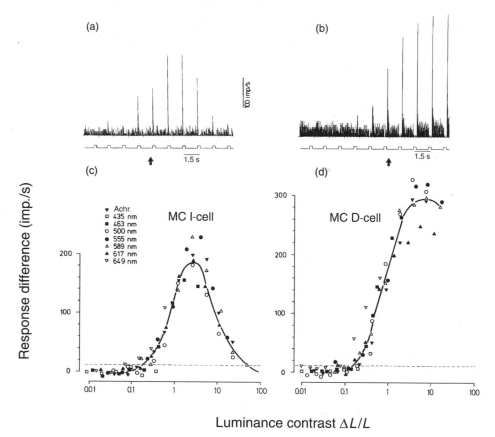

Figure 7.5 The response histograms in (a) and (b) for transient Increment and Decrement MC cells, using incremental stimuli as in Figure 7.4 with a test-light wavelength of 589 nm. Here, the diameter of the stimulus field was 0.5° in diameter. In (c) and (d) the transient response within the first 45 ms is plotted as a function of luminance contrast (after subtracting the response to the white adapting background). Different symbols represent different wavelengths. In contrast to what was the case for opponent cells in Figure 7.4, responses and threshold are independent of wavelength (from Valberg and Lee, 1989). The horizontal dashed line represents a threshold response criterion (10 impulses/s) that is applied in Figure 7.6.

earlier (e.g. Figure 6.10) to calculate the response obtained under a particular adaptation condition (exemplified by the solid lines in Figure 7.4) and directly compare the responses and sensitivities of cells in this situation with psychophysical thresholds obtained under the same conditions.

In Figure 7.6, spectral threshold sensitivities of the increment responses in a situation like that of Figure 7.4 are shown for the four most sensitive cells encountered with 'S–L', 'M–S', 'M–L' and 'L–M' opponencies, and one typical non-opponent MC cell. D-cells have not been included in this figure because their sensitivity for this kind of stimulation is much lower than for I-cells, and therefore it is not likely that

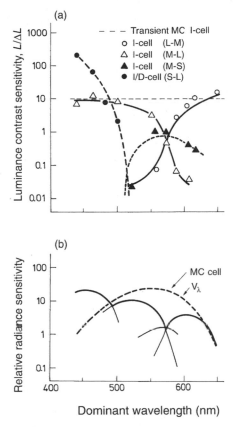

Figure 7.6 (a) Spectral contrast sensitivity for responses of four opponent I-cells and one MC I-cell to incremental test lights. The 'S–L' and 'M–L' cells are the same as those presented in Figure 7.4. Symbols represent experimental data, and the curves are computed by the neural model mentioned earlier. The horizontal dashed line represents the luminance sensitivity of the MC cell. (b) This figure shows the same curves as in (a), but using energy units on the ordinate instead of the photometric unit luminance. In this plot, the dashed curve for the MC cell is equal to the spectral luminous efficiency function $V(\lambda)$ (from Valberg and Lee, 1989).

they contribute to the psychophysical increment threshold in this situation. In the figure, sensitivity is defined as the inverse of the Weber ratio at threshold, as derived from response curves such as those shown in Figures 7.4 and 7.5. The threshold criterion was 10 impulses/s. Sensitivities to luminance contrasts are shown in Figure 7.6(a), and the same data are re-plotted in terms of radiance contrast in Figure 7.6(b). As a consequence of the choice of units for light intensity, the sensitivity of the MC cell with its $V(\lambda)$ spectral sensitivity is a straight line in Figure 7.6(a) and equal to the $10° V(\lambda)$ curve in Figure 7.6(b). Clearly, the cell with the highest increment luminance sensitivity in this situation is the blue-sensitive 'S–L' cell.

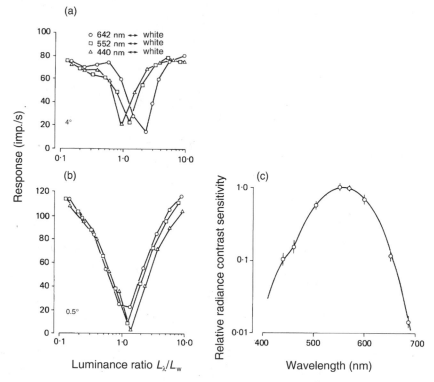

Figure 7.7   The response magnitudes of an MC cell in an experiment where three wavelengths of varying luminance were sinusoidaly alternated with white light, in either a 4° (a) or a 0.5° (b) field. With the smaller spot at high or low luminance ratios responses tended to be more vigorous, and minima better defined in (b). The average spectral sensitivity for eight MC cells for a 0.5° stimulus is plotted by the data points in (c). The MC cell sensitivity is the same as the CIE 10° $V(\lambda)$. Eccentricity between 3 and 10°. See also Figure 4.19 (adapted from Lee *et al.*, 1988).

The spectral sensitivity of MC cells

Throughout this book, we have referred to magnocellular cells as having a spectral sensitivity that corresponds well with $V(\lambda)$, the luminous efficiency function of the human eye. The evidence for this is given in Figure 7.7. For stimuli of two different sizes, parts (a) and (b) illustrate how the response of an MC cell depends on luminance ratio, $Y = L_\lambda/L_W$, between two alternating stimuli, one monochromatic with variable luminance and the other white with constant luminance (as in heterochromatic flicker photometry, HFP, see Figure 4.17). For every wavelength, there is a clear minimum for every wavelength close to equal luminance $(L_\lambda/L_W = 1)$. The average spectral sensitivity of eight retinal cells is shown in Figure 7.7(c). These experiments, extended to 26 MC-cells and performed at an eccentricity larger than 2°, gave very good correspondence with the 10° curve of the

CIE. In addition, our own psychophysical studies using the same apparatus gave similar results (Figure 4.19). These and other results are strong indications that it is the MC cells that determine the spectral sensitivity in psychophysical experiments using either flicker photometry or the minimally distinct border method.

From Figure 7.6 it is evident that MC cells have the same luminance contrast sensitivity throughout the spectrum, and a much higher sensitivity in the middle of the spectrum than any of the opponent cell types. The difference is largest around 570 nm, where about 10 times less light is needed to activate non-opponent MC cells than opponent cells. For PC cells lacking S-cone inputs, white light has the same threshold as mid-spectral yellow light.

This difference in the responsiveness of PC and MC cells to achromatic light is also evident when using achromatic sinusoidal gratings as stimuli. Responses to such stimuli are given in Figure 7.8(a) and sensitivity in Figure 7.8(b). Below 10 percent contrast, in Figure 7.8(a), the difference in response magnitude between MC and PC cells is about a factor 10. For higher contrasts, the MC cells show response saturation whereas the PC cell response continues to increase in a linear manner, reducing the difference between their response magnitudes. Figure 7.8(b) reproduces the results of Hicks *et al.* (1983), who measured spatial contrast sensitivity of MC and PC cells in the geniculate of the macaque monkey. Up to 2–3 cycles/deg MC cells are the most sensitive by a factor of about 10. Both figures show that MC cells start responding to luminance contrast long before the PC cells, and that low contrasts and low spatial frequency therefore can be used to isolate and study MC activity without intrusion from opponent cells.

A difference between detection and identification

Provided MC cells do not contribute to chromatic vision, and opponent cells do, from the different sensitivities of MC cell and opponent cells shown in Figure 7.6, we conclude that for mid-spectral light there is a difference between the *detection* of a light and the *identification* of its color. At the spectral ends, however, detection and identification thresholds should coincide, i.e. color is observed simultaneously with detection. As we shall see, this corresponds well with experimental observations.

Several psychophysical studies have demonstrated that, in the middle of the spectrum, stimuli are devoid of color at detection thresholds; in order to tell which color the stimulus has, it is necessary to increase its intensity above threshold. This difference has been called the 'achromatic interval'. In Figure 7.9 we use the data of Graham and Hsia (1969) to illustrate this relationship. The ratio between the threshold for detection (without color discrimination) and the threshold for identification of the color is plotted as a function of wavelength. As expected from Figure 7.6, the difference is largest for 570 nm. At this wavelength it is necessary to increase stimulus luminance by a factor of about 10 in order to recognize its color. At the spectral ends the color is seen simultaneously with detection. The fully drawn

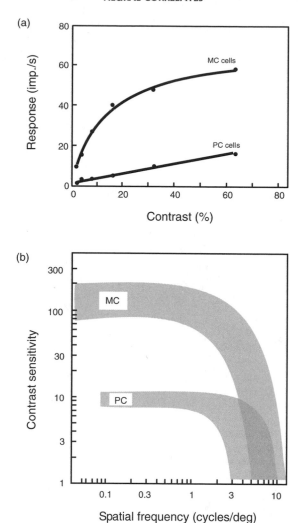

Figure 7.8 (a) A typical example of the responses of MC and PC cells to increasing Michelson contrast of achromatic stimuli. MC cells have the higher sensitivity and a response that begins to saturate at relatively low contrast values. For PC cells, the relationship between contrast and response is more linear (redrawn after Lee *et al.*, 1990). (b) Spatial contrast sensitivity for achromatic light for some MC and PC cells. Below 2–3 cycles/deg the MC cells are about a factor 10 more sensitive than the PC cells. The measurements are from ON and OFF cells of the geniculate nucleus of the macaque monkey (redrawn from Hicks *et al.*, 1983).

curve in Figure 7.9 reproduces the difference in the sensitivity of MC and PC cells of Figure 7.6(a). In terms of relative thresholds, the form of this curve corresponds well with the data points for psychophysical observations. We are therefore inclined to conclude that detection threshold is determined by whichever system is the more sensitive, either the MC system for mid-spectral lights or the opponent systems at the

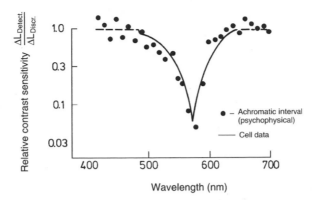

Figure 7.9 The achromatic interval. The data points represent the luminance difference between detection of a stimulus and identification of its color (after Graham and Hsia, 1969). The solid curve shows the difference between the sensitivity of MC cells and opponent cells from Figure 7.6(a) within the wavelength range where the MC cells detect the stimulus at lower contrasts than the opponent cells. In this interval one might expect the psychophysical threshold for detection to be lower than that for chromatic identification. Towards the spectral ends (dashed curve) the opponent cells are the more sensitive, and the threshold for identification and detection is the same (from Valberg and Lee, 1989).

spectral ends. Color perception requires activation of the opponent system, independently of the response of the MC system.

Other measurements of detection and identification thresholds confirm the hypothesis above. The results of one such experiment are shown in Figure 7.10. Here the detection and identification thresholds have been determined for stimuli of different spectral purity, in an experiment where the stimuli were projected on a steady white background. In a psychophysical experiment, human subjects were first asked to determine threshold for detection. In another experiment they were asked to increase the intensity of the light above detection threshold until they could identify its color. For high purity stimuli at 531 and 649 nm, Figure 7.10 shows that the thresholds for identification (solid circles) were about the same as for detection (open circles). However, as purity decreased, the two thresholds separated, with the sensitivity for identification falling below that for detection. With decreasing purity, the curve for detection approached an asymptote that resembled the luminance thresholds for MC cells in the same situation, being constant and independent of purity. In Figure 7.10 the thresholds for color identification run parallel to that of opponent cells, again supporting the notion that color perception requires activation of the opponent system.

Classical and global receptive fields

Wiesel and Hubel (1966) divided the receptive fields they found in the geniculate of the macaque monkey into several subgroups called types I, II and III (Figure 7.11).

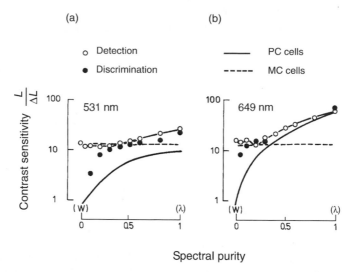

Figure 7.10 The threshold sensitivities for PC and MC cells are here compared with the human psychophysical sensitivity for detection and for identification of color as a function of colorimetric purity (saturation). The lower curve in (a) shows average sensitivity of two opponent, 'M–L', I-cells to 531 nm light while the lower curve of (b) gives the average sensitivity of two 'L–M', I-cells to 649 nm. The horizontal dashed curve represents the threshold luminance sensitivity of MC cells in this situation. Open circles represent psychophysical detection thresholds and solid circles discrimination (identification) thresholds. For maximum purity, detection and identification thresholds were equal, but discrimination (identification) sensitivity decreased more rapidly as purity decreased, in parallel with the diminished sensitivity of the PC cells. Detection thresholds reached an asymptote for low purity values and flattened out at about the same level as MC cell sensitivity. We conclude that detection threshold is determined by whichever system is the more sensitive, but color identification requires opponent cell responses. The stimulus was 4° in diameter projected upon a white steady background at 10° eccentricity (from Valberg and Lee, 1989).

Type I had either an excitatory cone input to its center and another cone type inhibiting the surround, or the opposite arrangement (see Figures 3.22 and 6.12). A type I field shows a spatially antagonistic center-surround structure for most wavelengths and cone opponency for large stimulus fields. In type II receptive fields excitatory and inhibitory regions are roughly co-extensive and spectral opponency is found for all stimulus sizes. Cone opponent cells in the retina and LGN have receptive fields of either type I or type II. Type I is most common among Increment PC cells with opponent L- and M-cone inputs.

Type II opponent cells, with about the same extent of excitatory and inhibitory cone inputs, are common for KC cells with S-cone inputs, for instance yellow-sensitive 'M–S' cells and the blue sensitive 'S–L' cells. Such receptive fields also seem to be relatively frequent among Decrement cells with L- and M-cone inputs. Magnocellular cells seem to have type III receptive fields, with a sum of L- and M-cones in the center (either excitatory or inhibitory) and the same cone combination in the surround, but

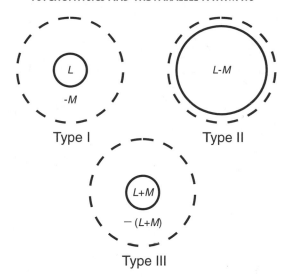

Figure 7.11 Possible receptive field center/surround structures of retinal and geniculate cells. Type I cells have concentric center and surround with a different cone input to each area while type II cells have coextensive cone-opponent fields. Type III cells have concentric center and surround fields with antagonistic, but non-opponent cone input.

with the opposite sign. In our own recordings from single cells in the retina and geniculate of *Macaca fascicularis*, we encountered PC cells with receptive fields that seemed to be somewhere between type I and type II, and many authors now believe that, at least for PC cells, types I and II represent the extremes of a continuum rather than separate classes. Since there is a connection between the receptive field structure (Figure 6.12) and the degree of inhibition, one might expect also a continuum of the strength of inhibition between weakly and strongly inhibited cells. A receptive field structure of most of the PC, KC and MC cells according to Zrenner *et al.* (1990) is shown in Figure 7.12.

The description above is limited to the so-called 'classical receptive fields' with the size of a few minutes of arc in the central fovea to about 5° further out in the retinal periphery. However, the response of cells is influenced by illumination of the retina outside the classical field (Valberg *et al.*, 1983, 1985a, 1991a). For instance, a steady peripheral annulus surrounding a stimulus that activates the classical receptive field of LGN opponent cells has the effect of altering the responses in such a way that the intensity–response curves are moved towards higher luminance ratios, as in Figure 7.13. Therefore, it seems that a greater 'global receptive field' may exert an effect on the smaller classical field through lateral adaptation. There has been a steady accumulation of reports that confirm the existence of such larger non-classical field surrounds at all levels in the visual pathway (Valberg *et al.*, 1985a; Allman *et al.*, 1985; Creutzfeldt *et al.*, 1991; Solomon *et al.*, 2002). Not only can the orientation sensitivity of a cell in V1 be altered by the orientation of a grating projected outside

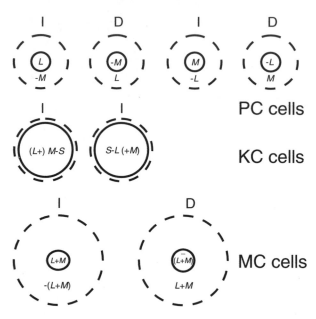

Figure 7.12 Center/surround structures for six types of opponent cells (PC and KC cells) and two types of non-opponent MC cells. KC cells with excitatory or inhibitory S-cone inputs seem to have relatively large and coextensive receptive fields. Sometimes this also seems to be the case for PC-Decrement cells. MC cells summate inputs from L- and M-cones in the center and are inhibited by the same sum in the surround.

the classical receptive field (as in Figure 1.10(b)), but steady surround illumination leads to an adaptive shift of the intensity–response curves. In neither case could the surround effects be explained by stray light.

The intensity–response curves of Figure 7.13 are well modeled by the intensity–response functions used to account for the responses of the opponent cells of Figure 6.10. If, without a surround (solid circles), the response of the cell is given by

$$N = A_L L / (L + \sigma_L) - A_M M / (M + \sigma_M) + N_o$$

which is the same as

$$N = A_L V_L - A_M V_M + N_o$$

then the shifted response curve in Figure 7.13 (open circles), resulting from adding a steady white surround, is mathematically described by the same response equation as before, but with adaptation constants σ'_L and σ'_M higher than without the surround, where $\sigma'_L = c \sigma_L$, and $\sigma'_M = c \sigma_M$. The parallel shift in Figure 7.13 is about 1 log unit on the x-axis, corresponding roughly to $c = 10$ for both cone mechanisms. Stray light would not lead to the observed parallel shift. Stray light would have had a greater effect

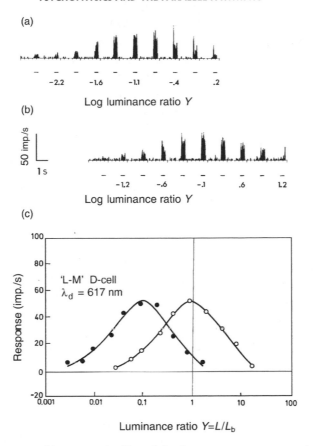

Figure 7.13 Response histograms (a, b) and luminance–response curves (c) for an 'L–M' Decrement cell. In (a) the stimulus was a $4 \times 5°$, 617 nm field of increasing luminance, L, alternating with a white adaptation field of $110 \, \text{cd/m}^2$ luminance, L_b. The numbers give the logarithm of the luminance ratio, $Y = L/L_b$. In (b) the situation was the same as in (a) but now with a static $110 \, \text{cd/m}^2$ white surround projected around the stimulus field. The inner and outer dimensions of the surround were $4.5 \times 5.5°$ and $20 \times 30°$, respectively. In (b) the threshold and maximum response are shifted to a higher luminance ratio compared with (a). In (c) the responses of (a) and (b) are plotted as a function of the luminance ratio Y. Solid circles refer to the no-surround stimulus (a), and open circles to the same stimulus with a white surround (b). The curve drawn through the data points is identical in both cases, only shifted 1 log unit along the x-axis. The static surrounding field has an adaptive effect on the cell that moves the luminance–response curve along the x-axis towards higher luminance ratios. This adaptation effect corresponds to that described for the cones in Figure 4.15(b). (Reproduced from Valberg *et al.*, 1985a, *Experimental Brain Research* **58**, 604–608, by permission of Springer)

on responses to low luminance increments and luminance ratios than to high ratios. Further evidence can be found in Valberg *et al.* (1985a, 1991a).

Another demonstration of a global receptive field is the influence of rapid movements of a contrast-rich pattern in the peripheral retina on the sensitivity to small spots projected onto the central retina. Breitmeyer and Valberg (1979) showed that sudden movements of a black/white grating in the periphery lead to reduced contrast sensitivity in the fovea. This strange phenomenon, called the 'jerk effect', seemed analogous to the shift or periphery effects observed in the cat retina (MacIlwain, 1966; Fischer and Krüger, 1974). Recent recordings in macaque ganglion cells (Lee, personal communication) have demonstrated that in MC cells a strong jerk effect is present which does not evoke a response *per se* (silent surround), but does modulate sensitivity to a central test spot.

Temporal sensitivity and response

In the following we shall take a closer look at the behavior of magnocellular and opponent cells to temporal variations of the visual stimuli. Many experiments relevant to this topic were performed during the early 1990s in Barry B. Lee's laboratory at the Max–Planck Institute of Biophysical Chemistry in Göttingen, Germany, in collaboration with Vivianne Smith and Joel Pokorng. Most of these experiments were performed using red and green diodes that made it possible to achieve high retinal illuminance. The formation of an image on the retina was made through a Maxwellian view optical system. With the red and green diodes modulated sinusoidally in phase, a pure luminance modulation resulted of the yellow mixture color. With the diodes modulated in counter-phase (180° out of phase), the result was a pure red–green chromatic modulation about a yellow average (Smith *et al.*, 1992).

Figure 7.14(a) shows a diagram that combines the effect of the diode contrast modulations on the L- and M-cones. The Michelson contrast in L- and M-cones is plotted along the x- and y-axes. The excitation in each cone type is calculated from tabulated values for spectral absorption in the cones and the spectral emission of the diodes. The figure shows the vector for pure luminance modulation when both cones are modulated in phase, and the direction of isoluminant chromatic modulation (of chrominance) when they are modulated in counter-phase. When the stimulus modulates the L- and M-cones in different proportions, the resultant stimulus vectors in Figure 7.14(a) will point in different directions, as indicated by the angle α.

Figure 7.14(b) shows the response histograms for a D-center, magnocellular cell and a I_{L-M}-parvocellular cell (Lee *et al.*, 1993). To begin with, the response of the phasic MC cell follows the luminance modulation and responds with short trains of impulses whenever there is a luminance decrement in the stimulus. However, with isoluminant red–green exchange, the main MC cells response is reduced to a minimum, but a new component turns up in between each decrement response. The cell now fires twice within a stimulus period, whenever the chromaticity changes in the directions of red and green increments or decrements. This frequency doubling is the so-called 'second harmonic response' of the MC cell, and it arises whenever the red–green color difference dominates over the luminance change. It is as if the cell signals the rectified or absolute difference $|L-M|$ between the cone excitations. The

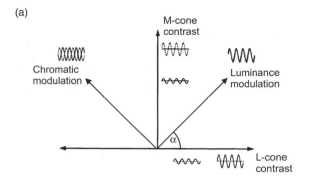

(a)

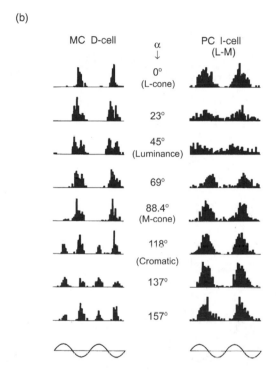

(b)

Figure 7.14 (a) Modulations of L- and M-cone absorptions represented in a cone contrast space. Fully drawn and dashed sinusoidal curves indicate the temporal modulation of L- and M-cones, respectively. The cones are either modulated with the same phase (right) or in anti-phase (left). When the cones are modulated in-phase equally strongly ($+45°$), the result is pure luminance modulation. Pure isoluminant, chromatic modulation is achieved by modulating the cones in anti-phase and adjusting the relative amplitudes of modulation. (b) Examples of the responses of an MC Decrement cell and an 'L–M', PC Increment cell to a combination of luminance and chromatic modulations. The cells were tested for many combinations of luminance and chrominance, corresponding to different radial angles, α, in the contrast diagram of (a). The responses for modulation in eight of these directions with a vector length of about 40% cone contrast are shown here. The histograms span two periods of sinusoidal modulation. We see that the response of the MC cells is maximum for in-phase, luminance modulations, with a frequency doubling for anti-phase, isoluminant, red–green modulations. The response of the PC cell was largest for anti-phase, chromatic modulations and smallest for pure luminance modulation. The modulation frequency was 10 Hz (redrawn from Lee *et al.*, 1993).

same applies to MC I-cells. Below we shall see how this might relate to psychophysical results on border perception.

The firing of the particular PC I-cell of Figure 7.14(b) follows the excitatory L-cone input, with a roughly sinusoidal response for the different combinations of chrominance and luminance. The response is attenuated on approaching pure

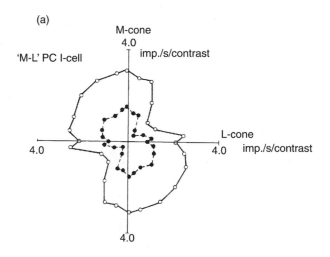

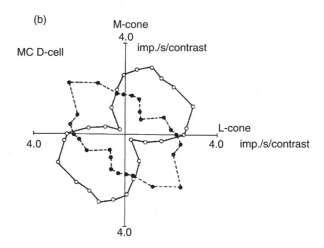

Figure 7.15 The sensitivity of MC and PC cells to 13 combinations of relative L- and M-cone modulation, plotted in the diagram of Figure 7.14(a). The radial angle indicates the relative weights of L- and M-cone modulation and the distance from the origin is proportional to sensitivity, measured as firing rate/percentage contrast. The open and solid symbols refer to the first and second harmonic response components, respectively. The PC cell in (a) has the highest sensitivity along the red–green direction, and it has only a small second harmonic component. The MC cell in (b) has a high first harmonic sensitivity in the luminance direction and low sensitivity in the red–green direction. In contrast, the second harmonic is significant in the chromatic direction but small in the luminance direction. (Courtesy of B.B. Lee.)

luminance modulation of a yellow light, since the difference of L- and M-cone excitations is a minimum when the two cones are in phase (L–M is minimum). The response is maximum near the isoluminant red–green modulation.

The polar diagrams of Figure 7.15 show examples of a PC and a MC cell's responses for different combinations of L- and M-cone excitations at a modulation frequency of 10 Hz. The amplitude and the polarity of the red and green diode modulations were adjusted to give different L and M excitation values while maintaining a constant combined cone contrast, i.e. while keeping the root mean square of the L- and M-cone Michelson contrasts at a fixed value [yielding a circle in the diagram of Figure 7.14(a)]. The response of the PC cell is plotted as open circles in Figure 7.15(a). Response magnitude is given as the length of a vector from the origin with the unit (impulses/s)/percentage contrast. This is an expression of sensitivity or gain. For the PC cell the highest gain is in the chrominance direction M–L, whereas for this cell the response is a minimum in the luminance direction M+L (about $+45°$ and $-135°$). The second harmonic response (inner black dots) was small.

In Figure 7.15(b) we see that the first harmonic response of MC cells (open circles) is largest in the luminance direction (about $+45°$ and $-135°$) and smallest in the chrominance direction. The second harmonic response of the MC cell is plotted as solid circles. For this cell, the second harmonic is largest in the red–green chrominance direction and smallest in the luminance direction. Further experiments have shown that a second harmonic is present in most phasic MC cells, albeit to a variable degree. If the MC cells had only had a linear sum of inputs from the L- and M-cones, as is suggested by the cone sensitivities adding up to $V(\lambda)$ in Figure 7.7, then they would not have shown a second harmonic component. The result of Figure 7.15 therefore indicates additional rectification of a small opponent signal from L- and M-cones (see also p. 375).

Dependency on temporal frequency

The magnitude of cellular responses depends on how a stimulus is presented over time. Figure 7.16 compares the contrast sensitivity of retinal MC and PC cells in the macaque monkey, for periodic, sinisoidal luminance and chrominance modulations of spatially homogeneous stimuli. MC cells exhibit the highest sensitivity for luminance changes, i.e. they require the least contrast of a temporally varying stimulus to elicit a criterion response. These cells show a gradual increase of sensitivity from 1 Hz up to a maximum around 10–20 Hz, dependent on adaptation conditions. The sensitivity of PC cells is about a factor of 10 lower at low and medium frequencies, rising to a maximum at around 20 Hz and approaching the sensitivity of MC cells for higher frequencies. Generally, MC cells display the highest contrast sensitivity for all temporal frequencies.

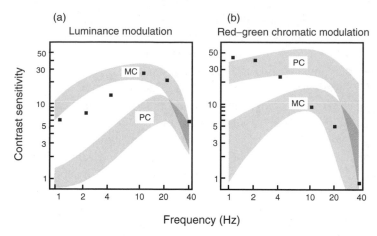

Figure 7.16   The hatched areas illustrate the sensitivity to temporal modulation of luminance and red–green chrominance for some of the most sensitive retinal MC and PC cells. The sensitivity measure is based on cone modulations. The black squares plot the mean contrast sensitivity of six human subjects for the same stimuli at the same eccentricity (10° to the side of the fovea). MC cells and human psychophysics give about the same results for luminance modulation (a). For red–green chromatic modulation (b), PC cells are about as sensitive as the human subjects at low temporal frequencies, but above 3–4 Hz the cells do better. A similar difference between psychophysics and single cell responses was found for blue–green flicker; only 'S–L' cells responded for frequencies above 10–15 Hz. Stimuli were 4° in diameter and the retinal illuminance was 1400 td (redrawn after Lee *et al*., 1989).

In Figure 7.16(b), for isoluminant red–green temporal changes, MC and PC cells have changed roles. For this stimulus, the PC cells are the more sensitive. Their sensitivity does not change much with frequency from 1 to 10 Hz, but decreases for higher frequencies. The second harmonic response of MC cells now has the lower contrast sensitivity by a factor of between 5 and 10. PC cells behave as low-pass filters for chrominance and as band-pass filters for luminance, whereas MC cells are band-pass for both.

The black squares in Figure 7.16 represent human psychophysical sensitivity to the same stimuli under the same conditions as those used during the electrophysiological recordings described above. In Figure 7.16(a), the correspondence is relatively good between psychophysical results and the luminance sensitivity of MC cells, supporting the notion that luminance thresholds are determined by the MC system (the higher number of PC cells by a factor of 6 and the possibility of summation between cells would not give the necessary sensitivity). For red–green chrominance, the correspondence with PC cells is better for low than for medium and high temporal frequencies. The psychophysical chrominance sensitivity decreases much more rapidly with high frequencies than does PC cell sensitivity, and at the medium and higher frequencies it compares best with the sensitivity of MC cells. These psychophysical results correlate with the psychophysical results of Swanson *et al*.

(1987), and those of Noorlander and Koenderink (1983) that are reproduced in Figure 5.35.

In exploring the luminance and chrominance dimensions, and the combination of both, Kremers *et al.* (1992) used light-emitting diodes to measure the sensitivity of human subjects and of macaque retinal ganglion cells. They found that, when luminance and chrominance were combined, the psychophysical sensitivity could be described by a convolution consisting of separate luminance and chrominance components, with physiological correlates in MC and PC cells, respectively. Modulation sensitivity was determined by the chrominance mechanism below 3–4 Hz and by the luminance mechanism above this frequency.

It is remarkable that PC cells respond to much higher temporal frequencies than are perceived. For instance, at the flicker null in a heterochromatic flicker experiment, PC cells are responding vigorously. It seems necessary to postulate some low-pass filtering of their signals in the cortex, as discussed below. This is an interesting example of a neural response that apparently is not used behaviorally.

Dependency on adaptation luminance

The dependency on adaptation luminance is somewhat different for MC luminance contrast sensitivity and PC chrominance contrast sensitivity, as shown in Figure 7.17(a) and (b) (Lee *et al.*, 1990). For MC cells and low temporal frequencies, luminance contrast sensitivity does not change much with light adaptation. MC cells display a Weber-like behavior, with constant contrast sensitivity. This result is similar to the behavior of humans in psychophysical experiments [see Figure 7.18(a)]. Since PC cells do not show Weber behavior for luminance contrast (not shown), this is an additional indication that, in primates, it is the phasic MC cells that are responsible for luminance threshold sensitivity at low temporal frequencies. The fusion frequency (the high frequency for which temporal changes are no longer observed) for luminance contrast increases with retinal illuminance for both cell types, but it is always higher for MC cells. In Figure 7.17(a), MC cells could follow up to about 90 Hz for 2000 td, whereas the PC cells reach about 55 Hz for luminance modulation at the same retinal illumination. The results for chrominance will be dealt with below.

Figure 7.18 presents human psychophysical sensitivities in the same experimental situation, and using the same apparatus as that applied to obtain the results of Figure 7.17 for macaque ganglion cells. The correspondence between psychophysics and electrophysiology is relatively good for luminance modulation at low retinal illuminances. For higher adaptation levels, there is still reasonably good correspondence at low frequencies, but there is a significant deviation for frequencies above about 10 Hz (see also Figure 4.28). Whereas one can measure a response modulation in MC cells up to about 80–90 Hz, the human subjects could not see 100 percent modulated stimuli at frequencies above 50–60 Hz in the same situation. It is not clear what causes this difference. In the literature it has been interpreted as some sort of

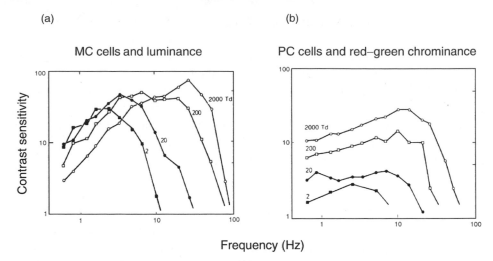

Figure 7.17 (a) The modulation sensitivity (contrast sensitivity) of MC cells for different levels of retinal illuminance (light adaptation). The ability of MC cells to follow high-frequency modulations improves as retinal illuminance increases. At 2 td the cut-off is about 10 Hz and increases to about 90 Hz at 2000 td. At the lower frequencies contrast sensitivity is relatively unaffected by the light level (following Weber's law). The frequency at which sensitivity peaks shifts from a few Hz to about 30 Hz with increasing adaptation level. (b) The sensitivity of PC cells for isoluminant, temporal red–green chromatic modulation. At all frequencies as light level increases, sensitivity improves (it does not follow Weber's law). The highest modulation frequency that can be followed by PC cells is between 50 and 60 Hz (from Lee *et al.*, 1990).

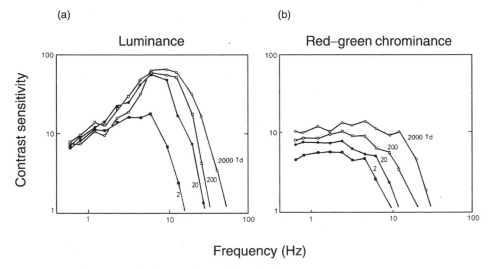

Figure 7.18 Human psychophysical sensitivity for the same stimuli as in Figure 7.17. Luminance sensitivity in (a) compares well with that of MC cells for low temporal frequencies, but sensitivity is lower for the higher frequencies. Sensitivity to isoluminant red–green modulation as a function of frequency (b) also decreases faster than is the case for PC cells in Figure 2.17 (from Lee *et al.*, 1990).

'cortical filter' that removes the high frequency signals coming from the retina (Lee *et al.*, 1990).

The similarity between physiology and psychophysics is less good for chrominance modulation, as illustrated by Figures 7.17(b) and 7.18(b) At 2000 td and 200 td, psychophysical sensitivity is less than PC cells' sensitivity for frequencies above 4–5 Hz. Again one may explain this in terms of a cortical filter. For the lowest adaptation levels, the relationships are reversed: at 2 and 20 td, the chrominance sensitivity is greater for the psychophysical task. There are no data for the sensitivity of MC cells to chromatic stimulation in the same situation. Such data would be of interest since there is a possibility that the second harmonic response of MC cells, such as that shown in Figure 7.14(b), may contribute to detection of chromatic differences.

Dynamic processing of contours

MC cells have about 10 times higher contrast sensitivity for white light than do PC cells (Figures 7.6 and 7.8). Figure 7.8 shows spatial contrast sensitivity for some MC and PC cells in the geniculate of the macaque monkey, and suggests that the mentioned sensitivity difference holds quite well up to about 3 c/deg. The convergence at the high end of the frequency scale suggests that there is little difference in spatial resolution between the two types of cells.

Figures 7.14 and 7.15 show examples of the frequency doubling in the responses of MC cells to an isoluminant red–green contrast. This behavior has not yet been traced back to a property of the receptive field. The fact that MC cells respond to chromatic contours (Figure 7.3) may, however, help understand this strange response property. Figure 7.19 shows the activity of an MC I-cell in response to borders of different luminance contrasts moving across its receptive field [as shown in Figure 7.19(c)]. In one case (a) the contour was between two achromatic stimuli, and in case (b) it was between white and red (642 nm). The contour moved with a speed of 4 deg/s. With a high luminance contrast across the contour (to the far left and to the far right in the figure), the cell gave a transient increment response (only to the direction of movement that corresponded to an incremental stimulus). However, at equal luminance ($Y = 1$) the white–red contour elicited a response for both directions of movement (corresponding to the second harmonic of Figure 7.14 when using homogeneous fields).

The situation of Figure 7.19 is similar to that used in the psychophysical task of trying to match two colors in luminance by means of the 'minimally distinct border' method (MDB method). Such studies have shown that the strength of the residual distinctness of the border between two adjacent colors at equal luminance (at the MDB point) is directly proportional to the absolute difference $\Delta|M–L|$ across the border, between the excitations in the M- and L-cones (excitations being normalized and equal for the white stimulus; Valberg and Tansley, 1977). The excitation difference that correlates with border distinctness for wavelengths compared with

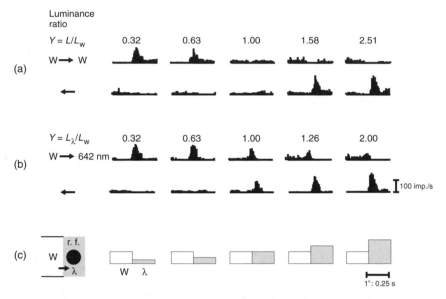

Figure 7.19   The responses of an MC Increment cell to a contour moving across its receptive field (r.f.). The stimulus was white on the left side and white or red (642 nm) on the right side. For each new response histogram in the same row, the luminance of the right field was changed as shown at the bottom of the figure. In the upper recordings of (a) and (b) the movement is towards the right whereas in the lower recordings it is towards the left. For white light, the cell responded only to an incremental luminance step, whereas for an isoluminant white/red difference ($Y = 1.0$), it responded for both directions of movement. This latter response is proportional to $|M–L|$, the absolute value of the difference between the L- and M-cone excitations across the border (adapted from Kaiser *et al.*, 1990).

white at equal luminance, has been called 'tritanopic purity'. It would correspond to spectral purity for a tritanope lacking S-cones.

Figure 7.20(a) shows tritanopic purity, p_t, as a function of wavelength. The corresponding measure for the actual white light, relative to equal energy white, is represented by the dashed horizontal line. Figure 7.20(b) gives the absolute value of the difference relative to white light, Δp_t. Figure 7.21 plots the responses of two cells, an MC and a PC cell, as a function of the absolute purity difference $|\Delta p_t|$ between two chromatic stimuli and white. Proportionality between response and purity applies to the response of MC cells only. PC cells reach saturation for relatively low values of tritanopic purity. For opponent cells with S-cone inputs the response shows no correlation at all with tritanopic purity, and it follows that only cells with some kind of opponency between L- and M-cones can contribute to the processing of a residual contour between colors at isoluminance.

The red–green opponent response of MC cells seems in some way to be related to this residual contour. The MC cells' response to isoluminant contours and a human subject's psychophysical scaling of this contour distinctness are relatively alike, as is

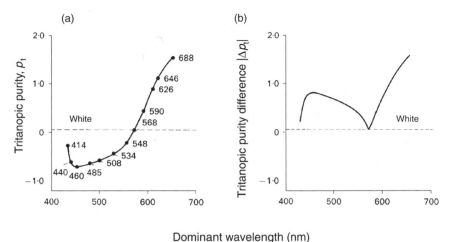

Dominant wavelength (nm)

Figure 7.20 (a) Tritanopic purity, p_t, as a function of dominant wavelength of near-spectral lights. The dashed line represents tritanopic purity of the white light used in the experiment relative to equal energy white of zero purity. The difference in tritanopic purity, Δp_t, for spectral lights and our white light is simply the distance between two points on the y-axis. (b) The absolute tritanopic purity difference, $|\Delta p_t|$ for different dominant wavelengths relative to the white stimulus used in the neurophysiological recordings. (Reproduced from Valberg *et al.*, 1992, *J. Physiol.* **458**, 579–602 by permission of The Physiological Society.)

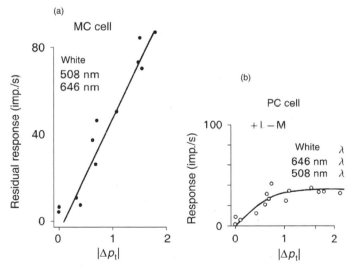

Figure 7.21 In (a), the amplitude of the residual response of an MC cell to the contour between isoluminant colors is plotted as a function of |M–L|, the absolute difference between the cone excitations L and M across the border, i.e. as a function of $|\Delta p_t|$. Regardless which wavelengths are combined with a white reference, the response is proportional to the linear difference of excitations. A linear relationship was found for all MC cells, but with somewhat different slopes. The response of a PC cell did not show the same linear relationship, as demonstrated by the example in (b) of an 'L–M' Increment cell. (Reproduced from Valberg *et al.*, 1992, *J. Physiol.* **458**, 579–602 by permission of The Physiological Society.)

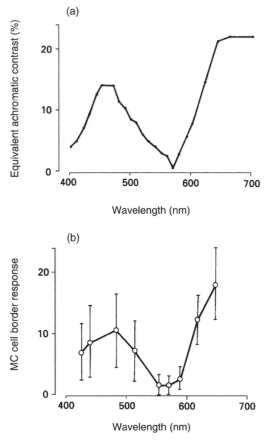

Figure 7.22   The graph in (a) plots the psychophysically rated distinctness of a contour when monochromatic lights of different wavelengths form a border with a white stimulus of the same luminance. The scale on the y-axis is the contrast of an achromatic luminance border that produces the same border distinctness. This relationship should be compared with the plot in (b) showing the residual response at isoluminance of 21 MC cells, scaled relative to their response at 20% achromatic contrast. The cell responses resemble the psychophysical data in form and absolute magnitude, indicating that the residual, second harmonic response of MC cells is proportional to border distinctness between adjoining isoluminant colors. (Reproduced from Kaiser *et al.*, 1990, *J. Physiol.* **422**, 153–183 by permission of The Physiological Society.)

demonstrated in Figure 7.22. Figure 7.22(a) presents the result of a psychophysical experiment where the task was to select an achromatic border contrast that matched the distinctness of a border between a white stimulus and a juxtaposed isoluminant test wavelength. In Figure 7.22(b) are shown the residual responses (normalized second harmonic) of MC cells to isoluminant borders between white and chromatic fields. The wavelength dependence of this response is quite similar to the magnitude of the matching equivalent achromatic contrasts of Figure 7.22(a), suggesting a common origin. The origin and mechanism of the rectified |M–L| signal, reflecting the difference in excitations of M- and L-cones, is still unknown. There are indications that opponency may arise as a result of a relative small substructure (since the chromatic border response is relatively sharp) in MC cells that combines L- and M-cones differently from the sum giving rise to $V(\lambda)$.

Hyperacuity

Despite the fact that MC cells make up only about 10 percent of all retinal cells, we have seen that they play a role in contour and form vision. Recent studies indicate that MC cells are able to account for *hyperacuity*, a visual capacity that makes it possible to read off a vernier scale (e.g. on a slide rule) with an accuracy that corresponds to an angle of only a few arcsec between the marks (Lee *et al.*, 1995). Hyperacuity operates at luminance contrast levels that are so low that it is not likely that PC cells are involved in the task. The finding that chromatic contrast is as effective as luminance contrast in Vernier performance when stimuli are normalized to detection thresholds (Krauskopf and Farell, 1991), can probably be explained by the frequency-doubled response of MC cells to red–green chromatic modulation.

This result and its interpretation imply that there must be some interpolation between MC cells within the neural net. The location of a contour can be established with relative spatial coarse detection mechanisms (Watt and Morgan, 1984; Shapley and Victor, 1986), thus providing a capacity for fine spatial discrimination. Such interpolation may also explain some odd results when color differences are used to produce Kanizsa's illusory triangle in Figure 1.12. The appearance of a masking triangle in this illusion requires luminance contrast in order to work (Livingstone and Hubel, 1988). If one uses color to bring forward the illusion, the illusory triangle is likely to disappear at isoluminance. Thus, MC cells may be more strongly involved in contour perception than their relative number seems to imply.

Defocus

Defocusing an image on the retina, for instance by putting an additional +5 diopter lens in front of the eye, reduces perceived luminance contrast by a significant amount, whereas the strength of chrominance contrast is hardly affected (Seim and Valberg, 1988, 1993). This surprising difference in the perception of grays and chromatic color (which can be deduced also from the spatial contrast sensitivity curves of Figure 4.29) may be important for persons in whom the retinal image is diffuse, whether it be from cataract or an uncorrected refraction error. For them, chrominance contrast may allow objects that would otherwise disappear to remain visible, provided they are not too small.

In situations where the detection of low luminance contrast of achromatic stimuli is accomplished by MC cells, one would expect from the above results that the

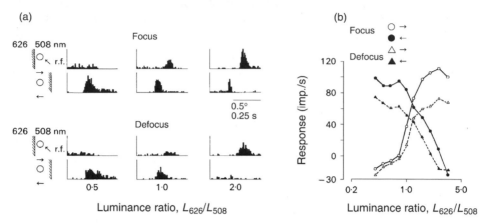

Figure 7.23 The defocusing of a contour by a +5 diopter lens in front of the eye leads to a reduced response in retinal MC cells. In this case a border between 626 and 508 nm stimulus was moved back and forth over the receptive field (r.f.). The original response histograms for both directions of movement are shown in (a) for three luminance ratios between the two wavelengths. After defocus, the transient response is reduced and smeared out in time. Nevertheless, the diagram in (b) shows that in both conditions, the response curves for the two direction of movement intersect at the same luminance ratio, indicating that the cell's isoluminance point remains the same (Reproduced from Valberg, *et al.*, 1992, *J. Physiol.* **458**, 579–602 by permission of The Physiological Society.)

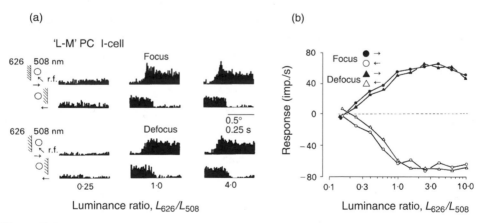

Figure 7.24 Defocusing a contour by +5 diopters in front of the eye does not significantly affect the response of an opponent PC cell. As in Figure 7.23, a border between 626 and 508 nm light was here moved across the receptive field (r.f.) of an 'L–M' Increment cell. In (a) we see the typical sustained response histograms for the sharp image, and the somewhat less brisk response change with defocus. In (b) the firing rate for the same cell is plotted as a function of the luminance ratio between the wavelengths that form the moving contour. There is no significant difference in the response to focus and defocus. (Reproduced from Valberg *et al.*, 1992, *J. Physiol.* **458**, 579–602 by permission of The Physiological Society.)

responses of these cells would be diminished by defocus, whereas the chromatic responses of opponent cells would be largely unaffected. To test this, we measured how defocus actually affects the responses of MC and PC cells. Figures 7.23 and 7.24 show the results of such an experiment, using a +5 diopter lens in front of the animal's eye. Figure 7.23 shows the border response of a retinal MC cell, with and without defocus, for equal luminance between the two colors as well as for luminance contrasts. Firing rate is plotted as a function of luminance ratio in Figure 7.23(b), with the two curves being for movement of the border in opposite directions. The response amplitude, measured in a time window of 30 ms before and after the contour moved over the receptive field, is reduced by 30–40 percent after defocus. The responses are smeared out in time, but the curves cross for the same luminance ratio (at equal luminance) whether the contour is sharp or blurred. This means that defocus has not changed the equal luminance ratio for the cell.

The effect of defocus on the response of an opponent PC, I_{L-M} cell is shown in Figure 7.24. This response, too, is less sharp in time after defocus. The firing rate, measured in the same time window as in Figure 7.23, is plotted as a function of contrast in Figure 7.24(b). One sees that the response is virtually unaffected by defocus. The effect of defocus on contrast perception is large for luminance contrasts, attenuating the response of MC cells. However, defocus has only a minor influence on the perception of chromatic differences, a fact that seems to be reflected in the preserved responses of PC cells.

8 Brain processes

Cortical organization and vision

The neural processing of hearing, vision and motor activities (e.g. speech) is localized to specific, highly organized brain structures. In vision, image features such as lightness, orientation, color and movement are treated by separate neural populations. In the retina, these and other aspects of the visual image are implicit in the responses of the same receptors and ganglion cells (often called multiplexing), whereas in the higher brain centers, the processing of different features appears to be distributed over separate cortical areas. A particular region or object in the field of view is represented by the activity of cells in several distinct functional units, with movement being represented in one place and color in another.

This idea has a precursor in *phrenology*. About 200 years ago, the Austrian F.J. Gall claimed to have identified between 30 and 40 cortical centers specialized for different emotional and intellectual capacities simply by studying the correlation of skull topography with different mental faculties. Pierre Flourens, a Frenchman, criticized this view at the beginning of the nineteenth century. Based on experimental studies, he concluded that mental functions are not localized, but rather that all areas of the brain participate in all mental activity.

The British neurologist J. Hughlings Jackson challenged this latter hypothesis in the middle of the nineteenth century. His clinical studies of epilepsy pointed to the possibility that different motor activities, as well as the processing of different sensory modalities, might be localized to particular cortical areas. Later, the German neurologist and psychologist Carl Wernicke and the Spanish histologist Ramond y Cajal systematically extended these studies towards a theory identifying the

Light Vision Color. Arne Valberg
© 2005 John Wiley & Sons Ltd

individual neurons as the signal elements of the brain. These neurons form connections with each other and can be grouped into functional units.

The scientific dispute did not end here. At the beginning of the twentieth century the importance of brain areas, neurons and their connections was not completely accepted. Only after about 1930 did it become clear that physiological functions, in particular, were localized to specific regions of the brain. This conclusion was reached after observing patients with brain injury as a result of war, accidents, stroke or poisoning from carbon monoxide. Today, there is fair agreement that specialization of different regions is an important principle in the organization of the brain (Kandel *et al.*, 2000).

Visual centers and areas

The retina is a part of the brain, and we have in previous sections dealt extensively with its various properties. Although retinal cells, and cells in the LGN, are not very specialized, we have seen several examples of cell sensitivity being correlated with psychophysical sensitivity.

The cortex is a thin layer of gray matter densely packed with active neurons. It is between 1.5 and 4.5 mm thick, and exhibits extensive folding. In humans, about one-third of the cortex is hidden in these folds or sulci. The brain contains roughly 10^{11} neurons, with about 10^{15} synapses and more than 3000 km of nerves. Neurons with similar tasks are gathered in specialized, interconnected areas. In the human cortex, about 40 areas have been found to be specialized for vision (Van Essen *et al.*, 1991). Many of these areas are subdivided into smaller units. For instance, *primary visual cortex*, also called area 17, striate cortex or V1, is organized in columns of cells that, for a given position in the visual field, respond to certain attributes of an object, such as the orientation of borders (horizontal, vertical or somewhere in between). Neurons that deal with color also seem to be assembled in specific columns. Most areas of visual specialization are represented in both halves of the brain. There are, for instance, two V1 areas, where V1 of the left brain receives information from the right visual field, and V1 of the right brain half receives information from the left visual field (see Figure 8.1). These separate representations are unified by a commissure, the *corpus callosum*, that links the two hemispheres.

Beyond area V1 there is some discussion as to the proper identification and partitioning of visual areas and which criteria should be used. Obvious criteria would be that the visual areas should be activated by visual stimuli and that each area should represent an independent and more or less complete map of the contralateral visual field. Visual areas can be differentiated by features such as a characteristic pattern of anatomical connections, identifiable and unique functional properties, and a distinctive architecture (Zeki, 2003).

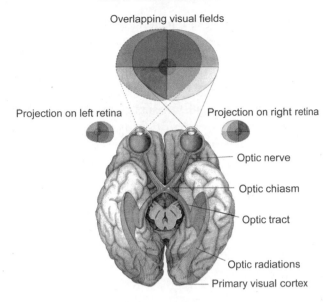

Overlapping visual fields

Projection on left retina

Projection on right retina

Optic nerve

Optic chiasm

Optic tract

Optic radiations

Primary visual cortex

Figure 8.1 The visual pathways from eye to higher brain centers. The visual image on the retina is inverted relative to that of the visual field in such a way that up is down and left is right (shown by the color code in the figure). Signals from the right half of the retina are sent to the right half of the brain by way of the optic chiasm and the LGN. From the geniculate the signals are led to the visual cortex via the optic radiation (from Posner and Raichle, 1994). (See also color plate section.)

Lateral geniculate nucleus

After the retina, the first center with visual neurons is the LGN. Figure 3.24 displays the location of these nuclei midway between the eyes and the visual cortex, one for each half of the visual field. Figure 3.25 shows the layers and Figure 3.26 the projections to primary visual cortex. In current brain theories, the LGN has sometimes been called the 'gatekeeper to the brain' (Crick, 1994). The reason for this is that the LGN seems to operate as some sort of filter, to a large extent deciding what visual information should be passed on to higher brain centers. Crick speculated that, because of this control function, the thalamus and the LGN might play a role in consciousness (Crick, 1994).

The geniculate body is about the size of a large pea, and has six main layers of cells. Every cell receives input from one eye only, there are no binocular cells in the LGN. The top four layers containing parvocellular cells receive inputs from the midget ganglion cells of the retina. Two layers receiving input from one eye are interleaved with a layer with input from the other eye. The bottom two layers are reserved for the magnocellular cells and receive inputs from the retinal parasol cells. Between the lamina of PC and MC cells, and between the PC cell layers one finds the KC cells with S-cone inputs which receive projections from the bi-stratified 'S–L' cells in the retina (and probably from monostratified 'M–S' cells; see Figure 3.19). Cells in the

LGN exhibit many of the same response patterns as the corresponding ganglion cells in the retina, differing only in the level of spontaneous firing, which is somewhat lower in the LGN.

Geniculate cells project to primary visual cortex, V1, predominantly to subdivisions of layer 4 (Figures 3.26). Layer 4 contains sublayers A, B and C, with the latter subdivided into $4C\alpha$ and $4C\beta$. In monkey visual cortex, PC cells project to layers 4A and $4C\beta$ and the upper part of layer 6. MC cells supply cells in $4C\alpha$ that in turn project to 4B. KC cells seem to send their axons to layer 1 or directly to the cytochrome oxidase blobs of layer 3 of V1, and not to a subdivision of layer 4, as do geniculate MC and PC cells. Although cortical neurons have long been thought to receive segregated inputs from geniculate magnocellular and parvocellular cells (in layers $4C\alpha$ and $4C\beta$, respectively), some recent data suggest that convergent excitatory inputs from both PC and MC types are also present (Vidyasagar *et al.*, 2002).

Primary visual cortex

The visual cortex is found at the back of the brain, on either side of the cleft that separates the two brain halves. Following the mapping of brain areas by K. Brodmann, the visual center can be subdivided into areas 17, 18 and 19. The fibers of the *optic radiation* from LGN project to area 17. Areas 18 and 19 receive inputs from area 17, but may also receive some direct input from the geniculate nucleus. Semir Zeki has revised Brodmann's map of the visual areas, and for the macaque monkey he advocated a subdivision of the old areas. V1 corresponds to area 17 (V stands for Visual). Area 18 is subdivided into V2, V3 and V4, whereas area 19 corresponds to V5. Figure 8.2 shows this partitioning of the monkey brain areas after Zeki. Area V1 is surrounded by V2. In humans, the topography of infoldings, of *gyri* ('hills') and *sulci* ('valleys'), will be a little different.

The colors in Figure 8.3 shows the projections of the visual field onto V1. One point in the field (and on the retina) corresponds to one point in V1. The fovea projects to a disproportionately large area in V1, a disparity that is referred to as 'cortical magnification'. About half of the neurons in LGN and in V1 represent the fovea and the area immediately surrounding it.

From Figures 8.2 and 8.3 one can see that the fovea is relatively well exposed on the surface of V1. This allows us to measure temporal changes in the aggregate electric potential at the scalp (visually evoked potentials, VEP). These measurements are typically made using skin electrodes to detect the difference in electric potential between a location just over V1 and some reference location.

Cortical magnification

The cortical magnification factor, M, given in mm/deg, is a useful magnitude for the linear magnification in the projection of a retinal area on to visual cortex. The unit is

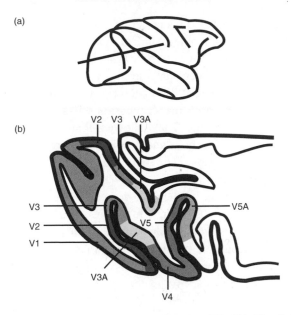

Figure 8.2 Zeki's divisions of the visual cortex into areas V1–V5. The division (b) refers to a horizontal cut through the brain of the macaque monkey as shown in (a) (after Zeki, 1993). (See also color plate section.)

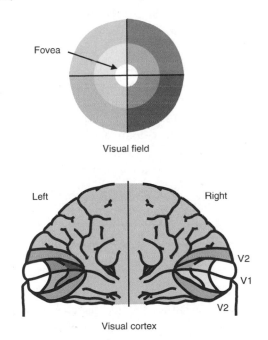

Figure 8.3 Visual field projections to primary visual cortex, V1, of the two brain halves. The foveal projection area, which is exposed at the tip of each lobe, has the greatest area, whereas the more peripheral parts of the visual field project to smaller areas and are hidden more deeply in between the two halves. Area V2 encloses V1. (See also color plate section.)

in millimeters of cortex per degree visual angle and gives the ratio between a length in the retinal image of a small object and the extent of its projection in V1. The magnification is highest in the fovea, where it is 7.75 mm/deg (Virsu and Rovamo, 1979), and decreases for greater eccentricities. At 1.5° eccentricity M equals 5.6 mm/deg, at 14° it is 1.2 mm/deg, and at 30° eccentricity it is down to 0.5 mm/deg.

Virsu and Rovamo have used the expression 'M-scaling' to describe spatial frequency in cortical projections. They showed that the contrast sensitivity for sinusoidal gratings was close to constant when spatial frequency was computed in periods per millimeter of cortex. The maximum contrast sensitivity was found between 0.5 and 1 periods/mm. Thus, when a grating is M-scaled (that is, the whole grating and *not* its spatial frequency), detection threshold is about the same at all eccentricities on the retina. It has also been claimed that there is direct proportionality between visual acuity and the cortical magnification factor. It has been a matter of some debate as to whether magnification factor is proportional to ganglion cell density over the visual field. This is approximately the case, but there may be extra magnification in the fovea.

Eye dominance and orientation selectivity

The visual centers of the brain are highly structured into what seems to be functional subunits or modules that organize neurons according to their selectivity for a wide variety of stimulus features. These include position on the retina, orientation, eye dominance, direction of motion, binocular disparity, chromaticity and spatial and temporal frequency. Figure 8.4(a) shows a schematic drawing of the orderly sequence of eye dominance columns and orientation modules in primary visual cortex. These columnar structures contain cells that treat information about eye, orientation and color of a stimulus. In Figure 8.4(a) eye dominance columns and orientation columns are represented as being perpendicular to each other. More recent measurements have shown that the cells selective to a particular orientation radiate from a non-orientation selective center in the middle of a column [Figure 8.4(b)], the center cells being selective for the color (spectral distribution) of a stimulus. Cells of different orientation selectivity are positioned on circles around this center, and the iso-orientation contours meet the border of the ocular dominance columns roughly at a right angle. One module of cells that treats all orientations receives information from one point in the retina. This represents a *retinotopic* order in that neighboring modules treat the same information from neighboring retinal areas. In short, the striate cortex is composed of repeating modules that contain all the cells necessary to analyze small regions of visual space for a variety of stimulus attributes. Thus, the same orientation for different retinal locations is represented in V1 by a regular and patchy arrangement of neurons with a spacing of less than 1 mm.

The so-called 'simple cells' (see for instance Heggelund, 1985) in V1 have elongated receptive fields and are orientation-selective. Hubel and Wiesel (1963)

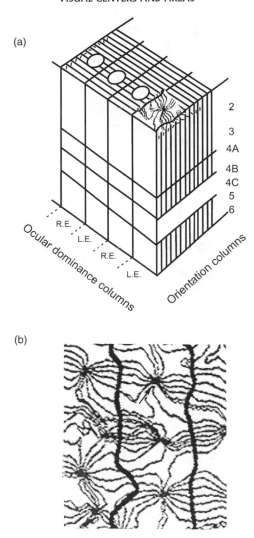

Figure 8.4 (a) The traditional view of V1 functional organization is one of ocular dominance columns and orientation columns crossing each other at right angles. For each position on the retina, an orientation column contains cells selective for 180° in about 10° steps. More recent findings, illustrated in (b), indicate that cells selective to the same orientation are positioned on 'spokes' of a wheel centered on a non-orientation selective cytochrome oxidase 'blob'. The cortical depth layers are numbered from 1 to 6. Color coding cells are abundant in the blob regions 2, 3 and 6 outside layer 4C of the eye dominance columns.

thought that the receptive fields of these cells arose from a convergence of inputs from cells with concentric, circular fields arranged in a row with a particular orientation on the retina, but the exact mechanism of orientation specificity is still a matter of dispute. For the simple cortical cell shown in Figure 8.5, the most effective stimulus (the stimulus giving the largest response) would be a bar with

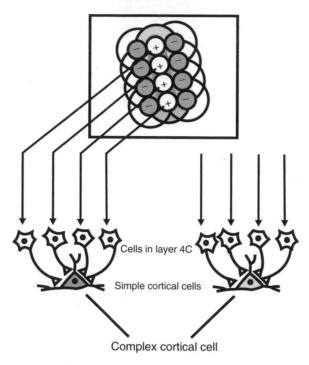

Cells in layer 4C

Simple cortical cells

Complex cortical cell

Figure 8.5 Convergence of input to a cortical simple cell from cells in layer 4C has been a popular model for orientation selectivity. The receptive field of a simple cell receives input from cells with circular receptive fields positioned after one another along a line of a particular orientation. This cortical cell responds best to a bright rectangle that fills the elongated receptive field center. A stimulus oriented at right angles to the long axis gives little or no response. In a hierarchical model, simple cells in turn converge on complex cells.

positive contrast, and with an orientation corresponding to that of the receptive field. A stimulus oriented at 90° relative to the long axis would be the least effective. Other neighboring cortical cells would have similarly shaped receptive fields, but with the long axis oriented at a slightly different angle. Every retinal position has cortical cells with receptive fields of all orientations.

In a plane through V1, parallel to its surface, eye dominance columns for the right eye alternate with those for the left. Using special radioactive marking techniques to mark cells showing high metabolism during responses, these alternations will pop out as zebra-stripes (see Figure 8.6). Color-sensitive cells would, by the same technique, show up as small islands or 'blobs' inside the weave of orientation and eye dominance columns. A similar regular arrangement to that known for orientation-selective cells in V1 has been reported for the cortical organization of color-selective cells in V2, where cells selective for the hues on a color circle are positioned in successive locations (Xiao *et al.*, 2003).

Surface of
ocular dominance columns

Figure 8.6   Radioactive tracing of nerve fibers from one eye leads to a faint radioactive radiation from cortical cells that receive information from that eye. When this cortical tissue is placed on a photographic plate, a zebra-stripe pattern develops as the plate is exposed to radioactive and non-radioactive cortical cells receiving information from either the left or the right eye.

Cells in V1 also respond to extended contours, and such contour integration is probably achieved by interconnections between orientation selective cells in V1. The length of these horizontal interconnections covers about 4° of visual angle at 4° eccentricity (corresponding to about 7 mm in the cortex), a length that agrees with psychophysical contour integration (Stettler *et al.*, 2002). Experiments on humans and monkeys have demonstrated strong attentional modulation of V1 activity, including contour integration, indicating a possible top-down influence of feedback from higher centers to V1.

Color processing

Color not only aids the discrimination and recognition of objects, it also adds quality and aesthetic values to our visual world. What do we know about the contribution of higher brain centers to the perception of color and brightness? In view of the surprisingly many correlates to perceptual attributes of color that have been found at retinal and geniculate levels (see the previous chapter), one might ask whether there are (object) color properties that have not been sufficiently explained at the lower levels.

Color matching experiments have demonstrated important linear properties of the cone receptors that have led to the quantification of color stimuli and to a flourishing color technology. Cone opponency already appears as a property of the bipolar and ganglion cells in the retina, and of the parvo- and koniocellular relay cells of the geniculate. Threshold sensitivities of MC and opponent pathways correspond to psychophysical detection and discrimination thresholds in a number of temporal and spatial tasks. Relatively simple linear rules, applied to the supra-threshold inputs of geniculate opponent cells to V1, can account for the transformation to magnitudes that correspond to color perception (see the chapter on color vision). The physiological substrate for luminance (but not for lightness and brightness) has been found to reside in the magnocellular cells. However, we need a deeper understanding of peripheral and central processes in order to explain adaptation, color constancy, simultaneous contrast, color scaling, and temporal and spatial tuning.

Despite the considerable success in demonstrating correlates between early processing and an abundance of psychophysical data, higher level processing must be invoked to account for perceptual qualities. The transition from physical chromaticity to color attributes such as the unique hues, and the perception of white and black, for example, cannot be explained by early-level processes alone.

Color and luminance

The term 'color cells' has often been reserved for cells that differentiate between isoluminant chromatic stimuli. This has probably led to a misrepresentation of the number of cells participating in color coding. It has been found that the opponent chromatic information provided by LGN is largely retained in V1 (Ts'o and Gilbert, 1988; Lennie et al., 1990; Engel et al., 1997; Cottaris and DeValois, 1998), but it seems to be mixed with luminance in a way that also supports form information (Johnson et al., 2001; Gegenfurtner, 2003). As we have seen, opponent cells in the retina and the geniculate typically have a significant luminance response. In fact, pre-cortical opponent cells without a luminance response to iso-chromatic and achromatic stimuli are extremely rare (Lee et al., 1987; Valberg et al., 1987; Lankheet et al., 1998). In V1, a combination of chromatic sensitivity and luminance responses also seems to be the rule (Lennie et al., 1990; Johnson et al., 2001). The fact that some opponent cells in V1 are more spatially selective than those of the LGN has given rise to the notion that color-sensitive cells are involved in spatial coding and form vision as well as color. Multiple transformations of the signals received from the LGN seem to be the rule, giving rise to simple and complex cells, and orientation-selective luminance and color detectors.

> In view of the fact that opponent cells on many levels respond to achromatic colors and to luminance, one may ask what constitutes an early level 'chromatic response'. A simple requirement is the response differentiation with respect to wavelengths that comes with opponency, resulting in a biphasic spectral response curve with a 'zero crossing' between activation and inhibition at some neutral wavelength. A prerequisite for a 'chromatic response' is the discrimination between an achromatic stimulus and a specific range of spectral stimuli and chromaticities of the same luminance, regardless of the luminance response. To find a correlate with hue discrimination, however, it is

necessary to determine the relative strength of responses across different opponent cell types. The magnitude of a 'chromatic signal' resulting from combining the outputs of cells responding optimally along the cardinal axes is usually taken to correlate with color strength (Figure 6.23). In the physiological model of color vision presented earlier, we proposed that an achromatic response is subtracted at some level after the geniculate, by a comparison across opponent cells. It would therefore be interesting to know more about how cells in the cortex behave to white and pure achromatic stimuli in comparison to chromatic stimuli. However, such data are largely missing.

Wavelength tuning vs color

Tuning to color rather than to wavelength is a property that for a long time was believed not to occur before area V4 of the macaque, an area that today is considered a multimodal sensory area. According to Zeki (1983a, b), this was a true color area since the responses of cells in V4 accounted quite nicely for color constancy, V1 being selective only for differences in the wavelength composition (i.e. the spectral distribution) of a stimulus. This view called into question early findings of double-opponent cells in V1 (Michael, 1978a, b) and later evidence of the same (Thorell et al., 1984; Johnson et al., 2001) since such cells would support color constancy. Recently, responses of cells in V1 have been shown to be modulated by remote stimuli in such a way that they neutralize the influence of the overall chromaticity of the illumination (Wachtler et al., 2003) thus indicating that a fair part of color constancy may indeed be accounted for in V1, or even earlier.

Double-opponent cells

There are two major functional cell types in monkey V1, the simple cells and the complex cells. Simple cells respond selectivity to the position of a stimulus within the receptive field, while complex cells respond selectively to spatial patterns regardless of their position with respect to the receptive field. The spatial and chromatic tuning of the latter cells seem to differ in a complex manner.

As already mentioned, double-opponent, simple cells that exhibit cone opponency in the center as well as in the surround of their receptive field have been reported in the cortex of primates. Cells with an 'L–M' cone combination in the center and the opposite excitatory combination 'M–L' in the surround were commonly found. Other cells had the opposite arrangement. One also found cells with S-cone inputs, for instance 'S–(L+M)' in the center and '(L+M)–S' in the surround (here too the opposite configuration existed, although it was encountered less frequently). Double-opponent cells were first observed in the retina of the goldfish (Daw, 1968), but in the primates they do not show up before area V1. They may exist in at least two variants, the oriented types being more frequent in V1 than the non-oriented types. However, it is still a matter of dispute how common these cells are (Thorell *et al.*, 1984; Ts'o and Gilbert, 1988; Lennie *et al.*, 1990; Johnson *et al.*, 2001).

What would be the function of such cells? A cell with an 'L–M' center and an 'M–L' surround' is activated not only by long wavelength stimuli in their centers, but also by intermediate wavelengths in the surround. This corresponds to *lateral excitation*, a concept used by Valberg and Seim (1983) to describe color induction as a physiological process. A white spot surrounded by green would activate such a cell in the same way as red in its enter. Replacing the white central spot with a red spot would increase the firing rate to a higher level (Conway, 2001). It therefore seems reasonable to assume that such cells are involved in simultaneous color contrast.

When a white or a chromatic stimulus is covering the whole receptive field of these cells, the excitation of the center will be opposed and largely cancelled by the inhibition of the surround. A reddish illumination that excites the 'L–M' center of such cells will simultaneously inhibit it through the 'M–L' surround. Under a greenish illumination, the cell will be activated by the surround, and inhibited by the center. If we assume a high degree of symmetry between mirror opponencies, and balance between the associated spatial antagonisms in each double-opponent cell, the change in firing rate of such a cell to changing the illumination might well be negligible. Double-opponent cells therefore have been taken to be involved in chromatic adaptation and color constancy. Convergence of signal pathways from nearby simple I- and D-type opponent cells with spatially overlapping activating and inhibitory cone mechanisms in the receptive field is a possible mechanism for the field structure of double-opponent cells.

Cytochrome oxidase blobs

The cells in the blobs of the upper layers 2 and 3 of V1, and in layers 5 and 6, are rendered visible by staining with the mitochondrial enzyme cytochrome oxidase, an enzyme associated with high metabolic activity. The patches of cytochrome oxidase stain are arranged in parallel rows about 0.5 mm apart, corresponding to the centers of the ocular dominance columns. The blobs contain cells that are non-selective for orientation, a high proportion of which are opponent cells that respond to differences in cone activation. The functional specialization of the continuation of PC and KC cell pathways into the blob regions is still a matter of debate, and no clear picture has emerged from the data accumulated to date. Here we want to draw attention to some properties that are relevant for modeling color processing beyond LGN.

The findings of Ts'o and Gilbert (1988) suggest that 'blue–yellow opponency' is segregated from 'red–green opponency', with individual blob regions being dedicated to either of these two systems. Different parts of a blob may contain cells with either L- or M- center input with either an ON or an OFF sign. The recordings by Ts'o and Gilbert showed that cells in layer 4C often had identical 'color specificity' to the cells in the blobs located directly above, a finding that is consistent with direct projections from layer 4C to the next, more superficial layer, layer 3. It also appeared that cells in a particular blob interacted only with neighboring blobs having the same opponency (in a 'like talks to like' fashion). Similar connections were found outside the blob regions, between spectrally non-selective cells of the same orientation sensitivity, suggesting contour integration. Anatomical studies suggest a segregation of the intrinsic connections between the blob color system and the oriented, non-color selective inter-blob regions (Livingstone and Hubel, 1984). While this seems to support the idea of form and color being processed by different parvocellular cell populations, the existence of spectrally selective, oriented cells indicates a convergence of form and color information. Single blob cells respond to a wide range of contrast, over several log units, as if they combine inputs of cells operating over different contrast ranges. This higher level response may thus be accomplished by inputs from an ensemble of low level cells, each with different threshold sensitivity, as suggested by the data of Edwards *et al.* (1995). Close-range connections between cells in the same blob region may be involved in constructing new receptive field types, such as spectrally selective, oriented cells and units that combine Increment and Decrement cells of the same opponency. Long-range horizontal connections between cells with the same function may laterally adapt and optimize a cell's response and allow it to be influenced by context.

Modeling color coding

One obvious question to ask is: what happens in V1 to the relatively simple opponent inputs from the geniculate? There is a general consensus that the behavior of neurons in all layers of striate cortex is described relatively well by the model also used in the retina and the LGN, in which a cell responds to the weighted sum of signals from the three classes of cones. This implies nonlinear intensity–response curves also for cortical cells, with a hue selectivity of the combined response highly dependent on luminance ratio (Bezold–Brücke hue shifts; Figure 6.24). Cells in LGN lose their chromatic selectivity at very high intensity (Figures 6.10 and 6.24). In this model, an explanation of the Bezold–Brücke phenomenon needs no separate low-level achromatic channel (although achromatic responses need to be subtracted somewhere to achieve a pure chromatic response). The nonlinearity of intensity–response curves observed for cells in the retina and the LGN is a good enough explanation of psychophysical hue and chromatic changes with intensity. These hue and chroma shifts are also a convincing argument against the assumption that the response of 'color cells' must be independent of luminance. Addition and subtraction, and vector combination of input signals from LGN neurons to V1, constitute a general model that has the potential to explain many psychophysical data.

Most studies, if they explore relatively narrow ranges of intensity and cone-contrasts (limited by CRT monitors; Derrington *et al.*, 1984; Ts'o and Gilbert, 1988; Lennie *et al.*, 1990), or investigate larger ranges of stimulus intensity (Lee *et al.*,

1987; Valberg *et al.*, 1985b, 1987), agree that the LGN cells combine cone signals linearly. Is a linear transformation of the cardinal responses p_1 and p_2 in the LGN like those used to arrive at the (F_1, F_2) coordinate system of Figures 5.32 and 5.33 likely to reflect reality? Additional transformations is suggested by the fact that the wavelength tuning of cells in and after V1 appears to be different from that of the LGN; many more directions of color space are represented (Lennie *et al.*, 1990; Gegenfurtner, 2003). It also seems that other directions than the cardinal axes of the LGN are emphasized (Wachtler *et al.*, 2003). This may be accomplished by a broad range of LGN to V1 transformations in which a continuum of different weights are applied for the LGN cardinal axes p_1 and p_2 of Figure 6.19. Different relative weights of these cardinal directions would account for the maximum response being tuned to different wavelengths (or hues). For instance, a cell that integrates inputs from two orthogonal cardinal directions in the LGN (for instance S–L and L–M) would have a preferred hue at an intermediate direction, in this case a non-spectral purple hue depending on the relative weights of the inputs. If we add the requirement that equal color strength be represented by equal distances in the diagram, transformations similar to (F_1, F_2) would result. Such models predict responses that are highly dependent on the absolute and relative intensity levels (due to the initial nonlinearity in the intensity–responses of the cones). Earlier in this book we saw how important such low-level nonlinearities are in accounting for salient properties of color perception. These nonlinearities must, necessarily, be passed on to cortical units, via the LGN.

Separate I- and D-channels (ON- and OFF-channels) are required in order to preserve information about the direction of a stimulus change (lighter vs darker). In addition, I- and D-channels may also converge, as we have postulated in order to account for color processing for light and dark stimuli (see Figure 6.9 and 6.11). In this picture of early, retinal separation of some stimulus features, and the advantage of keeping them separated as well as combining them later, it seems that divergence and separation, as well as convergence and integration, run in parallel. Neurons in V1 appear to be more selective than cells at lower levels, but they too are multi-dimensional in that they respond to more than one of several features [orientation, disparity, spatial frequency (size), chromaticity, direction of movement, etc.]. Even with some degree of multiplexing present in the higher visual areas, serial selection of one or a few attributes from an object also appears to be an effective processing strategy. One example is separating the color and orientation of an object, with color opponency being established at an earlier stage and orientation selectivity later.

Another example of feature extraction might be edge or border detection between isoluminant chromatic areas. Red–green contour sensitive cells may be a combination of Increment and Decrement PC cells as in double-opponent cells, or represent a continuation of the MC pathway (as may complex cells). In MC cells, for instance, the red–green second harmonic edge responses (analogous to a red–green frequency doubling response at isoluminace) are well known [Figures 7.14(b) and 7.19; Lee *et al.*, 1989, 1993; Kaiser *et al.*, 1990; Valberg *et al.*, 1992).

If color perception is associated with the non-orientated, color-selective cells in the blob regions, synchrony, or coincidence detection between blobs and other locations, would appear to be required in order to bind together features and different parts of the same object. In other cases, the binding problem might be solved by a hierarchical system and response integration.

Image analysis techniques that minimize mutual information in natural, colored scenes have identified statistically independent image elements, or filters, that resemble the receptive field properties and color tuning of simple neurons in monkey V1. This suggests that the decomposition of spatio-chromatic information into statistically independent luminance, red–green, and yellow–blue orientation selective channels offers an optimum means of coding natural images (Buchsbaum and Tailor, 2001). At the very least, the analysis indicates that simple cells in and outside the blob regions can be viewed as filters that simultaneously reduce spatial and chromatic redundancy, and that the non-oriented blob cells are part of an independent and separate pathway (Caywood et al., 2001).

Higher visual areas

After V1, information processing appears even more complicated. The signal pathways of MC cells project onto the areas V2, V3 and V5 (also called the middle temporal area, MT). Whereas V3 analyzes form and depth, and probably also some aspects of motion perception, V5 (MT) seems specialized for processing movement. Some of these areas are likely to also receive input from PC cells. A simplified schematic route of signals from the retina to cortical areas and the tasks their activity is associated with is shown in Figure 8.7 (van Essen et al., 1991).

A body of accumulated data provides evidence for two parallel pathways from striate cortex (V1) to extra-striate areas, an organization originally suggested by Ungerleider and Mishkin (1982). Figure 8.7 reproduces a proposed scheme for the division of labor between the two visual streams. One system, called 'the ventral stream', leads from striate cortex via areas V2 and V4 to the *inferior temporal cortex* (ITC). For instance, the spectral selective cells in the blobs of V1 project to the thin stripes of V2 and from there to V4. The ventral pathway has been thought to deal with aspects that are important for discerning shapes, color and texture and for object recognition. Its neurons help decide *what* the stimulus might be. At the highest level of processing in this pathway, we find cells that respond preferentially to faces.

The other system, the *dorsal* or *parietal stream*, involves areas V2 and V3, and the movement-sensitive MT, also called V5. This pathway deals with the localization of a visual stimulus in space, i.e. determining *where* the stimulus might be. Neurons in MT are sensitive to the direction and speed of movement. It has been conjectured that these two streams are continuations of the retino-geniculate-cortical PC and MC

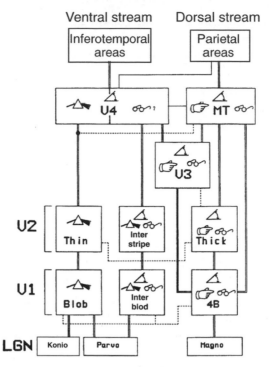

Figure 8.7 Parallel pathways from the LGN to the visual areas V1, V2, V3, V4 and V5 (MT) of the cortex. The symbols are meant to indicate the attributes of the stimuli for which the neurons in these pathways are functionally specialized (prism = wavelength; glasses = binocular disparity; angle = orientation of a contour; index finger = direction of movement; modified from van Essen *et al.*, 1991).

pathways, respectively. Even if there may be dominance of inputs from the PC cells over MC cells in the ventral stream, and a clear dominance of MC in the dorsal stream, interaction and cooperation between channels seem to appear in higher visual processing.

Cells associated with the ability to recognize faces of known persons seem to be located in a relatively small area in the ventral stream. People who suffer from *prosopagnosia* cannot even recognize their own face in the mirror, let alone the faces of friends or close family members, such as their own spouse or children. However, this does not prevent them from seeing that a face is a face and the attributes of a face, such as the nose, the mouth and eyes, and they are still able to interpret facial expressions. This would seem to demonstrate that the recognition of a familiar face and the analysis of facial expression as an expression of emotion are dealt with by different brain systems.

Place cells have been known to exist in the hippocampus of lower vertebrates (Moser and Paulsen, 2001), but until recently it was unclear whether this place coding had a homolog in humans. Recent recordings from hippocampus and the parahippo-campal region have now provided evidence for a human code for spatial navigation based on cells that respond at specific spatial locations and cells that respond to views and landmarks (Ekstrom *et al.*, 2003). A recently discovered area, human visuomotor area V6A, appears to analyze absolute position inside a room. The receptive fields of these 'position cells' do not change in space with gaze shift, thus encoding space independently of retinotopic coordinates.

Highly selective neural activity does not necessarily require conscious awareness. In anesthetized animals, cells in V1 and the next higher areas respond with their usual selectivity to visual stimuli presented within their receptive fields. What does the abundant data on alert and anesthetized animals tell us about the influence of consciousness on responses at different levels of visual processing? The proportion of neurons that are inactivated by anesthesia seems to increase at higher levels in the visual pathway. At the earlier stages of cortical processing only a few cells seem to be influenced by the level of awareness and linked to perception. In contrast to V1 and V2, it seems that a great majority of cells in inferior temporal cortex (ITC) of the ventral stream, and in area MT of the dorsal stream, are sensitive to the state of consciousness. Logothetis (2002) suggested that the small number of neurons whose behavior reflects conscious perception are distributed over the entire visual pathway rather than being localized to a single area in the brain. Other experiments, reviewed in Gazzaniga *et al.* (2002), have shown that the processing of sensory inputs is influenced not only by the state of awareness, but also by selective attention, imagery and expectations based on previous experience.

Brain injury from a stroke or trauma can occasionally result in cerebral *achroma-topsia*. This is a condition where the person has lost the ability to see colors. This color vision defect can manifest itself without any associated loss of visual field or visual acuity. To such people the world appears to be colored in shades of gray. Sacks (1995) has given a vivid description of such a case in the story of a painter who lost all chromatic vision after a car accident. He came to experience the world as if it were made from lead. Such cases support the idea that color is processed separately from other attributes of the visual image. According to Zeki (1993), this particular functional loss is usually associated with damage localized to the human color center, or human area V4. However, the deficits in patients with achromatopsia differ from those of monkeys with lesions in V4 (Kandel *et al.*, 2000) in that humans cannot discriminate hues but can differentiate shape and texture, the latter being difficult for the monkey. V4 in the monkey is probably not directly comparable with the human color area that is being affected in achromatopsia.

Recently, optical imaging techniques have revealed a spatially organized representation of hues in monkey area V2 much like the orderly sequence of orientation selective cells in V1 (Roe and Ts'o, 1995; Xiao *et al.*, 2003). The responses to different colors peaked at different locations in the thin stripes of V2, and the peak responses were spatially arranged in the order of the hues on a color circle.

Motion detection and direction sensitivity

The perception of form and movement probably arise from two independent neural networks. When something unknown approaches you at a great speed, you inevitably try to avoid it without stopping to determine whether it is a harmless piece of paper or something more dangerous. Physiological data indicate that the analysis of speed takes less time than the analysis of form or function. The neurons that deal with movement and speed are believed to project to other areas of cortex than those that handle form (Zeki, 1993). Damage to the 'movement channel' limits the ability to perceive movement, without affecting the perception of stationary objects. Damage to this channel may result in *movement blindness*.

The movement-sensitive neurons in V5 (MT) are clustered into columns of similar preferred directions. All movement directions are represented across the retinotopic map, and almost every cell in this area is direction-selective. In analogy with the organization of orientation-sensitive cells in V1, neighboring cells would be activated by a slightly different direction of movement, and as a consequence of retinotopic mapping, the same direction of movement for an adjacent position on the retina would be represented by other cells some small distance away. The perception of, for instance, the shape of a moving animal behind occluding trees would seem to depend on spatial interpolation of contours and temporal integration over the occluded body parts moving in synchrony (see Figure 1.4).

The strange phenomenon of movement blindness may be caused by damage to area V5. Zihl *et al.* (1983) reported a case of movement blindness in a woman who, after a stroke, had lost the ability to see moving objects, although she had no problem seeing them at rest. This caused significant problems in daily life. She would, for example, find it impossible to pour water into a glass. To her the water seemed frozen to ice, and she was unable to follow the water level as she filled the glass. She also had problems crossing streets. A car that was far away would suddenly be close without her having seen it approaching. This and other cases of cortical movement blindness are strong evidence that movement is, indeed, analyzed in a particular brain area.

In primates, a combination of orientation-sensitive cells and cells sensitive to the direction of movement is not encountered before the visual cortex. Neurons in the movement channel are selective to movement in a particular direction, for example in the direction of 2 o'clock, while they are unresponsive to movement in the opposite direction. Figure 8.8 gives a possible model for a possible structure of a neural network that is sensitive to the direction of movement. Owing to lateral inhibition operating in one direction only, a leading edge of bright light that moves from left to right on the retina sequentially activates the ON-cells (I-cells) in the intermediate layer that converge on the direction-sensitive cell at the bottom of the figure. Inhibition opposite to the direction of movement results in a transient increment response to the rightward passing of the dark/bright border and inhibition of the trailing bright/dark edge. A dark/bright border that moves towards the left inhibits the

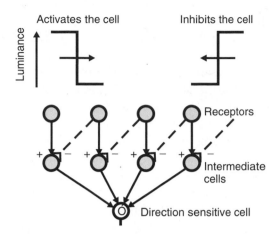

Figure 8.8 The structure of a simple model that explains a cell's selectivity for a certain direction of movement. The cell at the bottom of the figure receives signals from a set of cells with excitatory and inhibitory inputs; these intermediate cells are inhibited by a bright object (a luminance increment) that moves from the right towards the left (an inhibitory signal is projected ahead of the moving edge), and they are activated by a bright object (a luminance increment) approaching from the opposite direction.

intermediate cells just ahead of the border and keep them inhibited as long as the light is activating the receptors above. Thus the bottom I-cell, summing the inputs from the previous layer, will be activated maximally by one direction of movement and inhibited by movement in the opposite direction.

Vision in depth

Perception of depth in three-dimensional space requires cells that code for disparity, i.e. or the degree of non-correspondence between same scene images in the two eyes (Figure 3.7). Of course, this cannot happen before the signals from the two eyes are compared. This comparison is made by the binocular cells of the cortex. We find such cells in layer 4C in V1 and also in area V2. It seems that MC cells and the complex cells of V1 have an important role in stereopsis (see also Figures 1.2 and 3.9), but V3 and adjoining areas may also subserve depth vision.

The binding problem

Different aspects of vision, such as the perception of color, movement and the expression in a person's face would seem to arise in different, specialized parts of the brain. Damage to one of these areas, for instance by stroke, can result in bizarre perceptual phenomena such as, for instance, color without form, or form vision

without movement perception. Considering the reports of such and similar phenomena, one is led to conclude that, in order to perceive something as a meaningful
whole, many specialized areas of the brain must contribute in parallel to the percept.

After several feature dimensions have been multiplexed in the activity of cones and
retinal ganglion cells, is it reasonable to assume a separation at a later stage in which
each cortical area processes a different object feature, or a set of features? Several
aspects of the visual world need to be sorted out: first, there is the information about
the objects themselves and their surface properties (their size, shape, color, texture,
whether they are familiar or unfamiliar); second, the interrelations between the
objects making up the visual scene (constancy of size, color, etc., figure/ground,
grouping, movement speed and direction, etc.); and third, their relation to the
observer's space (spatial position, orientation, etc.). Are all these dimensions and
their sub-modalities separated out at some stage in the visual process? One might
argue that the large degree of selectivity for object features in neurons in the different
cortical areas and modules confers a certain identity on cells belonging to each
module. This identity defines one or several object-related properties. Do these
properties need to be physically integrated at some later stage, in order to correlate
and compare them with all other aspects of the visual scene, or are they somehow
autonomous? In a hierarchical model, object attributes may be processed at the lowest
possible level and the result passed on to higher neural levels (Lennie, 1998), such as
the information about light increments and decrements in I- and D-channels. If this
information were only to converge and be combined linearly at a later stage, the
feature identity would be lost. If each neuron in an area, say V1, were sensitive only
to a small region of a multidimensional feature space, a composite object might be
represented by an abstract multidimensional vector, as illustrated in Figure 8.9. These
vectors might be more or less fuzzy, ranging from sharply defined arrows in Figure 8.9(a) for segregated pathways to diffuse, overlapping distributions of multiplexed stimulus features in Figure 8.9(b).

The segregated pathway hypothesis that the perceived attributes of a composite
stimulus are associated with distributed neural activity of many separate modules and
areas of the brain, each dealing with a particular sub-modality or property (Livingstone and Hubel, 1988), leaves us with the problem of explaining how all the distinct
features are brought together to be associated with the same object. This is referred to
as the 'binding problem' (Singer, 1993). In recent years, several models for such
integrative processes in the brain have been proposed.

Properties such as color, movement, form and orientation are perceived as
belonging to an object. Since color is conveyed by the relatively slow opponent
channels while movement is signaled by the faster, phasic MC pathways, the
hypothesis of parallel processing might lead us to expect that color will be separated
from form with the color lingering behind the moving object. Such strange situations
have, in fact, been reported for patients with brain injuries. However, since this form
of separation does not occur normally, one may assume that visual cohesion relies on
coincidence detectors that can signal some sort of synchrony between the distributed

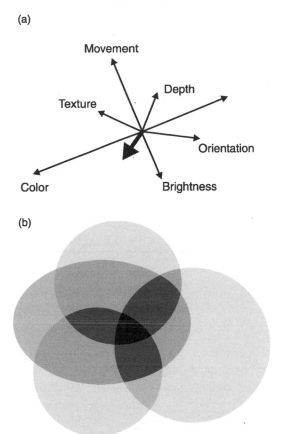

Figure 8.9 (a) A multidimensional vector space where each dimension represents a set of object features. Each vector represents the neural responses to one particular feature. The orientation vector represents the activity of all 20 or so orientation-selective cells, while the color vector represents the activity of Increment and Decrement cells responding along the different cardinal directions of cone space. A complex object is represented by a combination of these and other vectors to form an abstract resultant state vector represented by the thick arrow. This resultant vector represents an abstract response state of the system and need not necessarily be represented by a particular cell, as in the convergent structure of a hierarchical model. (b) serves to illustrate the possibility that the vectors representing feature dimensions need not be sharply defined, but that they can be relatively broadly tuned. A particular object may thus be represented by the response state of an ensemble of fuzzy vectors.

neural activities. The cortical processes in the separate areas, the correlates for form, color, movement and direction, must run in step.

The populations of nerve cells whose responses correlate with the object's various attributes must be activated simultaneously and run in synchrony. It has been suggested that for cells that respond to different aspects of the same retinal location, or of the same object, the responses oscillate in phase, and that the activity of cells responding to another object oscillates out of phase with the first. Perhaps cells that are spatially distributed but nevertheless 'bound together' by virtue of being associated with the same complex stimulus somehow achieve temporally synchronized firing. Synchrony of cell firing might bind distributed activity in a temporary functional unit. It has been speculated that such synchronization could be mediated by oscillatory brain activities in the 30–70 Hz range that have been discovered in many animals. However, since the response properties of cells become increasingly complex the further downstream post-retinal processing occurs, we cannot exclude the possibility that parallel and serial processing are followed by neural integration, leading to some degree of hierarchical convergence.

Mirror neurons

How would you go about constructing a neural system that would make it possible to interpret and understand the actions of other beings? The answer may lie in the discovery of so-called 'mirror cells' (Rizzolatti *et al.*, 1996). While recording from cells in the ventral premotor area of a monkey, it was found that some cells fired when the monkey saw the experimenter or another monkey taking food from a plate to put into their own mouth. The strange thing about this was that these were the same cells that responded when the observer monkey himself was allowed to make a grasping movement and feed itself some raisins. Further experimentation led to the notion that these cells were 'mirror cells', responding to the observation of a familiar action carried out by somebody else. How did these cells recognize a vicariously performed action? Might these findings be extrapolated to suggest that similar responses in mirror neurons elsewhere in the neural system can bring about a correspondence in state of mind and thereby constitute the neural correlate of recognizing emotions and even empathy? Functional MRI experiments on mental imagery with closed eyes have suggested that the same brain structures are indeed active when conciously viewing a visual event and when recalling and imaging the same event.

The 'split brain'

The right half of the body is governed by the left half of the brain, and the right field of view is projected (via the left retinal hemifield) to the left side of the visual cortex. In some interesting psychophysical experiments, R. Sperry (Nobel Laureate in 1981) analyzed the function of the two brain halves in patients with 'split brains'. These were patients suffering from epilepsy, and in whom the connections between the two halves had been cut in order to prevent the spread of epileptic seizures from one half of the brain to the other half. These 'split brain' patients behaved relatively normally in everyday situations, but in an experimental setting it was possible to study situations where each brain half was unaware of the sensory input presented to the other half. By presenting different, or even conflicting, information to each half, it was possible to study differences in left and right brain information processing.

If, for instance, the left hand was holding a cup behind a curtain so that it could not be seen, and the patient was asked what he was holding in his hand, he would be unable to answer. The explanation given was that the speech center, being located in the left brain, had insufficient information about what the left hand was doing. However, the patient was able, with his left hand, to write the correct answer 'cup' (since the visual information was passed through to the right half of the brain). When the word 'book' was presented only to his left field of view, the patient would be unable to convey orally what he had read although he was able to write the answer with his left hand. When the patient was forced to respond orally, the answer would be based on a guess. However, after he had answered (with his left brain), the right brain would hear the answer and decide whether it was right or wrong. In other words, communication within each brain half was intact, but the two halves could not communicate with each other internally. The two brain halves operated as individual, independent units.

There is an anecdote about another 'split brain' patient who could not manage to get dressed for a party. The reason, it was said, was to be found in a conflict between the two brain halves. One of them wished to go to the party, but the other did not like the idea, so while one hand was dressing the other hand was undressing.

Localization of brain activity: methods

Until now, the method that has given us the most information by far about the behavior of nerve cells is the use of microelectrode recordings from single units. A few other recent methods provide information about the functioning of cell groups, the selective activation of which is achieved through a specific design of the physical stimulation.

Optical imaging

In this method, the cortical surface is observed while it is illuminated with red light. Active cortical regions absorb more light than the less active ones. The temporal and spatial changes in light absorption can be recorded with a video camera as a means of mapping cortical activity. The distribution of orientation specific neurons of Figure 8.4(b) was obtained using this technique. The method can be applied to experimental animals and to human patients undergoing neurosurgery.

Positron emission tomography

Positron emission tomography (PET) is a method used to localize brain activity that is linked to a particular task. It takes advantage of the fact that cells that are active in performing a particular task have a greater energy uptake than when they are at rest. After injecting a small dose of radioactive sugar (glucose) into the blood, it is possible to register which brain areas are more radioactive (emitting gamma rays) and therefore have a higher concentration of glucose than the surroundings. Cortical areas that have been most active during the task will have consumed the most glucose and will 'lighten up' in the PET scan maps, regardless of whether the task was a motor activity, such as moving the index finger, a sensory task or a mental one. In this way it has been shown that emotions and cognitive (logical, lingual) functions are localized to certain areas in the brain. The spatial resolution in PET scans is a few millimeters. The time resolution is fairly low and rapid changes cannot be registered.

Magnetic resonance imaging and functional magnetic resonance imaging

Magnetic resonance imaging (MRI) is a means of imaging the anatomy of the brain that is often used in conjunction with a PET scan. MRI provides a clear outline of the brain, with markers that are needed for determining the precise position of PET activity. MRI is based on the fact that electrons are influenced by an external magnetic field; they are like small magnets that are forced to 'line up' in a certain direction in the magnetic field. This alignment can be registered by radio waves.

MRI today has a spatial resolution of less than 1 mm, which is considerably better than PET. The temporal resolution is less than 1 s and is continuously improving. Further developments of the MRI technique have made it possible to localize brain activity by *functional MRI* (fMRI, Raichle, 1994). Brain activity is visually mapped by subtracting the MRI image of an idle brain from the image of the brain actively engaged in a visual or mental task, for example. The method relies on the fact that the magnetic properties of blood depend on its oxygen content, and in fMRI it is the oxygen uptake in the blood that is monitored. In fMRI the blood oxygen level is the measure of neural activity, and changes in activity in a defined brain region can be registered by fMRI.

Sometimes the level of activity in an fMRI signal can be used to produce contours of equal activity when the stimulus is changed. By recording a contrast–response function, it is possible to compare a threshold activity with psychophysical sensitivity for the same stimulus and the same subjects. However, because the resting brain has some baseline activity, only the activity change induced by changing stimulus contrast or by introducing a new attribute is of interest. A typical difference in the fMRI signal is only about 5 percent of the total signal, and fMRI sensitivity thus lags far behind psychophysical sensitivity.

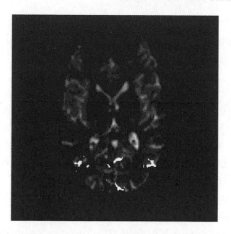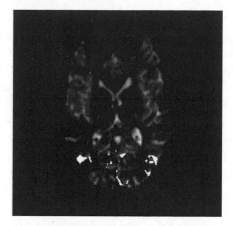

Figure 8.10 The two visual areas V5 are activated by movement of a visual stimulus. Here, areas are localized by means of fMRI during central fixation. Movement-dependent activity is indicated by the symmetrically located white areas half away from the midline. The smaller white spots at the back of the brain, closer to the midline, are from V1/V2 (Freitag *et al.*, 1998).

Figure 8.10 shows fMRI images of two subjects looking at moving targets. The white areas are those that were particularly active during this task. The two large areas on each side of the brain are areas V5, where the cells are highly specialized for movement detection.

A shortcoming of PET and fMRI is their poor temporal resolution. Better temporal resolution can be obtained by measuring electromagnetic activity in the brain. However, this can be done non-invasively only at the cost of spatial resolution.

Visual evoked potentials

A Visual evoked potential (VEP) recording is a form of *electroencephalogram* (EEG) for measuring the potential differences that occur over the brain during visual processing. In the simplest arrangement, this is done by placing one or more electrodes on the skin above the visual cortex and another at some reference position elsewhere on the head. With this arrangement it is possible to measure the fluctuations in electric potential that correlates with activity in area V1 and adjacent areas. With a large number of electrodes (more than 100) distributed over the scalp in a systematic fashion, it is possible to record the topographical distribution of the fluctuating electric potential arising from visual stimulation, or from other sensory input. Using a sophisticated model of the electric properties of the head, and an MRI image of its anatomy, the data can be analyzed mathematically to determine the location of cell groups generating that particular topography of electric potential across the skull (brain mapping). Area V1 is relatively accessible for VEP recordings, although part of the fovea is hidden between the two brain halves (see Figure 8.3). Cortical processing of visual information is distributed over large parts of the brain, and the different attributes of a stimulus may activate several brain areas simultaneously.

One advantage with the multielectrode VEP method over PET is that it does not interfere with the 'inner life' of the brain and that it does not require radioactive marking. The spatial resolution in VEP brain-mapping depends on the number of electrodes and can be better than with PET. However, it is less reliable for sources of activity that are situated in the deeper layers of the cortex. VEP has the best time resolution of all techniques. Resolution is determined by sampling frequency and can be set to values below 1 ms. Various sources of noise (external noise from the mains, electrical instruments and machines, and internal noise from global EEG and muscle activity) must be minimized when recording VEP. Because cortical activity unrelated to visual stimulation (EEG) is significant, even in differential recordings using a scalp reference, a VEP recording is usually built up of many repeated responses to identical presentations of the same stimulus. Since the timing of the response contribution from non-visual brain activity is random with respect to the stimulus, the electric noise from different sources will largely cancel out when a large number of synchronized responses are added together. The visual responses, however, are time-locked to the stimulus, and will survive this averaging procedure.

Figure 8.11 shows a conventional experimental setup for recording VEPs by using three electrodes. Potential differences between the forehead and the V1 region at the back of the head are registered and stored on a computer; the ear is connected to ground. A combination of the methods mentioned above, each of which is continuously being developed and improved upon, is a powerful tool in brain research.

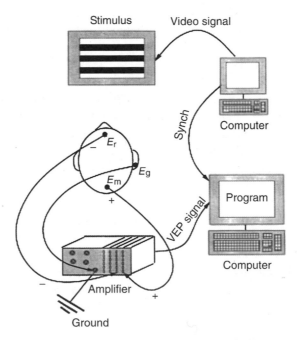

Figure 8.11 An experimental setup for recording VEP. The potentials are, for example, measured between the electrodes E_r and E_m.

Superconducting quantum interference device

A superconducting quantum interference device (SQUID) is one of the most sensitive sensors for measuring extremely weak magnetic fields, such as those that are generated by electric neural activity in the brain. The technique has found increasing use as a research tool in recent years. It is complementary to VEP, but being much more laborious and expensive, it is available only at some major physics research facilities.

Visual pathways and clinical investigation

Although we have described examples of striking co-variation between psychophysical behavior and neural processes, demonstrating this convincingly is usually a time-consuming process. Therefore, it is not always straightforward to exploit such correlates in clinical investigation of visual function. Clinical tests must be simple, easy to administer, and fast.

In some forms of reading impairment, called dyslexia, there have been repeated suggestions that there is a correspondence between the disturbed visual function and neural deficits in the magnocellular pathways. Dyslexia is mainly a reading disability, and those who suffer from it often report that they see letters moving and interchanging positions, or experience other visual disturbances that indicate impaired temporal processing. The suggested connection to a deficit in the MC pathway has led to the application of moving visual stimuli for probing the function of the MC system. Other tests can be envisaged as well, taking advantage of the difference between MC and opponent cells. As we have repeatedly seen, exclusive activation of MC cells is likely to require achromatic stimuli of low spatial frequency and low luminance contrasts. This applies to a broad range of temporal frequencies. However, as yet, there is no conclusive data on how people who suffer from dyslexia perform under such tests.

Isolation of PC cell responses can be achieved by using low red–green contrasts of low spatial and temporal frequencies, since high red–green contrasts may elicit the MC cells' second-harmonic response. Moreover, isoluminant yellow–blue color contrasts along a tritanopic confusion direction activate only S-cones of the KC pathway. It is also possible to calculate pairs of light stimuli that will silence the activity of one or more receptor type. Abnormal differences in the temporal behavior of different photoreceptor classes have been found in ERG recordings of retinitis pigmentosa using cone-isolating stimuli, a fact that can be utilized for early diagnosis of the disease (Kremers, 2003).

Several studies have concluded that the larger-bodied MC cells and their neural pathways are the first to be affected when the ocular pressure increases above normal, a condition known as glaucoma (Quigely *et al.*, 1988). It is expected that the larger size of MC cells and their thick axons render them more vulnerable to the mechanical effects of elevated intraocular pressure. Reduced functionality in the MC pathways is likely to reduce achromatic luminance contrast sensitivity and sensitivity to movement for low contrasts. However, KC cells with excitatory S-cone inputs also appear to be large (DeMonasterio and Gouras, 1975), and can therefore be injured before the smaller PC cells. Psychophysical tests, such as measuring the extent of the visual field by blue–yellow perimetry (which determines the threshold for a short-wavelength light on a high luminance yellow background), seem to indicate that this sensitivity decreases in glaucoma. Such tests may help diagnose glaucoma in its early stages of development.

The electroretinogram

The electroretinogram (ERG) is a tool for monitoring activity in the retina. Flashing bright light onto the retina modifies the electric currents through the retinal layers and, as a result, brings about a change in the electric potential at the cornea, starting with an early negative change at the back of the eye relative to the cornea, followed by a positivity. The Swedish physiologist Ragnar Granit (Nobel Prize winner in 1954) analyzed the ERG into three components P-I, P-II and P-III (Granit, 1933), and although this component analysis has been slightly modified over the years, it remains a basis for our understanding of the ERG.

ERG are typically recorded in response to a bright flash that floods the entire retina. The earliest negative waveform in the response is attributed to the receptors themselves (the a-wave). This potential is generated by hyperpolarization of the photoreceptor inner segments. An a-wave may still be recorded even when the inner retina is destroyed but the outer retina is intact. Using cone-isolating techniques (silent substitution), it is possible to measure ERG to single cone types, a method that has been proven useful in the early diagnosis of retinitis pigmentosa (Kremers, 2003), an eye disease that damages rods and leaves the patient with foveal tunnel vision before he finally becomes blind. The subsequent positive b-wave is thought to be generated by the activity of bipolar cells.

A pattern ERG can be recorded to reversing checkerboard patterns of black and white squares, and this ERG response is thought to reflect the activity of ganglion cells. More recently, a method has been developed which makes it possible to record many simultaneous ERGs from a large number of small areas distributed over the retina. This method, referred to as *multifocal ERG* (mfERG), can be used for locating retinal scotomas – retina areas with severely compromised vision – or retinal areas in which the response to light modulation deviates in some way from that of the surrounding retina. This method promises to be highly informative in future studies of retinal function.

Visual evoked potentials

In light of the proposed link between dyslexia and impaired function of the MC-pathway, it might be of clinical relevance to be able to distinguish the responses of the magnocellular and the opponent systems in evoked cortical potentials. VEPs provide an objective method for monitoring brain activity, but the recorded waveforms are not easy to interpret. There is, however, some indication of a different VEP response of MC and PC systems that depends on stimulus contrast (Nakayama and Mackeben, 1982; Valberg and Rudvin, 1997; Rudvin *et al.*, 2000). VEP responses to small uniform stimuli of increasing contrast show intensity–response curves with two branches. The potentials to the stimuli of lowest contrasts, up to 20 percent Michelson contrast, may be associated with MC activity, whereas potentials for the higher contrasts may be due to combined MC and opponent inputs to visual cortex (Figures 8.12 and 8.13). This interpretation is supported by the parallel development of delay times to peak amplitude. Delays decrease with increasing contrast up to

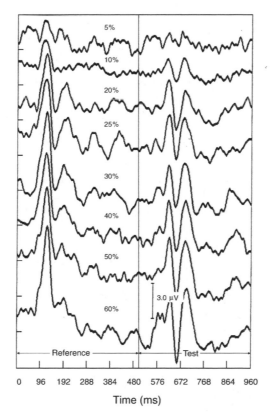

Figure 8.12 An example of the waveforms of VEP measured in an experimental setup like that of Figure 8.11. The voltage fluctuations generated over the visual cortex are measured as a voltage difference between the electrodes E_m and E_r when a 3°, achromatic circular test light is exchanged with a coextensive reference light of 25 cd/m². The reference field is switched on at 0 ms, and the luminance drops to a lower value at 500 ms. Contrast is defined by $\Delta L/L_b$, or Weber contrast. The first main waveform (P120) appears with a peak latency of about 120 ms after a luminance increment, the delay decreasing as the contrast increases. The stimulus conditions are like those in the recordings from single cells in Figures 4.14 and 6.7.

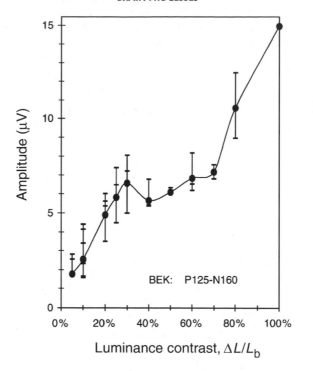

Figure 8.13 The P120 waveform in Figure 8.12 is here plotted as a function of Weber contrast. The response increases with increasing contrast up to a plateau, flattens out and increases again after 50–60% stimulus contrast. This behavior has been interpreted in terms of the contribution of MC and opponent cells to the response: the more sensitive MC cells are thought to be responsible for the initial, low-contrast part of the curve, with PC cells dominating the response for the higher contrasts.

about 20 percent contrast and increases thereafter, as we would expect if the slower opponent cells took over at higher contrast.

A similar distinction between responses of MC cells and the less contrast-sensitive opponent cells can be made for VEP waveforms recorded to the ON/OFF presentation of chromatic gratings (Nygård *et al.*, 2002).

Cortical visual impairment

Cortical visual impairment (CVI) is a term used for injuries or developmental problems in or after the optic chiasm, for instance in the geniculate nucleus, the optic radiation, or in the visual cortex. Sometimes these injuries result in symptoms that are similar to those of defects in the peripheral visual apparatus, such as visual field defects and low acuity. The foveal retina is greatly over-represented in primary visual cortex and low acuity may result from injuries to this area. In other cases acuity is not reduced, but the individual has perceptual or cognitive problems, such as difficulty in interpreting visual inputs and integrating them with other brain functions. *Visual agnosia* refers to an impairment of processes leading to object detection and recognition, although other vital visual functions are preserved. This may be described as 'seeing without meaning'. This description may be apply to deficiencies in the perception of space (depth, distance, perspective, etc.) that is found in people who lost their sight at a very early age but regained it much later in life (von Senden, 1960; Sacks, 1995; and the case of Virgil mentioned in the Introduction). *Prosopagnosia* is a failure to recognize faces, despite correct identification of the eyes, the nose, the mouth, etc.

The failure to perceive movement (*akinetopsia*) or to determine direction of movement is another example of CVI. *Simultanagnosia* refers to problems with the perception of more than one object at a time. Brain injuries following accidents can also lead to loss of color perception (*achromatopsia*). *Hemianopia* is characterized by the loss of half of the visual field, either to the right or to the left of the mid-line, in both eyes. This reflects the loss of visual function in half of the brain, as a result of stroke, head trauma or a tumor. A person who is unable to see objects in his blind field may have problems orientating himself in space. A common cause of CVI is head trauma from car accidents or an oxygen deficit resulting from diminished blood supply to the brain. The circulation of blood may be interrupted by cardiac arrest or at birth. The effects on the brain of neonates can be diagnosed fairly early by ultrasound or MRI.

Other causes of CVI may be prenatal or postnatal exposure to infections or toxic agents, or to substances that affect the central nervous system, such as alcohol and narcotics. CVI is seldom seen during a regular eye examination; it often requires an interdisciplinary approach. One reason for studying the impairments that follow brain injuries is that it provides us with insight into the functioning of the normal brain. For instance, we may learn more about the organization of the brain in the specialized subsystems, or modules that are required for us to recognize and be aware of different attributes of visual stimuli.

People who have lost a limb often report pain in some part of the missing limb, in a toe, for example. This 'phantom pain', which usually subsides in a matter of months or years, is usually explained in terms of nerves in the brain associated with the lost organ, and with pain, still being active and 'projecting' the pain into the brain's body space, an external map completed early in life. Imagine a similar situation in vision: contact with some visual center in the brain is disconnected because of injury, e.g. after stroke or the removal of a tumor, or even after a visual loss due to retinal damage, such as in AMD. Although the normal visual input signals to a visual center are missing, the center may still have some level of spontaneous activity. In such a case, the analogy to phantom pain might be that the person experiences phantom images in terms of color patterns, movements or other attributes of vision that bear no relation to objects in the external world. This may explain why people with maculopathy occasionally report seeing 'strange things'.

Appendix

A physiologically based system for color measurements

As a supplement to the existing *XYZ* system for color measurements, committee TC 1-36 of the CIE has for some time been engaged in the development of a physiological based system for colorimetry. The committee has already adopted as cone fundamentals (fundamental color matching functions) the spectral sensitivities L_λ, M_λ, and S_λ for a 2° field of Stockman and Sharpe (2000). In Table A1 these cone sensitivities (here called specific tristimulus values) are tabulated from 390 to 830 nm in 5 nm intervals.

Figure A1 shows the chromaticity diagram derived from the Stockman–Sharpe$_{2000}$ fundamentals of Table A1, where the integrated values are normalized to equal values of 20.0000 for CIE Illuminant E (the equal energy spectrum) such that $(l_E, m_E) = (1/3, 1/3)$. The chromaticity coordinates (l, m) for the spectrum locus are:

$$l_\lambda = L_\lambda/(L_\lambda + M_\lambda + S_\lambda),$$
$$m_\lambda = M_\lambda/(L_\lambda + M_\lambda + S_\lambda),$$
$$s_\lambda = S_\lambda/(L_\lambda + M_\lambda + S_\lambda).$$

Light Vision Color. Arne Valberg
© 2005 John Wiley & Sons Ltd

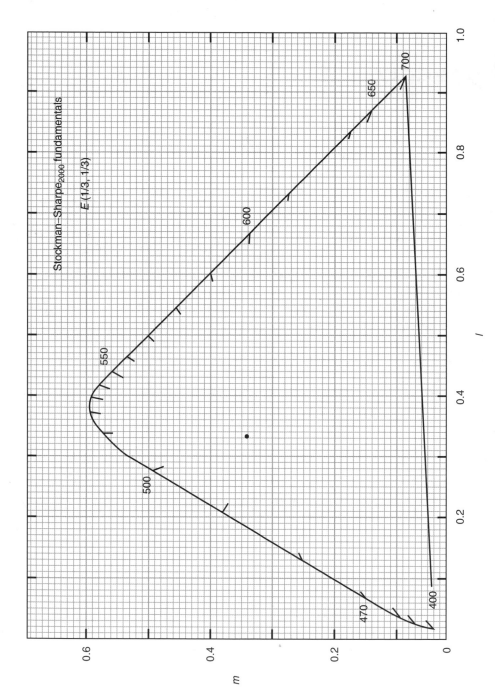

Figure A1 A chromaticity diagram (l, m) derived from the Stockman–Sharpe$_{2000}$ fundamentals L, M, and S of Table A1 (courtesy of J. H. Wold).

Table A1 Specific tristimulus values and spectral chromaticity coordinates based on the Stockman–Sharpe$_{2000}$ cone fundamentals for 2° fields (energy units) at wavelengths from 390 to 830 nm and 5-nm intervals[a,b]

Wavelength (nm)	Specific Tristimulus Values (1/nm)			Chromaticity Coordinates	
λ	L_λ	M_λ	S_λ	l_λ	m_λ
390	3.57825E-04	3.88464E-04	1.63368E-02	0.02095	0.02274
395	9.06989E-04	1.01101E-03	4.07681E-02	0.02125	0.02368
400	2.07654E-03	2.39386E-03	9.69362E-02	0.02048	0.02361
405	4.16746E-03	4.95676E-03	2.09532E-01	0.01906	0.02267
410	7.51968E-03	9.27389E-03	3.98711E-01	0.01810	0.02232
415	1.15397E-02	1.53210E-02	6.52569E-01	0.01698	0.02255
420	1.59063E-02	2.28480E-02	9.30211E-01	0.01642	0.02358
425	1.97722E-02	3.11862E-02	1.15413E00	0.01641	0.02588
430	2.43041E-02	4.16112E-02	1.37329E00	0.01689	0.02891
435	2.94064E-02	5.46497E-02	1.54615E00	0.01804	0.03352
440	3.47099E-02	6.83156E-02	1.69578E00	0.01930	0.03798
445	3.87466E-02	8.00249E-02	1.69663E00	0.02134	0.04408
450	4.29938E-02	9.18061E-02	1.63482E00	0.02430	0.05188
455	4.77169E-02	1.03555E-01	1.47200E00	0.02940	0.06379
460	5.57999E-02	1.22621E-01	1.34617E00	0.03660	0.08043
465	6.95722E-02	1.52434E-01	1.26329E00	0.04684	0.10263
470	8.57700E-02	1.85498E-01	1.10602E00	0.06227	0.13468
475	1.02434E-01	2.16614E-01	8.83656E-01	0.08517	0.18011
480	1.20836E-01	2.48628E-01	6.67918E-01	0.11648	0.23967
485	1.41363E-01	2.82701E-01	4.96784E-01	0.15351	0.30700
490	1.65164E-01	3.20210E-01	3.62536E-01	0.19479	0.37765
495	2.00834E-01	3.76559E-01	2.74684E-01	0.23570	0.44193
500	2.49147E-01	4.51123E-01	2.10196E-01	0.27365	0.49549
505	3.10155E-01	5.43742E-01	1.52115E-01	0.30830	0.54049
510	3.82553E-01	6.49132E-01	1.04074E-01	0.33683	0.57154
515	4.62577E-01	7.58425E-01	7.32582E-02	0.35741	0.58599
520	5.41959E-01	8.61203E-01	4.99712E-02	0.37296	0.59265
525	6.07625E-01	9.33908E-01	3.31812E-02	0.38586	0.59307
530	6.64455E-01	9.86782E-01	2.15627E-02	0.39721	0.58990
535	7.11947E-01	1.02176E00	1.38509E-02	0.40740	0.58468
540	7.59627E-01	1.04956E00	8.70803E-03	0.41786	0.57735
545	7.92440E-01	1.05165E00	5.42251E-03	0.42846	0.56861
550	8.10660E-01	1.03056E00	3.35207E-03	0.43948	0.55870
555	8.32677E-01	1.00882E00	2.05812E-03	0.45167	0.54721
560	8.46224E-01	9.67866E-01	1.26655E-03	0.46615	0.53315
565	8.57468E-01	9.20888E-01	7.80248E-04	0.48196	0.51760
570	8.62216E-01	8.57933E-01	4.82202E-04	0.50110	0.49862
575	8.55592E-01	7.80716E-01	2.99518E-04	0.52278	0.47703
580	8.35863E-01	6.88948E-01	1.87292E-04	0.54811	0.45177
585	8.23942E-01	6.03865E-01	1.18068E-04	0.57702	0.42290
590	7.99861E-01	5.19499E-01	7.51235E-05	0.60621	0.39373
595	7.63902E-01	4.33703E-01	4.82934E-05	0.63783	0.36213
600	7.19078E-01	3.52691E-01	3.13926E-05	0.67091	0.32906
605	6.68311E-01	2.79336E-01	2.06479E-05	0.70522	0.29476
610	6.08482E-01	2.16482E-01	1.37489E-05	0.73757	0.26241
615	5.43867E-01	1.64775E-01	9.27174E-06	0.76747	0.23252
620	4.77864E-01	1.23010E-01	0.00000[c]	0.79528	0.20472
625	4.13816E-01	9.02609E-02	0.00000[c]	0.82094	0.17906
630	3.45502E-01	6.55038E-02	0.00000[c]	0.84063	0.15937

(continued)

Wavelength (nm)	Specific Tristimulus Values (1/nm)			Chromaticity Coordinates	
λ	L_λ	M_λ	S_λ	l_λ	m_λ
635	2.82692E-01	4.69173E-02	0.00000^c	0.85766	0.14234
640	2.29165E-01	3.31444E-02	0.00000^c	0.87364	0.12636
645	1.83898E-01	2.29943E-02	0.00000^c	0.88886	0.11114
650	1.42388E-01	1.62916E-02	0.00000^c	0.89733	0.10267
655	1.07561E-01	1.12970E-02	0.00000^c	0.90495	0.09505
660	8.01940E-02	7.70132E-03	0.00000^c	0.91238	0.08762
665	5.90709E-02	5.24329E-03	0.00000^c	0.91847	0.08153
670	4.29957E-02	3.62434E-03	0.00000^c	0.92226	0.07774
675	3.08877E-02	2.50593E-03	0.00000^c	0.92496	0.07504
680	2.18823E-02	1.72675E-03	0.00000^c	0.92686	0.07314
685	1.52787E-02	1.18251E-03	0.00000^c	0.92816	0.07184
690	1.04933E-02	8.02610E-04	0.00000^c	0.92895	0.07105
695	7.30449E-03	5.54151E-04	0.00000^c	0.92949	0.07051
700	5.08495E-03	3.85266E-04	0.00000^c	0.92957	0.07043
705	3.52760E-03	2.67255E-04	0.00000^c	0.92957	0.07043
710	2.41808E-03	1.83926E-04	0.00000^c	0.92931	0.07069
715	1.65597E-03	1.27194E-04	0.00000^c	0.92867	0.07133
720	1.14406E-03	8.87680E-05	0.00000^c	0.92800	0.07200
725	7.91328E-04	6.21532E-05	0.00000^c	0.92718	0.07282
730	5.51282E-04	4.38768E-05	0.00000^c	0.92628	0.07372
735	3.84581E-04	3.10428E-05	0.00000^c	0.92531	0.07469
740	2.68038E-04	2.20265E-05	0.00000^c	0.92406	0.07594
745	1.89110E-04	1.58674E-05	0.00000^c	0.92259	0.07741
750	1.33256E-04	1.14109E-05	0.00000^c	0.92112	0.07888
755	9.44203E-05	8.24989E-06	0.00000^c	0.91965	0.08035
760	6.72458E-05	6.00170E-06	0.00000^c	0.91806	0.08194
765	4.79623E-05	4.36605E-06	0.00000^c	0.91656	0.08344
770	3.44281E-05	3.19212E-06	0.00000^c	0.91515	0.08485
775	2.46736E-05	2.33174E-06	0.00000^c	0.91366	0.08634
780	1.78757E-05	1.72358E-06	0.00000^c	0.91206	0.08794
785	1.29706E-05	1.27664E-06	0.00000^c	0.91039	0.08961
790	9.43668E-06	9.48271E-07	0.00000^c	0.90869	0.09131
795	6.87838E-06	7.06159E-07	0.00000^c	0.90689	0.09311
800	5.04449E-06	5.30665E-07	0.00000^c	0.90482	0.09518
805	3.71706E-06	4.00799E-07	0.00000^c	0.90267	0.09733
810	2.73332E-06	3.01965E-07	0.00000^c	0.90052	0.09948
815	2.02164E-06	2.28721E-07	0.00000^c	0.89836	0.10164
820	1.50601E-06	1.74177E-07	0.00000^c	0.89633	0.10367
825	1.12297E-06	1.32362E-07	0.00000^c	0.89456	0.10544
830	8.40069E-07	1.00547E-07	0.00000^c	0.89311	0.10689
Totals	2.00000E01	2.00000E01	2.00000E01		

^{a} Denoting the wavelengths 390, 395, ..., 830 by $\lambda_1, \lambda_2, ..., \lambda_{89}$, the specific tristimulus values are normalized so that $\sum_{i=1}^{89} L_{\lambda_i} \Delta\lambda = \sum_{i=1}^{89} M_{\lambda_i} \Delta\lambda = \sum_{i=1}^{89} S_{\lambda_i} \Delta\lambda = 100$, where $\Delta\lambda = 5$ nm.

^{b} The chromaticity coordinates l and m of a stimulus with relative spectral energy distribution $P(\lambda)$ are defined by the equations $l = L/(L+M+S)$ and $m = M/(L+M+S)$, where $L = \sum_{i=1}^{89} P(\lambda_i)L_{\lambda_i} \Delta\lambda$, $M = \sum_{i=1}^{89} P(\lambda_i)M_{\lambda_i} \Delta\lambda$ and $S = \sum_{i=1}^{89} P(\lambda_i)S_{\lambda_i} \Delta\lambda$. For the equal-energy stimulus, E, this gives $l_E = 0.33333$ and $m_E = 0.33333$.

^{c} The zero-values of S_λ from 620 to 830 nm do not have a physiological meaning, but only signify that the values are so small, that these values can be neglected. Without introducing the zero-values here, the chromaticity coordinates l_λ and m_λ would be undetermined for wavelengths beyond 615 nm.

Glossary

This Glossary does not provide accurate definitions, it only explains some central concepts and words used in visual science. A word in the explanation that is in *italics* is a key word that is itself explained in the list.

Abney effect	In an additive color mixture, the hue of a chromatic stimulus changes when white is added to it (while luminance is kept constant).
Absolute threshold to light	The smallest number of light quanta incident on the cornea, or absorbed in the rod receptors, required to evoke a sensation of light when the eye is completely dark-adapted.
Absorption spectrum	The spectral distribution of the amount of light absorbed, I_a, by a pigment relative to the amount of incident light, I_i, plotted as the wavelength dependency of the fraction I_a/I_i (I_a being equal to $I_i - I_t$, where I_t is the amount of transmitted light). See *principle of univariance*.
Accommodation	Changing the eye's focus from far to near by decreasing the radius of curvature of the lens. Expressed in diopters. The ability to accommodate is more critical at low light levels than in bright light (due to the reduced depth of field when the pupil expands).
Accommodation range	The eye can decrease the radius of curvature of the lens to increase its power. In younger years this change corresponds to about 12–16 diopters. The ability to accommodate decreases with age, and after 50 the accommodation range is usually less than 2 diopters.
Achromatic colors	Colors with no chroma, i.e. black, gray and white.
Achromatizing lens	A lens combination where the *chromatic aberration* (color deviation) in one lens is compensated for by another lens.
Achomatopsia	Total absence of chromatic color vision. This may be due to lack of cone receptors in the retina or to brain injury.
Action potentials	Nerve impulses ('spikes') of short duration (approximately 1 ms), by which nerve cells transmit information. The number of impulses per second is a measure of the cell's activity.
Action spectrum	The spectral distribution of energy (or radiant power) required to obtain a criterion response at different wavelenghts (e.g. a threshold perception of light).

Light Vision Color. Arne Valberg
© 2005 John Wiley & Sons Ltd

Acuity	See *visual acuity*.
Adaptation	The ability of the visual organ to adjust its sensitivity and function to the prevailing light level and color. The term can be used for the process itself or for the final state. The retina is said to be light adapted (corresponding to *photopic vision*) or dark-adapted (*scotopic vision*). The size of the *pupil* plays only a minor role in adaptation.
Additive color mixture	When two or more colored lights are added (overlaid), either on a screen or otherwise superimposed on the retina.
Agnosia	The inability to perceive objects through otherwise normally functioning sensory pathways (e.g. depth agnosia, movement agnosia, color agnosia, *prosopagnosia*). Loss of knowledge.
Amacrine cells	Cells that convey information laterally to bipolar and ganglion cell terminals in the inner plexiform layer of the retina.
Amblyopia	Reduced vision without clear signs of a disease. May occur in children who squint. Should be treated at a young age to prevent permanent loss of coordinated vision of the two eyes (e.g. depth vision).
AMD	Age-related macular degeneration is a disease that usually attacks the receptors in or near the *fovea* and therefore typically has a severe impact on visual resolution and contrast sensitivity. This is the most common cause of *low vision* in elderly people in the western world.
Angular frquency	$\omega = 360f = 2\pi/\lambda$.
Angular magnification, M	For a magnifying glass $M = 25/f = 0.25P$ when the object is at a distance equal to the focal length, f, from the lens and the image is seen at infinity. P is the *lens power* in diopters.
Annulus	Ring-shaped visual stimulus.
Anterior chamber	The fluid-filled space between cornea and iris/lens. With age, the depth of the anterior chamber is reduced as the lens grows larger.
Aqueous humor	A fluid produced by the ciliary body that serves to eliminate waste products from the eye. It passes from the posterior chamber to the anterior chamber before flowing out through Schlemm's canal.
Astigmatism	Deviation of the curvature of the cornea from a spherical form towards a cylindrical form. This causes the image planes for some contours (e.g. vertical and horizontal contours) to be different. Can be optically corrected.
Autostereogram	Because of the distance between the two eyes, the left and right eye will see the same detail on a three-dimensional object at a different angle, and thus somewhat displaced relative to the other details that are behind or in front of the one in focus. The texture of an autostereogram contains two separate images of the same three-dimensional object, each as the object would appear for one eye alone. The spatial relations of object details in each image are the same as when the scene is viewed with one eye. By squinting, the images are brought into register with each other, and the sensory impression of depth arises.

Axon	The outward going nerve fiber from a cell. An axon transmits signals to presynaptic terminals where the cell has established synaptic contacts with other nerve cells.
Bandwidth	Range of wavelengths (or frequencies) represented in a stimulus, or to which a system is sensitive.
Bezold–Brücke phenomenon	The hue of a color stimulus of constant chromaticity changes when luminance changes.
Binding problem	The problem of how different attributes, such as color, movement, shape, depth, etc., can be simultaneously attached to an object. The processing of these attributes takes place in different, separated brain regions and it takes time for signals to travel between the areas.
Binocular sector	The part of the frontal visual field that is viewed by both eyes.
Binocular vision	Coordinated vision with both eyes.
Bipolar cells	Cells in the retina that convey information from the *photoreceptors* to the *ganglion cells*.
Blindsight	The ability of some cortical blind people to point to the location of light that they cannot see.
Blind spot	A spot in the *monocular* visual field that is blind because it corresponds to the position on the retina where the nerve fibers leave the eye, and where there are no *photoreceptors*.
Border colors	Colors that arise at black and white borders due to dispersion, e.g. by a prism or an imaging lens. When the beam of light is too wide to produce a *spectrum* of monochromatic light upon refraction, a partial spectrum of *optimal colors* may still be seen at black and white borders.
Brightness	Apparent amount of light emitted by a surface.
B-system	Brightness system.
Cataract	An ailment due to opacities of the optic media of the eye. The eye lens becomes less clear and the imaging on the retina more diffuse. Removing the eye lens and replacing it with an artificial lens is now a routine operation in this situation.
CIE	Commission Internationale de l'Eclairage.
cGMP	Cyclic guanosine monophosphate. A substance that conveys information to the cell membrane about light absorption in the pigment disks of the receptors. CGMP controls the ion current accross the cell membrane by opening special ion channels (cGMP-gated channels) in darkness and closing them in light.
Chromatic aberration	Dispersion in the optical media of the eye gives rise to chromatic aberrations. The eyes focus for wavelengths around 580 nm, and they are therefore nearsighted (myopic) for short-wavelength blue light and farsighted (hyperopic) for long-wavelength red light. The retinal images formed by short and long wavelengths are therefore not in focus. When chromatic aberration of a lens is severe, *border colors* can be seen at black and white borders. Transversal aberration leads to differences in image size, dependent on wavelength, and longitudinal aberration gives differences in focal length.

Chromatic adaptation	The self-adjustment of the visual system to the color of the prevailing illumination in such a way that object surfaces appear to have the same color for all daylight phases and for most artificial lights. The visual system works towards neutralizing (in an unknown way) the effect of the color of the illumination. For example, a white surface appears white even if the illumination changes from bluish daylight to yellowish incandescent light. See *color constancy*.
Chromatic colors	Colors are divided into achromatic colors (black, grays and white) and chromatic colors (yellow, red, blue, green and their transitions).
Chromaticity	Two-dimensional color coordinates (r, g) in a unit color triangle $R + G + B = 1$, or the (x, y)-coordinates in the CIE system for color measurement, in a plane where the sum of tristimulus values $X + Y + Z = 1$.
Chrominance	An isoluminant chromatic color stimulus has a chrominance proportional to its chromatic difference from an achromatic color of the same luminance (as defined in a colorimetric color space). Color stimuli can be described by chrominance and luminance coordinates (or by chrominance and luminance ratio for object colors). Achromatic color stimuli have zero chrominance.
Color	Can be used both for a physical color stimulus and for the qualitative, subjective experience.
Color constancy	A tendency for objects and reflecting surfaces to not change their color appearance much when the color of the illumination changes, e.g. from daylight to incandescent light.
Complementary colors	Pairs of colors that yield white in an additive color mixture.
Cone	Receptor cell in the retina that operates at *photopic* light conditions.
Contrast	Michelson contrast, $C_{\text{Mich}} = (L_{\text{max}} - L_{\text{min}})/(L_{\text{max}} + L_{\text{min}})$, is commonly used for periodic stimuli. Weber contrast, $C_{\text{Web}} = (L - L_b)/L_b = \Delta L/L_b$, where L stands for stimulus luminance and L_b for background luminance. Combined cone contrast, $C_{\text{LMS}} = [(1/3)(C_L^2 + C_M^2 + C_S^2)]^{1/2}$, where C_L, C_M and C_S are the individual cone contrasts for the absorptions (excitations) in L-, M- and S-cones.
Contrast rendering/contrast rendering factor, CRF	The ratio T between the contrast of an image and the imaged object; $T = $ *(image contrast)/(object contrast)*. For all modes of optical reproduction, increasing *spatial frequency* (see *gratings*) implies a smaller T. In the retinal image, contrasts are reduced by *dispersion*, *diffraction*, light scatter from the eye media and lens aberrations.
Contrast sensitivity curves for luminance and chrominance	• At photopic levels, curves of contrast sensitivity ($1/C_{\text{threshold}}$) plotted as a function of spatial and temporal frequency, have an inverse U-form. For all but the lowest frequencies, contrast sensitivity increases when luminance level increases, giving curves their inverse U-form. The higher the adaptation luminance, the smaller the stimulus contrasts that can be detected (increased contrast sensitivity), and the finer the details that can be resolved (increasing *acuity*).

- Contrast sensitivity and resolution decrease with retinal eccentricity (distance from the *fovea*). With increasing age, sensitivity decreases at middle and high spatial frequencies.
- According to one theory, the primary visual cortex has several separate mechanisms or cell groups each of which is sensitive to a narrow band of spatial frequencies and orientations. It has been assumed that, at any location in the retina, there are at least six such mechanisms, whose sensitivities overlap along the spatial frequency dimension.
- *Flicker sensitivity* ($1/C_{\text{threshold}}$) varies as a function of temporal frequency, and depends on spatial frequency.
- Spatial contrast sensitivity curves for *chrominance* [i.e. for chromatic contrast detection (color discrimination) at equal luminance] differ from the curves for luminance contrast. Chrominance sensitivity increases steadily towards lower spatial frequencies, and sensitivity decreases rapidly at higher spatial frequencies, with a cut-off frequency between 10 and 20 cycles/deg, depending on the color combination.

Cornea	Transparent frontal surface of the eye.
Corresponding points	Points on the retinae of the two eyes that have the same angular distance from the fovea. See *horopter*.
Cortex	Highly convoluted outer layers of cells that surround the top and sides of the brain.
D-cells (Decrement cells)	Cells that are activated by luminance decrements; their firing rate increases when luminance decreases (also called OFF-cells).
Decrement response	Response to a decrement in stimulus luminance (response to negative contrast).
Dendrite	Portion of a cell specialized for receiving inputs from other cells via a great number of contact points; the *synapses*.
Depolarization	Change in membrane potential towards amore positive value inside the cell relative to the negative resting potential.
Deuteranopia	Color vision deficiency resulting from the absence of M-cones. Deuteranopes cannot distinguish between red, yellow and yellow-green, or between purple, white and green.
Diffraction	*Monochromatic light* waves that are generated at the rim of an aperture (such as the pupil) give rise to a pattern of superimposed waves at a distance from the pupil. Analogous to waves on water, the amplitudes of these waves add and subtract on the retina depending on their relative phase (interference). In the eye diffraction limits resolution and *visual acuity*.
Dichromacy	Color vision defect caused by the absence of one of the three cone types. *Protanopes* lack L-cones while *deuteranopes* lack M-cones. Both are unable to distinguish reddish colors from yellow, white and green, although with different nuances. *Tritanopes* lack S-cones and do not distinguish between yellow, white and blue.
Difference sensitivity	A measure of the ability to *discriminate* between two suprathreshold stimuli along a certain stimulus dimension.

Diopter	See *refractive power*.
Direct signal route	The pathway from the *photoreceptors* through bipolar cells to *ganglion cells*. These signals can be modified laterally (sideways) by *horizontal* and *amacrine cells*, i.e. from cells that integrate information laterally in the retina.
Discrimination	The ability to identify an object or an image after distinguishing it from the background. Discrimination usually requires a larger contrast than detection. One talks about color discrimination when one sees a qualitative difference between two color stimuli.
Dispersion	The index of refraction, *n*, of a medium (e.g. glass, water) varies with the wavelength of light, and the refraction at the boundary between two translucent media (glass and air, for example) is therefore wavelength-dependent. White light that is refracted by a glass prism gives rise to a spectrum where the short wavelengths are refracted more than the long ones. One consequence is that images of white/black borders produced by a simple lens have colored borders (see *border colors* and *spectrum*).
Dorsal	In the direction towards the top of the head.
Double opponent cell	A cell with cone opponency at every point in its receptive field. Such cells are activated by light within a particular spectral range (color) impinging on the center of their receptive field as well as by the complementary color in their receptive field surround.
D-system	Darkness system; should not be confused with D-cells or Decrement cells, although an older theory assumes that they are the same.
Duplicity theory	We depend on different types of *photoreceptors* for vision in daylight and at night. The *cones* are the receptors for daylight vision (*photopic vision*) and the *rods* are the receptors for vision at low light levels (*scotopic vision*).
Electroretinogram (ERG)	Gross potential reflecting the electric activity of retinal cells.
Elementary colors	Six particularly simple color qualities (approximate wavelength in brackets): black, white, yellow (ca. 570 nm), blue (ca. 470 nm), green (ca. 500 nm) and red (approximately complementary wavelength 495 nm). Also called unique colors.
Excitation	The linear effect of absorption of light quanta in photoreceptors, a process giving rise to changes in the membrane potential (polarization) of receptors. It also refers to an increased activity of nerve cells as opposed to the decrease of activity during inhibition.
Eye lens	The eye's lens is elastic and can change its radius of curvature to adjust its refractive power (*accommodation*). The range of accommodation is 12–16 diopters before the age of 25, but less than 2 diopters after the age of 50. Usually, the closest point of fixation increases to over 1 m around 70 years of age. The eye absorbs UV light and becomes more yellow (transmits less blue light) and rigid with age.

Fechner's law	This law states that the smallest sensory threshold differences, Δr, are proportional to the *Weber ratio*, $\Delta I/I$, and that they can be integrated (summated) to a final sensory difference, R. This leads to Fechner's law: $R = c \log(I/I_t)$, where I_t is the threshold intensity. The dB scale for loudness of sound is based on this law.
Flicker	The impression of temporal variation in luminance or color.
Flicker sensitivity	The inverse of the threshold contrast for the perception of flicker in a temporally modulated stimulus.
fMRI	Functional magnetic resonance imaging is a means of imaging brain activity. The technique is based on recording differences in regional blood flow and in the oxygen uptake of active brain cells between baseline and stimulus conditions.
Focal point	The point at which a lens focuses parallel light rays.
Fovea	The foveal pit is an area of about 1.5 mm in diameter (corresponding to ca 5°) which is free of capillaries and where the nerve cells are bent aside, leaving the receptors more directly exposed to light. The fovea is specialized for resolving fine detail.
Fovea centralis	A smaller part of the fovea of about 1° in diameter that does not contain rods; also called the 'foveola'.
Fundus	The part of the retina (the back of the eye) that is visible when looking into the eye with an ophthalmoscope.
Ganglion cells	The cells representing the final stage of neural processing in the retina. They generate nerve pulses that are led to *the lateral geniculate nucleus* (LGN; or corpus geniculatum laterale in Latin) via the optic nerve.
Gestalt psychology	Theory that emphasizes properties of figures that transcend their components.
Glaucoma	An eye disease. If the transport of fluid out of the eye is somehow impeded, this results in elevated pressure inside the eye, which, if maintained, may eventually lead to visual loss (because the pressure on the optic nerve becomes too high).
Grating	A pattern of alternating dark and light bars (see *sinusoidal grating*).
Hering contrast	A lateral contrast that extends over large areas, as opposed to the local effects in border contrast (see also *Mach bands*).
Horizontal cells	Cells in *the outer plexiform layer* that convey information across the retina (laterally) and modulate signal transmission from the *receptors* to the *bipolar cells*. Horizontal cells may play a role in the formation of the *receptive field* surround and in *adaptation*.
Horopter	An imaginary circle that can be drawn through a fixated point (an object) and the nodal point (close to the entrance pupil) of the two eyes. Every point on this horopter is imaged in corresponding retinal points, i.e. in points that have the same position relative to the fovea in the two eyes (no binocular disparity). Objects that are situated outside this circle give rise to double images that serve as cues for depth perception.

Hue	The hue of a color can be characterized by relative proportions of the closest elementary hues yellow, red, blue and green. They can be ordered in a hue circle.
Hyperpolarization	A change in the potential across the cell membrane from the (negative) resting potential towards an even greater negative value relative to the outside. Hyperpolarization is the result of excitation (light absorption) in primate photoreceptors.
I-cells (Increment cells)	Cells that are activated by luminance increments; their firing rate increases when luminance increases (also called ON-cells).
Illuminance, E	See *radiometry* and *photometry*.
Increment and decrement thresholds	The smallest detectable increment or decrement change, ΔI. The magnitude of this threshold depends on relative sensitivity for different background intensity levels, I, above absolute threshold. The inverse *Weber ratio, $I/\Delta I$*, is a measure for the contrast sensitivity relative to the background.
Increment response	Response to an increase in stimulus luminance (response to a positive contrast).
Interference filter	Color filter with a relatively narrow wavelength range of light transmission. Characterized by spectral half-width in nm at 50 percent transmission.
IR	Infrared electromagnetic radiation of wavelengths above 780 nm.
Iris	A smooth muscle ring controlling the size of the pupil. The color of the eye. Eye color depends of the amount of pigment in the iris. Blue eyes have little pigment and brown eyes have more pigment.
Isoluminance	A situation where different color stimuli have the same luminance.
Koniocellular cells, KC cells	Cells with S-cone input that are found in the koniocellular layers of the *LGN* receiving inputs from corresponding retinal cells.
Lateral geniculate nucleus (LGN)	Part of thalamus that relays visual signals to the cortex. A layered structure between retina and visual cortex, with the parvocellular, magnocellular and koniocellular layers receiving inputs from corresponding retinal cells.
Lens power (thin lenses); P	$P = 1/f$ (m^{-1}), where f represents the focal length of the lens.
Light	A stimulus that gives a visual sensation. The visible range of electromagnetic radiation from 380 to 760 nm.
Lightness	The visual attribute by which one can tell if a surface reflects more or less light. A visual impression of intensity that increases with increasing reflectance factor.
Low vision/visual disability	A person with visual acuity between 0.1 and 0.33 is defined as visually disabled by the World Health Organization. Vision loss is generally a result of organic defects that lead to loss or impairment of one or more visual functions. When a person is unable to master normal visual tasks, he or she has a disability in relation to his/her and other people's expectations. There are different degrees of low vision and blindness.
Luminance; L_v	See *radiometry* and *photometry*.
Luxmeter	An instrument used to measure *illuminance*.

Mach bands	The (illusory) enhancement of lightness contrast across contours arising from a luminance difference between two adjacent areas.
Macula lutea	The central region of the retina that contains the fovea. The yellow spot, ca. 2–3 mm in diameter, corresponding to about 10°.
Magnocellular cells, MC cells	A relatively large type of retinal ganglion cell (also called parasol cells) that projects to one of the magnocellular layers of the *LGN*. MC cells are phasic, responding transiently to a change in stimulus.
Maximum spectral luminous efficacy	Maximum ratio, K_m, between light flux and corresponding radiant flux (for monochromatic light of 555 nm). At 555 nm, where V_λ reaches its maximum of 1.0, 1 W corresponds to 683 lm, and $K_m = 683$ lm/W.
Melatonin	A hormone that regulates the circadian rythm.
Mesopic vision	Twilight vision in which both rods and cones are active; between *scotopic* and *photopic vision*.
Metameric colors	Color stimuli that match in color appearance, but have different spectral distributions.
Microelectrode	A thin metal tip (usually tungsten) that serves as an antenna for the electrical activity of a cell when placed close to it (or inside it).
Minutes of arc	1 arcmin $= 1' = 1/60°$ $(60' = 1°)$.
Modulation transfer function, MTF	An MTF is a plot of the *contrast rendering factor* as a function of *spatial frequency*. It characterizes the imaging quality of an optical system (lenses, cameras, etc.).
Monochromat	A person totally lacking chromatic color vision, e.g. a rod monocromat.
Monochromatic light	Radiation with a narrow wavelength distribution, ideally a single wavelength.
Monocular sector	A sector about 30° to the left and right in the field of view that can only be seen by one eye (since the nose blocks the view from the other eye). See also *binocular sector* and *nasal*.
Movement blindness	Persons with this defect cannot see moving objects, for example, water flowing from a mug into a glass, nor the rising water level in the glass. They can see that an object has shifted position from one place to another, but not the movement in between.
Multiplicative color mixture	Color mixture where light is absorbed in successive pigment layers. For instance, when light passes through a yellow filter and then through a blue filter, energy is removed from the light by every new filter, and the resulting spectral distribution is obtained by multiplying the spectral distribution of the incident light by the spectral transmission factors of each new pigment. See *subtractive color mixture*.
Munsell system	A color atlas that approximates the ideal of all neighboring color chips having the same percieved color difference. The coordinates of the system are chroma (color strength), hue and value (lightness).

Myelin	Electrically insulating layers of fatty tissue that surround nerve fibers.
Myopia	Nearsightedness. The image plane of a distant object is in front of the retina, the reason being that the eye is too long for its optics. Myopia can be corrected with concave lenses or by flattening a part of the cornea with a laser.
Nanometer (nm)	$1\,\text{nm} = 10^{-9}\text{m}$.
Nasal	The visual field and the retina are divided vertically into the nasal and the temporal fields. Objects in the temporal visual field are imaged on the nasal retina, and vice versa.
Natural Color System, NCS	The Natural Color System uses a perceptual scaling based on the relative proportions of unique colors. Color differences are therefore not equal everywhere in the NCS color space (as in the Munsell system). The cordinates are *hue*, *chromaticness* and black content.
Nerve impulse	See *action potentials*.
Neural network	Collection of interconnected neural elements (neurons) that form a functional unit.
Neuron doctrine	The hypothesis that the neurons are the fundamental signaling elements in the nervous system, and that their activity is directly linked to perception.
Neurotransmitter	A signaling substance that conveys signals from one cell to the next across the synaptic cleft. The substance can have an excitatory or inhibitory influence on the next cell. See *transmitter*.
Noise	Random variation of the activity of a sensory unit or in the stimulus itself.
Nyquist criterion	Used here as a criterion for visual resolution, i.e. when two nearby points of light are being resolved and not seen as one. Their separation requires that at least one *photoreceptor* detects the intensity minimum between the two intensity maxima.
Object colors	Colors of reflecting surfaces (or of related colors viewed in lighter surroundings). In contrast to the colors of light sources, object colors posess a black component induced by a surround of higher luminance (e.g. a white reflecting surface).
ON- or OFF-cells	Cells that respond with an increased firing rate for light increments (ON-cells) or for light decrements (OFF-cells) at the center of the receptive field. In this book we call them *I- and D-cells* (*Increment and Decrement cells*).
Ophthalmoscope	Instrument for looking at the retina through the optics of the eye.
Opponent colors	Opposite *elementary colors*. For example yellow and blue (there are no colors that are both yellow and blue at the same time). Red and green is another pair.
Optimal colors	*Object colors* that have a spectral distribution with only one step from 0 to 1.0 (or from 1.0 to 0) in their spectral distribution.

Ora serrata	A sawtooth-like seam behind the eye lens. This is one of the two places where the retina is fastened to the eye, the other place being where the optic nerve leaves the eye.
Outer plexiform layer	The retinal layer containing the horizontal cells and the synaptic contacts between *photoreceptors* and *bipolar cells*. 'Outer' means away from the center of the eye.
Outer segment (of a photoreceptor)	The light-sensitive part of a photoreceptor containing the pigmented membrane disks.
Parvocellular cells, PC-cells	A type of small, cone-opponent, retinal ganglion cells that project to cells in the parvocellular layers of the *LGN*, and cells within these layers. PC cells are tonic, with a sustained response to a change in the stimulus.
Penumbra	Half-shadow.
Perimeter	An instrument that is used to determine visual sensitivity for different parts of the monocular visual field.
Perception	Subjective qualitative experience or impression of some sensory input (internal representation). Can also be used for our understanding, comprehension and ideas, and is therefore sometimes linked to hypotheses and interpretations of sensory information about the environment.
PET	Positron emission tomography. A method for imaging brain activity.
Phase	Relative position (in time or space) of a sinusoidal function.
Photochemical adaptation	Changes in concentration of photopigment in the receptors, mainly in rods. These changes (bleaching of pigment at high intensity and regeneration at lower intensities) alter the optical density of the receptors and thereby their ability to absorb light quanta. In cones, photochemical adaptation is only significant at very high *retinal illuminances*, above about 10 000 troland.
Photometry	'The art of light measurement'. Photometry is based on radiometric measurements, combined with the spectral luminous efficiency function, V_λ, of the human eye (an action spectrum that specifies the spectral weighting function for the spectral power distribution of the stimulus). See *radiometry*.
Photon	See *quantum*.
Photopic vision	Daylight cone vision at light levels where the rods do not contribute (higher than ca. 10 cd/m^2).
Photoreceptors	Rods and cones. Sensory receptor cells in the retina that absorb electromagnetic energy (light quanta) within the visible range of the spectrum.
Pigment epithelium	Dark layer behind the retina.
Presbyopia	Decrease in the accommodation ability with advancing age.
Primary colors	This is a term often used for the colors in a color mixture (in aditive mixtures it is for practical reasons often red, green and blue, but in principle other combinations can be used as well).

Principle of univariance	For each of the pigment systems in rods and cone receptors, an absorbed light quantum contributes equally to vision, regardless of frequency. A *photoreceptor's* excitation depends only on the number of absorbed light quanta and not on their energy $E = h\nu$. This means that the same effect (the same excitation) can be achieved with different frequencies (or wavelengths) of light, providing the intensity is adjusted in accordance with the receptor's frequency-dependent probability of absorption.
Prosopagnosia	An inability to recognize faces, even when features such as mouth, nose, eyes, etc. are readily identified. Even close family members may have to be identified by their voices rather than their faces.
Protanopia	Color vision deficiency resulting from the absence of L-cones. Protanopes cannot distinguish between red, yellow and yellow-green, or between red, white and blue-green.
Pupil	The round aperture of the eye that is limited by the *iris*. In younger years, its diameter ranges from about 2 mm in strong daylight to about 8 mm in darkness. This change in size regulates the illumination of the retina only by a factor of 1:16. The pupil reflex is elicited by both rods and cones. A 3 mm pupil minimizes the negative effects of lens aberration and *diffraction*.
Purity (colorimetric)	Relative amounts of monochromatic light in an *additive color mixture* with white light.
Purkinje's phenomenon	The relative darkening of red and orange surface colors at dusk (as compared with blue and green). This is a consequence of the shift of maximum spectral sensitivity towards shorter wavelengths in the transition from cone vision in bright light to rod vision in the dark.
Qualia	The quality of conscious experience. The feeling or the perception of a quality such as the 'redness of red' or a pain.
Quantum	A small 'package' of energy $E = h\nu$. A quantum of light is called a photon.
Radians	$360° = 2\pi$ rad, and thus α (rad) $= 2\pi(\alpha°/360°)$. 1 rad $= 57.3°$.
Radiometry (e, energy units) and Photometry (v, visual)	Wavelength, λ (nm); radiant flux, ϕ_e (W); irradiance, E_e (W/m^2); radiance, L_e (W/sr m^2); light flux, ϕ_v (lumen; lm); illuminance, $E_v = \phi_v/A$ (lm/m^2; lux, (lx)); luminance $L_v = I_v/A$ (cd/m^2); light intensity $I_v = \phi_v/\omega$ (candela; cd).
Rayleigh criterion	Two small points of light very close to each other (e.g. two stars) can be separated only if their have an angular separation larger than the angular distance between the first intensity maximum and the first intensity minimum in the *diffraction* pattern from one of them. For imaging through a circular pupil, this angular distance is $\theta = 1.22\lambda/D$ rad, where λ is the wavelength and D the diameter of the pupil.

Receptive field	*Classical receptive field*: a small area of the retina that evokes a response in a nerve cell (e.g. a ganglion cell) when stimulated. If an excitation of the *photoreceptors* within the center of this area leads to activation, as in *I-cells*, stimulation of the surrounding receptors will usually result in inhibition (center-surround structure of the receptive field). *D-cells* are activated by light decrements in the receptive field center or increments in the surround, and they are inhibited by center increments. In the fovea the receptive field center of a ganglion cell collects information from only one or a few cones. *Global receptive field*: a larger retinal area, extending beyond the classical receptive field surround, through which the response of a cell can be modulated.
Receptor, photoreceptor	A neural element that transforms one type of energy to another. Most sensory cells respond selectively to specific physical stimuli, such as pressure, light, temperature, etc. *Photoreceptors* are excited by the absorption of light quanta and transform this energy into a change in the electric potential across the cell membrane.
Reflection factor, β. Spectral reflection factor, $\beta(\lambda)$	The ratio, β, between the light flux, Φ, that is reflected from a surface, and the flux, P, that impinges on it: $\beta = \Phi/P$. $\beta(\lambda)$ gives its spectral distribution.
Refractive index, n	Ratio between the speed of light, c, in vacuum and its speed, v, in the medium in question; $n = c/v$.
Refractive power, P	$P = n/f$, where n is refractive index and f is focal length. The unit is diopters [m^{-1}].
Related and unrelated colors	The appearance of chromatic stimuli undergoes changes in lightness, hue and chromatic strength depending on the luminance of their surroundings. Object or surface colors that are viewed in a well-illuminated, natural environment are examples of *related colors*. Since they are all darker than a white surface, they have some degree of induced blackness. Blackness is in itself a typical related color. *Unrelated color* refers to the appearance of light sources, of bright colors viewed in the dark, or of stimuli with higher luminance than their surroundings (they are sometimes called void colors); 99 percent of all natural colors we see during the day are related colors. The luminance ratio between a surface and white (the *reflection factor*) determines whether its color is related or unrelated. See also *Bezold–Brücke phenomenon*.
Retinal	A light-absorbing molecule in the cone and rod photoreceptors which, when bound to different opsins, makes different pigments with different spectral sensitivities.
Retinal illuminance	See *troland*
Rhodopsin	The light-absorbing pigment of the rods (visual purple).
Rod	Receptor cell in the retina that is active at low (*scotopic*) luminance levels.

Saccades	Abrupt eye movements from one fixation point to a new fixation point.
Saturation (color vision)	Apparent amount of chromatic color relative to achromatic color of the same lightness.
Schlemm's canal	Aqueous humor is excreted from (the anterior chamber of) the eye through Schlemm's canal. Obstruction of this passage can lead to a rise in intraocular pressure and to glaucoma.
Sclera	A tough, opaque, white membrane that encloses most of the eyeball (continuous with the cornea).
Scotoma	Blind area of the visual field due to damage or illness in the visual system.
Scotopic vision	Vision in darkness for light levels below the cone threshold of 0.1 troland, corresponding to about 0.005 photopic cd/m^2.
Sensitivity, s	Sensitivity (s) = criterion response (R)/physical stimulation (I), or $s = \Delta R/\Delta I$ at threshold (ΔR being constant).
Sensory space	Describes for instance the cone excitation space, but it is often also used for the spatial organization of the different qualities of a sensory experience, for example color (a perceptual color space). The dimensionality of such a space corresponds to the dimensions of that particular mode of perception. A three-dimensional space can seldom reflect all of these dimensions. For example, a color space can be defined in terms of pure perceptual qualities, such as, for instance, opponent unique color qualities (yellow–blue, red–green, white–black) and their transitions, but also according to the psychophysical dimensions hue, saturation and lightness, or simply in terms of equal distances between neigboring colors.
Simultaneous contrast	Mutual interactions between adjacent areas and figures within the visual field. Black and grey colors are typical examples of simultaneous contrast where bright surrounds affect areas of lower luminance. Simultaneous contrast is active within the lightness domain (*related and unrelated colors*) as well as within the chromatic domain (colored shadows, color induction). Within the spatial domain it gives rise to size illusions [see Figure 1.10(c) and (d)].
Sinusoidal grating	A pattern of many light and dark bars with 'soft' luminance transitions (no sharp edges). The luminance contrast across the bars varies as a sine function. Sine gratings are used for measuring the contrast rendering quality of a lens system. In recent years such gratings are increasingly being applied in the assessments of spatial contrast sensitivity in vision (see *contrast sensitivity*).
Soma	Cell body. The soma summates all signals from the *dendrites*.
Spatial frequency	Number of periods per degree of a repetitive pattern, for example a sinusoidal grating.
Spatial summation	Summation of some effect (excitation or inhibition) over an extended area (for example within the *receptive field* of a nerve cell).

Spectral luminous efficiacy, $K(\lambda)$	A function that converts radiometric units to photometric units for daylight vision (*photopic vision*). $K(\lambda) = 683V(\lambda)$ (lm/W). There exists a similar expression for night vision (*scotopic vision*) where $K'(\lambda) = 1700V'(\lambda)$.
Spectral luminous efficiency function for photopic vision; $V(\lambda)$	The relative spectral sensitivity of the light-adapted eye to electromagnetic radiation. For each wavelength, it specifies the relative efficiency of a light in evoking a criterion response (minimum flicker, for example). A similar function, $V'(\lambda)$, specifies the relative spectral sensitivity for *scotopic vision* (dark-adapted eye).
Spectral opponency	The response of a spectrally opponent cell is strongly influenced by the spectral composition of a stimulus. The cell is activated by increased absorption in one cone type and inhibited by increased absorption in another cone type. For a given cell, its spectral opponency is designated by appropriately signed labels, for example 'L–M', 'M–L', 'S–L' or 'M–S'. Owing to the different spectral sensitivities of the cones, opponent cells will be activated by some wavelengths, inhibited by others, and unresponsive to a particular neutral wavelength (zero crossing).
Spectral line	When heated, most gases emit light within several narrow spectral wavelengths. In fluorescent tubes a continuous spectrum overlies these spectral lines.
Spectral reflection curves	The spectral distribution of the reflection factor of a surface, $\beta_\lambda = \Phi_\lambda / P_\lambda$, can, to a good approximation, be written as a sum of three basic curves: $\beta_\lambda = aS_{1\lambda} + bS_{2\lambda} + cS_{3\lambda}$, where a, b and c are constants that must be determined for each surface.
Spectral reflection factor; β_λ	The property of a surface of reflecting different proportions of the incident light for different wavelengths. A white or gray surface reflects the same proportion of light at all wavelengths (they have a constant spectral reflection factor); $\beta_\lambda = \Phi_\lambda / P_\lambda$.
Spectral transmission factor; τ_λ	The property of transmitting different proportions of the incident light for different wavelengths. A neutral gray filter transmits the same proportion of light at all wavelengths; $\tau_\lambda = \phi_\lambda / P_\lambda$.
Spectrophotometer	An instrument for measuring the spectral distribution of light.
Spectrum	A spatial distribution of electromagnetic radiation of wavelengths. The visible spectrum ranges from 380 to 760 nm and is usually produced by dispersion in a prism or by interference in a regular grating pattern.
Spherical aberration	Light rays that pass through a lens near its rim are refracted more strongly than those that pass through its center. This is a prism effect, and in visual imaging the resulting distortions of the image are larger for a wide pupil than for a small one.
Spikes	See *action potentials*.
Square wave grating	Periodic sequence of dark and brighter bars with sharp contrast edges. This grating can be described mathematically as a sum of sinusoidal functions with different *spatial frequencies* (cycles/deg), amplitudes (contrast), and phase (spatial position).

Stevens' law	A scaling of a sensory magnitude, S, that follows a power law of stimulus magnitude, $I : S = $ const. I^n.
Stimulus/stimuli	Modulation of physical energy that elicits a sensory response. A stimulus can be described along several physical dimensions, such as size, radiant power, energy, luminance, wavelength, contrast, etc.
Subtractive color mixture	See *multiplicative color mixture*.
Synapse	Point of contact between the axon of the presynaptic cell and the dendrite of the postsynaptic cell where signals are transmitted from the first cell to the next (see also *dendrite* and *axon*). Synapses can be excitatory or inhibitory, and either chemical or electrical.
Synaptic layers	*Outer*: the retinal layer containing synaptic contacts between *photoreceptors, horizontal cells* and *bipolar cells*. *Innner*: the retinal layer with synaptic contacts between *bipolar cells, amacrine cells* and *ganglion cells*.
Temporal retina	The visual field and the retina is divided horizontally into two halves. The nasal retina is closest to the nose and receives input from the temporal visual field. The temporal retina is furthest away from the nose and receives input from the nasal visual field.
Thalamus	A subdivision of the brain between the retina and visual cortex where one finds the *LGN*. The thalamus has been considered to be a gateway to the cortex since all sensory information (except for olfaction) passes through this area before it reaches the cortex.
Threshold	The lowest intensity or energy of a stimulus that can be detected or discriminated in a given situation. The threshold value will depend on the task, the threshold criterion, the physical conditions, and on physiological and psychological states.
Tonic and phasic ganglion cells	Tonic is a term used to describe the sustained temporal response of PC and KC cells, whereas phasic describes the transient response of MC cells. The corresponding retinal cells are often called midget-, bistratified- and parasol cells.
Top-down processing	Higher brain centers influence the processing done at lower levels of the visual pathway.
Transduction	The process by which physical energy impinging on a receptor is converted into neural signals (e.g. to a change in the membrane potential of the receptor).
Transient response	Short lasting response, typical of MC cells.
Transmission factor, τ. Spectral transmission factor, $\tau(\lambda)$	The ratio, τ, between the light flux, Φ, that is transmitted through a material and the flux, P, that is incident on it: $\tau = \Phi/P$; $\tau(\lambda)$ denotes its spectral distribution.
Transmitter/transmitter substance	Molecules of a chemical signaling substance that is released in a synapse. The amount released depends on the strength of the incoming signal (rate of incoming action potentials, for example) that causes the release.
Trichromatic color vision; trichromat	Normal color vision with all cone types intact; a person with normal color vision.

Tristimulus values	Three numbers for the amount (vector length) of three basic color stimuli (often called primaries) that match a test color in an additive color mixture. In the CIE 1931 system the tristimulus values have the notation X, Y and Z. They apply to a specific triplet of (virtual) primaries.
Tritanopia	Color vision deficiency that is caused by the lack of S-cones. Tritanopes cannot distinguish between blue, white and yellow, or between blue and green.
Tritanopic purity	Purity in the M–L direction of cone color space, describing the chromatic purity for a tritanopic subject who lacks S-cones.
Troland; td	A unit for retinal illuminance. Retinal illuminance (td) = area of the pupil (mm^2) × Luminance of the stimulus surface (cd/m^2); troland $= A \cdot L$ (td).
Umbra	Core shadow in the presence of an extended light source or more than one light source.
Unique colors	See *Elementary colors.*
Univariance	See *principle of univariance.*
Unrelated color	The color appearance of a self-emitting light source or of light surfaces viewed in a dark environment. See *related colors.*
UV	Ultraviolet radiation with a wavelength shorter than 400 nm. UV is divided into three wavelength ranges; UV-A (315–400 nm), UV-B (280–315 nm), and UV-C (100–280 nm).
V1	Primary visual cortex, striate cortex, area 17 in Brodman's nomenclature.
Ventral stream	Visual areas on the lower part of the brain that get heavy PC cell inputs; concerned largely with identification of objects. Ventral: toward the bottom of the head.
Vergence, V	Vergence $V = 1/r$ (m^{-1}), where r is the radius of curvature of the wavefront. Converging light has a positive vergence, and diverging light has a negative vergence. Vergence is measured in *diopters.*
Vernier acuity	The ability to detect failures in (perfect) alignment (e.g. of two thin lines). Normally one can detect a failure in alignment corresponding to a few seconds of arc.
VIGRA	A specific 3×10 bit computer display system (videographic system) for the presentation and manipulation of a variety of visual stimuli on a color monitor.
Virtual	Not real. Often used for images that cannot be produced on a screen, such as, for instance images made by concave lenses.
Visual acuity	A measure for the ability to resolve of small details of maximum contrast, often assessed by means of a letter chart (e.g. the Snellen chart). Decimal acuity $= 1/\alpha$, where α is the minimum angle of resolution (MAR), expressed in minutes of arc (1 arcmin $= 1/60°$). Foveal acuity of 1.0 or better is regarded as normal. Another measure is log MAR $=$ log α. Resolution improves as luminance increases, and it decreases with distance from the fovea.

Visual cortex	Areas of cortex specialized for vision, with further subdivisions into functional units and modules. Receives input from the *LGN* via primary visual cortex, V1.
Visual Evoked Potential, VEP	Electrical brain activity in a collection of nerve cells provided by a visual stimulus; can be recorded by scalp electrodes.
Visual image	The image that is perceived consciously (as opposed to the physical image on the retina).
Vitreous humor	The eye is filled with a gel-like, transparent fluid (vitreous) that provides an inner pressure and thus maintains the spherical shape of the eye. Fine threads crossing the vitreous may break and form thin, curled, moving shadows that eventually sink to the bottom of the eye.
Weber–Fechner law	Mathematical relation between the *Weber ratio* at threshold and the subjective impression of lightness, s; $\Delta s = c \Delta L/L$. When integrated this leads to *Fechner's law*.
Weber ratio/Weber contrast	Weber ratio, or Weber contrast, is defined as $C_{Weber} = \Delta L/L$. For a wide range of photopic values of background luminance, L, one finds that this ratio is constant at threshold (*Weber's law*). At high luminance levels, *contrast sensitivity*, $1/C_{Weber}$ at threshold, is highest for fields larger than $12'$ or $0.2°$ (when they have sharp contours). For sinusoidal gratings with graded contours the sensitivity decreases for bars larger than 10 arcmin (e.g. for *spatial frequencies* below 3 cycles/deg). See *contrast sensitivity*, *Weber's law* and *Fechner's law*.
Weber's law	This law states that the *Weber ratio* is constant at threshold. The law is valid at photopic light levels from about 50 to 10 000 cd/m^2.

References

Abney, W. (1910) On the changes in hue of spectrum colours by dilution with white light. *Proceedings of the Royal Society of London. Series A* **83**, 120–127.

Abney, W. (1913) *Researches in Colour Vision and Trichromatic Theory.* Longmans: London.

Abney W. and Festing, E. R. (1886) Colour photometry. *Philosophical Transactions of the Royal Society, London* **177**, 423–456.

Adams, E. Q. (1942) X–Z-planes in the 1931 I.C.I. system of colorimertry. *Journal of the Optical Society of America* **32**, 168–173.

Allman, J., Miezin, F. and McGuinness, E. (1985) Stimulus specific responses from beyond the classical receptive field – neurophysiological mechanisms for local global comparison in visual neurons. *Annual Review of Neuroscience* **8**, 407–430.

ASTM (1991) *Standard Terminology for Appearance (E284-91b)* American Society for Testing and Materials: Philadelphia, PA.

Azzopardi, P. and Cowey, A. (1996) The overrepresentation of the fovea and adjacent retina in the striate cortex and dorsal lateral geniculate nucleus of the macaque monkey. *Neuroscience* **72**, 627–639.

Barlow, H. B. (1953) Summation and inhibition in the frog's retina. *Journal of Physiology* **119**, 69–88.

Barlow, H. (1972) Single units and sensation: a neuron doctrine for perceptual psychology? *Perception* **1**, 371–394.

Bastie, J. (1999) Light measurements before V(λ). In *CIE Symposium'99. 75 Years of CIE Photometry.* Hungarian Academy of Sciences: Budapest. Abstract book pp. 7–8.

Baumgartner, G. (1961) Die Reaktionen der Neurone des zentralen visuellen Systems der Katze im simultanen Helligkeitskontrast. In *Neurophysiologie und Psychophysik des visuellen Systems*, Jung, R. and Kornhuber, H. (eds). Springer: Berlin, pp. 298–311.

von Bezold, W. (1873) Über das Gesetz der Farbmischung und die physiologischen Grundfarben. *Poggendorffs Annalen der Physik, Leipzig* **150**, 221–247.

Birch, J. (1993) *Diagnosis of Defective Colour Vision.* Oxford University Press: Oxford.

Björnevik, L. R. (1992) Synspotensialer (Visual potentials). Master Thesis in Physics, University of Olso (in Norwegian).

Bjorvatn, B. and Holsten, F. (1997) Lysbehandling ved jet lag, nattarbeid og søvnlidelser. (Light treatment by jet lag, night work and sleep disorders.) *Tidskrift Den norske lægeforening* **17**, 2489–2492 (in Norwegian).

Blakemore, C. and Campbell, F. W. (1969) On the existence of neurones in the human visual system selectively sensitive to the orientation and size of retinal images. *Journal of Physiology* **203**, 237–260.

Bouma, P. J. (1946) *Physical Aspects of Colour.* N.V. Philips Gloeilampenfabrieken: Eindhoven.

Bowmaker, J. K. (1991) Visual pigments and colour vision in primates. In *From Pigments to Perception; Advances in Understanding the Visual Process*, Valberg, A. and Lee, B. B. (eds). Plenum Press: London, pp. 1–10.

Boynton, R. M. (1960) Theory of color vision. *Journal of the Optical Society of America* **50**, 929–944.

Boynton, R. M. and Kaiser, P. K. (1968) Vision: the additivity law made to work for heterochromatic photometry with bipartite fields. *Science* **161**, 366–368.

Boynton, R. M. and Whitten, D. N. (1970) Visual adaptation in monkey cones. Recordings of late receptor potentials. *Science* **170**, 1423–1426.

Brainard, D. H. (1996) *Human Color Vision*, Kaiser, P. K. and Boynton, R. M. (eds), 2nd edn. Optical Society of America: Washington, DC, Part IV, pp. 563–579.

Breitmeyer, B. and Valberg, A. (1979) Local foveal inhibitory effects of global peripheral excitation. *Science* **203**, 463–464.

Brill, T. B. (1980) *Light. Its Interaction with Art and Antiquities.* Plenum Press: New York.

Brindley, G. S. (1960) *Physiology of the Retina and the Visual Pathway.* Edward Arnold: London.

von Brücke E. T. (1878) Über einige Empfindungen im Gebiet der Sehnerven. *Sitzungsberichte der Akademie der Wissenschaften in Wien, Mathematisch-Naturwissenschaftliche Klasse, Abteilung 3* **77**, 39–71.

Buchsbaum, G. and Gottschalk, A. (1983) Trichromacy, opponent colours coding and optimum colour information transmission in the retina. *Proceedings of the Royal Society, London* B **220**, 89–113.

Buchsbaum, G. and Tailor, D. R. (2001) The elementary structure of natural color images and its possible neurophysiological correlates. *Journal of Vision* 1(3), 64a; available at: http://journalofvision.org/1/3/64, DOI 10.1167/1.3.64 (12 December 2001).

Burns, S. A., Elsner, A. E., Pokorny, J. and Smith, V. C. (1984) The Abney effect: chromatic coordinates of unique and other constant hues. *Vision Research* **24**, 479–489.

Burr, D. C., Morrone, M. C. and Ross, J. (1994) Selective depression of the magnocellular visual pathway during saccadic eye movements. *Nature* **371**, 511–513.

Campbell, F. W. and Robson, J. G. (1968) Application of Fourier analysis to the visibility of gratings. *Journal of Physiology* **197**, 551–566.

Carroll, J., Neitz, M., Hofer, H., Neitz, J. and Williams, D. R. (2004) Functional photoreceptor loss revealed with adaptive optics: an alternative cause of color blindness. *Proceedings of the National Academy of Sciences, USA* **101**, 8461–8466.

Cavanagh, P. MacLeod, D. I. A. and Anstis, S. M. (1987) Equiluminance: spatial and temporal factors and the contribution of blue-sensitive cones. *Journal of the Optical Society of America* **A4**, 1428–1438.

Caywood, M. S., Willmore, B. and Tolhurst, D. J. (2001) The color tuning of independent components of natural scenes matches Vλ simple cells. *Journal of Vision* 1(3), 65a; available at: http://journalofvision.org/1/3/65, DOI 10.1167/1.3.65.

Chalmers, D. J. (1995a) Facing up to the problem of consciousness. *Journal of Consciousness Studies* **2**, 200–219.

Chalmers, D. J. (1995b) The puzzle of conscious experience. *Scientific American December*, 62–68.

Chapanis, A. (1944) Spectral saturation and its relation to color vision defects. *Journal of Experimental Psychology* **34**, 24–44.

Chaparro, A., Stromeyer, C. F. III, Huang, E. P., Kronauer, R. E. and Eskew, R. T. Jr. (1993) Colour is what the eye sees best. *Nature* **361**, 348–350.

Chevreul, M. E. (1969) *De la loi du contraste simultane des coleurs.* Pitois-Levrault: Paris/ Leon Laget: Paris. [First published 1839.]

Chichilinsky, E. J. and Wandell, B. A. (1996) Seeing gray through the ON and Off pathways. *Visual Neuroscience* **13**, 591–596.

Churchland, P. S. (1986) *Neurophilosophy.* MIT Press: Cambridge, MA.

Churchland, P. M. (1994) *The Engine of Reason, the Seat of the Soul.* MIT Press: Cambridge, MA.

Churchland, P. S. and Sejnowski, T. J. (1992) *The Computational Brain.* MIT Press: Cambridge, MA.

Cicerone, C. M., Volbrecht, V. J., Donally, S. K. and Werner, J. S. (1986) Perception of blackness. *Journal of the Optical Society of America* **A3**, 432–436.

CIE (1970) *International Lighting Vocabulary*, Publication no. 17 (E-1.1). Central Bureau of CIE: Vienna.

CIE (1987) *International Lighting Vocabulary.* 4th edn. Publication no. 17. Central Bureau of CIE: Vienna.

CIE (1988) *Spectral Luminous Efficiency Functions based upon Brightness Matching for Monochromatic Point Sources 2° and 10° Fields.* Publication no. 75. Central Bureau of CIE: Vienna.

CIE (1990) *CIE 1988 2° Spectral Luminous Efficiency Function for Photopic Vision.* Technical Report. Publication no. 86. Central Bureau of CIE: Vienna.

CIE (1998) *Testing of Supplementary Systems of Photometry.* Technical report of TC 1–21. Draft. Central Bureau of CIE: Vienna.

Coblentz, W. W. and Emerson, W. B. (1918) Relative sensibility of the average eye to light of different colors and some practical applications to radiation problems. *Bulletin of the National Bureau of Standards* **14**, 167–236.

Conway, B. R. (2001) Spatial structure of cone inputs to color cells in alert macaque primary visual cortex (V1). *Journal of Neurophysiology* **21**: 2768–2783.

Cornsweet, T. N. (1978) The Bezold–Brucke effect and its complement, hue constancy. In *Visual Psychophysics and Physiology*, Armington, J. C. Krauskopf, J. and Wooten, B. R. (eds). Academic Press: New York, pp. 233–244.

Cottaris, N. P. and DeValois, R. L. (1998) Temporal dynamics of chromatic tuning in macaque primary visual cortex. *Nature* **395**, 896–900.

Creutzfeldt, O. D., Lee, B. B. and Elepfandt, O. D. (1979) A quantitative study of chromatic organisation and receptive fields of cells in the lateral geniculate body of the monkey. *Experimental Brain Research* **35**, 527–545.

Creutzfeldt, O. D., Lee, B. B. and Valberg, A. (1986) Colour and brightness signals of parvocellular lateral geniculate neurones. *Experimental Brain Research* **63**, 21–34.

Creutzfeldt, O. D., Kastner, S., Pei, X. and Valberg, A. (1991) The neurophysiological correlates of colour and brightness contrast in lateral geniculate neurones. *Experimental Brain Research* **87**, 22–45.

Crick, F. (1994) *The Astonishing Hypothesis. The Scientific Search for the Soul.* Scriber: New York.

Crick, F. and Koch, C. (1995) Why neuroscience may be able to explain consciousness. *Scientific American*, December: 66–67. [Comentor Chalmers (1995b).]

Dacey, M. D. (1993) The mosaic of midget ganglion cells in the human retina. *Journal of Neuroscience* **13**, 5334–5355.

Dacey, D. M. and Lee, B. B. (1994) The 'blue-on' opponent pathway in primate retina originates from a distinct bistratified ganglion cell type. *Nature* **367**, 731–735.

Dacey, M. D. and Packer, O. S. (2003) Colour coding in the primate retina: diverse cell types and cone-specific circuitry. *Current Opinion in Neurobiology* **13**, 421–427.

Dacey, M. D., Lee, B. B., Stafford, D. K., Smith, V. C. and Pokorny, J. (1996) Horizontal cells of the primate retina: Cone specificity without cone opponency. *Science* **271**, 656–658.

da Vinci, L. (1906). *A Treatise on Painting*. English translation by Rigaud, J. F. and Bell, G. London. [*Traktat von der Malerei*. German translation by Ludwig, H. (1882) New edition by Hetzfeld, M. (1925). Diederichs: Jena first published 1651.]

Daw, N. W. (1968) Colour-coded ganglion cells in the goldfish retina. Extension of their receptive fields by means of new stimuli. *Journal of Physiology* **1971**, 567–592.

De Lange, H. (1958) Research into the dynamic nature of the human fovea–cortex systems with intermittent and modulated light; II. Phase shift in brightness and delay in color perception. *Journal of the Optical Society of America* **48**, 784–789.

DeMonasterio. F. M. (1984) Electrophysiology of color vision. I. Cellular level. In *Colour Vision Deficiencies VII*, Verriest, G. (ed.) Junk: The Hague, pp. 9–28.

DeMonasterio, F. M. and Gouras, P. (1975) Functional properties of ganglion cells of the rhesus monkey retina. *Journal of Physiology* **251**, 167–195

Derefeldt, G. (1991) Colour appearance systems. In *The Perception of Colour, Vol. 6, Vision and Visual Dysfunctions*, Cronly-Dillon, J. R. and Gouras, P. (eds). Macmillan, London, 1991, pp. 218–261.

Derrington, A. M., Krauskopf, J. and Lennie, P. (1984) Chromatic mechanisms in lateral geniculate nucleus of macaque. *Journal of Physiology* **357**, 241–265.

Descartes, R. (1953) *La Dioptrique. I. Euvres et lettres*. Bibliothèque de laPléiade, Editions Gallimard: Paris. [First published 1637.]

De Valois, R. L. (1965) Analysis and coding of color vision in the primate visual system. *Cold Spring Harbor Symposia on Quantitative Biology* **30**, 567–579.

De Valois, R. L. (1969) Physiological basis of color vision. *Proceedings of the International Colour Meeting COLOR 69*, Stockholm, pp. 29–47.

De Valois, R. L. and DeValois, K. (1993) A multi-stage color model. *Vision Research* **33**, 1053–1065.

De Valois, R. L. and Jones, A. E. (1961) Single-cell analysis of the organization of the primate color-vision system. In *Neurophysiologie und Psychophysik des visuellen Systems*. Springer: Berlin, pp. 178–191 and 197–199.

De Valois, R. L., Abramov, I. and Jacobs, G. H. (1966) Analysis of response patterns of LGN celle. *Journal of the Optical Society of America* **56**, 966–977.

De Valois, R. L., DeValois, K., Switkes, F. and Mahon, L. (1997) Hue scaling of isoluminant and cone-specific lights. *Vision Research* **37**, 885–897.

Diller, L., Packer, O. S., Verweij, J., McMahon, M. J., Williams, D. R. and Dacey, D. M. (2004) L and M cone contributions to the midget and parasol ganglion cell receptive fields of macaque monkey retina. *Journal of Neuroscience* **24**(5), 1079–1088.

Du Buf, H. (1994) Responses of simple cells: events, interferences and ambiguities. *Biological Cybernetics* **68**, 321–333.

Eagleman, D. M. (2001) Visual illusions and neurobiology. *Nature Reviews Neuroscience* **2**, 920–926.

Edwards, D. P., Purpura, K. P. and Kaplan, E. (1995) Contrast sensitivity and spatial frequency response of primate cortical neurons in and around the cytochrome oxidase blobs. *Vision Research* **35**, 1501–1523.

Ekstrom, A. D., Kahana, M. J., Caplan, J. B., Fields, T. A., Isham, E. A., Newman, E. L. and Fried, I. (2003) Cellular networks underlying human spatial navigation. *Nature* **425**, 184–187.

Engel, S., Zhang, X. and Wandell, B. (1997) Color tuning in human visual cotex measured with functional magnetic resonance imaging. *Nature* **388**, 68–71.

Ernst, B. (1988) *M. C. Eschers tryllespejl.* Taco: Berlin.

van Essen, D. C., Felleman, D. J., DeYoe, E. A. and Knierim, J. J. (1991) Probing the primate visual cortex: pathways and perspectives. In *From Pigments to Perception. Advances in Understanding Visual Processes*, Valberg, A. and Lee, B. B. (eds). Plenum: London, pp. 227–237.

Estevez, O. (1979) On the fundamental data-base of normal and dichromatic colour vision. Ph.D. Dissertation, University of Amsterdam.

Evans, R. M. (1964) Variables of perceived color. *Journal of the Optical Society of America* **54**, 1467–1474.

Evans, R. M and Swenholt, B. K. (1967) Chromatic strength of colors; dominant wavelength and purity. *Journal of the Optical Society of America* **57**, 1319–1324.

Evans, R. M. and Swenholt, S. B. (1968) Chromatic strength of colors, part II. The Munsell-system. *Journal of the Optical Society of America* **58**, 580–584.

Farrell, B. A. (1962) Experience. In *The Philosophy of Mind*, Chappell, V. C. (ed.). Prentice-Hall (Spectrum Book), Engelwood Cliffs, NJ, pp. 23–48.

Fechner, G. T. (1966) *Elemente der Psychophysik.* Breitkopf and Härtel: Leipzig. *Elements of Psychophysics*, English translation. Holt, Rinehart & Winston: New York. [First published 1860.]

Fischer, B. and Krüger, J. (1974) The shift-effect in the cat's lateral geniculate nucleus. *Experimental Brain Research* **21**, 225–227.

Fosse, P. and Valberg, A. (2001) Contrast sensitivity and reading in subjects with age-related macular degeneration. *Visual Impairment Research* **3**, 111–124.

Fosse, P., Valberg, A. and Arnljot H. M. (2001) Retinal illuminance and the dissociation of letter and grating acuity in age-related macular degeneration. *Optometry and Vision Science* **78**, 162–168.

Fraunhofer, J. (1814) Denkschrift der bayr. *Akad.* **211**. (Cited in von Helmholtz, 1911.)

Freitag, P., Greenlee, M. W., Lacina, T., Scheffler, K. and Radii, E. W. (1998) Effect of eye movements on the magnitued of fMRI responses in the extrastriatal cortex during visual motion perception. *Experimental Brain Research* **4**, 409–414.

Friendly, D. S., Jafaar M. S. and Morillo, D. L. (1990) A comparative study of grating and recognition visual acuity in children with anisometropic amblyopia without strabismus. *American Journal of Ophthalmology* **110**, 293–299.

Frisby, J. P. (1979) *Seeing. Illusion, Brain and Mind.* Oxford University Press: Oxiford.

Fynn (1979) *Mister God, This is Anna.* Fount Paperbacks: London.

Gazzaniga,, M. S., Ivry, R. B. and Mangun, G. R. (1998) *Cognitive Neuroscience. The Biology of the Mind.* W. W. Norton: New York.

Gazzaniga, M. S., Ivry, R. B. and Mangun, G. R. (2002) *Cognitive Neuroscience. The Biology of the Mind*, 2nd edn. Norton: New York.

Gegenfurtner, K. R. (2003) Cortical mechanisms of colour vision. *Nature Reviews* **4**, 563–572.

Gibson, K. S. and Tyndall, E. P. T. (1923) Visibility of radiant energy. *Scientific Papers of the National Bureau of Standards* **19**(475), 131–191.

von Goethe, W. (1963) *Zur Farbenlehre*. Deutscher Taschenbuchverlag: München. [First published 1810.]

Gombrich, E. H. (1972) *The Story of Art*. Phaidon: Oxford.

Goodale, M. A. (2000) Perception and action in the human visual system. In *The New Cognitive Neurosciences*, Gazzaniga, M. S. (eds). MIT Press:, Cambridge, MA, pp. 365–377.

Gouras, P. (1968) Identification of cone mechanisms in monkey ganglion cells. *Journal of Physiology* **199**, 533–547.

Gouras, P. (1974) Opponent-colour cells in different layers of foveal striate cortex. *Journal of Physiology* **238**, 538–602.

Gouras, P. and Zrenner, E. (1981) Color vision: a review from a neurophysiological perspective. In *Progress in Sensory Physiology*, Vol. 1, Ottoson, E. (ed.). Springer: Berlin, pp. 139–179.

Graham, C. H. and Hsia, Y. (1969) Saturation and the foveal achromatic interval. *Journal of the Optical Society of America* **59**, 993–997.

Granger, E. M. and Heurtley, J. C. (1973) Visual chromaticity–modulation transfer function. *Journal of the Optical Society of America* **63**, 1173–1174.

Granit, R. (1933) The components of the retinal action potential in mammals and their relation to the discharge in the optic nerve. *Journal of Physiology* **77**, 207–239.

Grassmann, H. (1853) Zur Theorie der Farbmischung. *Poggendorffs Annalen Physik* **89**, 69.

Gregory, R. L. (1990) *Eye and Brain. The Psychology of Seeing*. Oxford University Press: Oxford.

Grenness, C. E. and Magnussen, S. (1989) Words and pictures. In *Basic Issues in Psychology*, Bjørgen, I. A. (ed.). Sigma: Berger, pp. 69–83.

Grini, F. (1997) *Light from the Pythagoreans to Quantum Theory*. Scandinavian University Press: Berger.

Grünert, U., Greferath, U., Boycott, B. B. and Wässle, H. (1993) Parasol (P_α) ganglion-cells of the primate fovea: immunocytochemical staining with antibodies against gaba$_A$-receptors. *Vision Research* **33**, 1–14.

Guild, J. (1931) The colorimetric properties of the spectrum. *Philosophical Transactions of the Royal Society of London, Ser. A* **230**, 149–187.

Gurnsey, R., Sharon, L. S., Potechin, C. and Mancini, S. (2002) Optimising the Pinna–Brelstaff illusion. *Perception* **31**, 1275–1280.

Guth, S. L. (1980) Vector model for normal and trichromatic color vision. *Journal of the Optical Society of America* **70**, 197–212.

Guth, S. L. (1991) Model for color vision and light adaptation. *Journal of the Optical Society of America* **A8**, 976–993.

Guth, S., Massof, R. W. and Benzschawel, T. (1980) Vector model for normal and dichromatic color vision. *Journal of the Optical Society of America* **70**, 450–462.

Hardin, C. L. (1988) *Color for Philosophers*. Hackett: Indianapolis, IN.

Hartline, H. K. (1938) The response of single optic nerve fibers of the vertebrate eye to illumination on the retina. *American Journal of Physiology* **121**, 400–415.

Hartline, H. K. (1940) The receptive fields of optic nerve fibers. *American Journal of Physiology* **130**, 690–699.

Hassenstein, B. (1968) Modellrechnung zur Datenverarbeitung beim Farbensehen des Menschen. *Kybernetik* **4**, 209–223.

Haupt, A. (1922) The selectiveness of the eye's response to wavelength and its change with change of intensity. *Journal of Experimental Psychology* **5**, 347–379.

Hebb, D. O. (1949) *The Organization of Behavior. A Neuropsychological Theory.* Wiley: New York.

Hecht, S., Shlaer, S. and Pirenne, M. H. (1942) Energy, quanta and vision. *Journal of General Physiology* **25**, 819–840.

Heggelund, P. (1985) Receptive field organization of simple and complex cells. In *Models of the Visual Cortex*, Rose, D. and Dobson, V. G. (eds). Wiley: New York, pp. 358–365.

Heggelund, P. (1991) On achromatic colors. In *From Pigments to Perception*, Valberg, A. and Lee, B. B. (eds). Plenum: London.

Heggelund, P. (1992) A bidimensional theory of achromatic color vision. *Vision Research* **32**, 2107–2119.

von Helmholtz, H. (1867) *Handbuch der Physiologischen Optik*, 1st. edn. Voss: Leipzig.

von Helmholtz, H. (1892) Versuch, das psychophysische Gesetz auf die Farbunterschiede trichromatischer Augen anzuwenden. *Zeitschrift für Psychologie und Physiologie der Sinnesorgane* **3**, 1–20.

von Helmholtz, H. (1896) *Handbuch der Physiologischen Optik*, 2nd edn. Voss: Hamburg.

von Helmholtz, H. (1911) *Handbuch der Physiologischen Optik*, Vol. 2. Voss: Hamburg, 3rd edn. [First published 1860.]

von Helmholtz, H. (1962) *Handbook of Physiological Optics.* Dover: New York. [English translation by Southall, J.P.C. for the Optical Society of America (1924) from the 3rd German edition, Vol 1, of *Handbuch der Physiologischen Optik*. Voss: Hamburg, 1909.]

Hendry, S. H. C. and Yoshioka, T. A. (1994) A neurochemically distinct third channel in the macaque dorsal lateral geniculate nucleus. *Science* **264**, 575–577.

Hensel, H. and Kenshalo, D. R. (1969) Warm receptors in the nasal region of cats. *Journal of Physiology* **204**, 99–112.

Hering, E. (1964) *Outlines of a Theory of the Light Sense.* Translated from German by Hurvich, L. M. and Jameson, D. Harvard University Press: Cambridge, MA. [Orginal title: *Grundzüge der Lehre vom Lichtsinne*. Springer: Berlin, 1920.]

Herse, P. R. and Bedell, H. E. (1989) Contrast sensitivity for letters and grating targets under various stimulus conditions. *Optometry and Vision Science* **66**, 774–781.

Hess, R. and Woo, G. (1978) Vision through cataracts. *Investigative Ophthalmology and Visual Science* **17**, 428–435.

Hess, R. F., Sharpe, L. T. and Nordby, K. (1990) *Night Vision. Basic, Clinical and Applied Aspects.* Cambridge University Press: Cambridge.

von der Heydt R., Peterhans, E. and Baumgartner, H. G. (1984) Illusory contours and cortical neuron responses. *Science* **224**, 1260–1262.

Hicks, T. P., Lee, B. B. and Vidyasagar, T. R. (1983) The responses of cells in macaque lateral geniculate nucleus to sinusoidal gratings. *Journal of Physiology* **337**, 183–200.

Hillebrand, F. (1888) Ueber die spezifische Helligkeit der Farben. Beiträge zur Physiologie der Gesichtsempfindungen. Mit Vorbemerkungen von E. Hering. *Sitzungsber. Akad. Wiss. Wien, Math.-Naturwiss. Kl.* **98**, 3 (Abt.).

Holsten, F. and Bjorvatn, B. (1997) Lysbehandling. Et alternativ ved psykiske lidelser med sesongvariasjon eller søvnforstyrrelser. *Tidskrift for Den norske lægeforening* **17**, 2484–2488 (in Norwegian).

Holtsmark, T. and Valberg, A. (1969) Colour discrimination and hue. *Nature* **224**, 366–367.

Hubel, D. H. and Wiesel, T. N. (1963) Receptive fields of cells in striate cortex of very young, visually inexperienced kittens. *Journal of Neurophysiology* **26**, 994–1002.

Hurvich, L. M. (1981) *Color Vision.* Sinauer: Sunderland, MA.

Hurvich, L. M. and Jameson, D. (1955) Some quantitative aspects of opponent-colors theory. II. Brightness, saturation and hue in normal and dichromatic vision. *Journal of the Optical Society of America* **45**, 602–616.

Hurvich, L. M. and Jameson, D. (1956) Some quantitative aspects of an opponent-colors theory. IV A psychological color specification system. *Journal of the Optical Society of America* **46**, 416–421.

Hyde, E. P. and Forsythe, W. E. (1915) The visibility of radiation in the red end of the visible spectrum. *The Astrophysical Journal* **XLII**, 285–293.

Hyde, E. P., Forsythe, W. E. and Cady, F. E. (1918) The visibility of radiation. *The Astrophysical Journal* **XLVIII**, 65–88.

Hyvärinen, L. (1992) *The LH Symbol Tests. A System for Vision Testing in Children.* Catalog no. C370, Precission Vision: Villa Park, IL.

Ikeda, M. and Nakano, Y. (1986) Spectral luminous-efficiency functions obtained by direct heterochromatic brightness matching for point sources and for 2° and 10° fields. *Journal of the Optical Society of America* **3**, 2105–2108.

Ikeda, M., Yaguchi, H. and Sagawa,K. (1982) Brightness luminous-efficiency functions for 2° and 10° fields. *Journal of the Optical Society of America* **72**, 1660–1665.

Ives, H. E. (1912) Studies in photometry of lights of different colours. I. Spectral luminosity curves obtained by the equality of brightness photometer and the flicker photometer under similar conditions. *Philosophical Magazine* **24**, 149–188.

Jackman, W. M. and Webster, J. D. (1886) On photographing the retina of the living human eye. *Philadelphia Photographer* **23**, 340–341.

Jameson, D. and Hurvich, L. M. (1955) Some quantitative aspects of an opponent-colors theory: I. Chromatic responses and spectral saturation. *Journal of the Optical Society of America* **45**, 546–552.

Johansson, T. (1952) *Färg. Den allmenna färglärans grunder.* Natur and Kultur: Stockholm (in Swedish).

Johnson, E. N., Hawken, M. J. and Shapley, R. (2001) The spatial transformation of color in the primary visual cortex of the macaque meonkey. *Nature Neuroscience* **4**, 409–416.

Judd, D. B. (1949) Response functions for types of vision according to Müller theory. *Journal of Research of the National Bureau of Standards (USA)* **42**, 1–16.

Judd, D. B. (1951a) CIE Technical Committee no. 7, 'Colorimetry and Artificial Daylight'. Report of Secretariat US Committee. In *Proceedings of the 12th Session of the CIE*, Stockholm, Vol. 9, Part 7. Central Bureau of the CIE: Paris, pp. 1–60.

Judd, D. B. (1951b) *Handbook of Experimental Psychology*, Stevens, S. S. (ed.). Wiley/Chapman and Hall: New York, pp. 811–867.

Judd, D. B. (1960) Appraisal of Land's work on two-primary color perceptions. *Journal of the Optical Society of America* **50**, 254–268.

Jung, R. (1973) Visual perception and neurophysiology. In *Handbook of Sensory Physiology Vol. VII/3. Central Processing of Visual Information*, Part A, Autrum, H., Jung, R., Loewenstein, W. R., MacKay, D. M. and Teuber, H. L. (eds). Springer: Berlin.

Jung, R., Baumgarten, R. and Baumgartner, G. (1952) Mikroableitungen von einzelnen Nervzellen in optischen Cortex der Katze: Die lichtaktivierten B-Neurone. Archiv für Psychiatrie und Nervenkrankheiten. *Zeitschrift für die gesamte Neurologie und Psychiatrie* **189**, 521–539.

Kaiser, P. (1988) Sensation luminance: a new name to distinguish CIE luminance from luminance dependent on an individual's spectral sensitivity. *Vision Research* **28**, 455–456.

Kaiser, P. K. and Boynton, R. M. (1996) *Human Color Vision*, 2nd edn. Optical Society of America, Washington, DC.

Kaiser, P. K., Lee, B. B., Martin, P. R. and Valberg, A. (1990) The physiological basis of the minimally distinct border demonstrated in the ganglion cells of the macaque retina. *Journal of Physiology* **422**, 153–183.

Kandel, E. R., Schwartz, J. H. and Jessell, T. M. (2000) *Principles of Neural Science*. McGraw-Hill: New York.

Kaplan, E. and Shapley, R. M. (1982) X and Y cells in the lateral geniculate nucleus of macaque monkeys. *Journal of Physiology* **330**, 125–143.

Kaplan, E., Lee, B. B. and Shapley, R. M. (1990) New views of primate retinal function. In *Progress in Retinal Research*, Osborne, N. and Chader, J. (eds). Pergamon Press: Oxford, pp. 273–336.

Kaufman, L. and Kaufman, J. H. (2000) Explaining the moon illusion. *Proceedings of the National Academy of Sciences* **97**, 500–505.

Kelly, D. H. (1994) Eye movements and contrast sensitivity. In *Visual Science and Engineering. Models and Applications*, Kelly, D. H. (ed.). Marcel Dekker: New York, pp. 93–114.

King-Smith, P. E. (1975) Visual detection analysed in terms of luminance and chromatic signals. *Nature* **255**, 69–70.

King-Smith, P. E. (1991) Chromatic and achromatic visual systems. In *The Perception of Colour*, Gouras, P. (ed.), Vol. 6, *Vision and VisualDysfunction*. Macmillan: London, pp. 22–42.

King-Smith, P. E. and Carden, D. (1976) Luminance- and opponent-color contributions to visual detection and adaptation and to temporal and spatial integration. *Journal of the Optical Society of America* **66**, 709–717.

Kiper, D. C., Fenstemacher, S. B. and Gegenfurtner, K. R. (1997) Chromatic properties of neurons in macaque area V2. *Visual Neuroscience* **14**, 1061–1072.

Kolb, H. and Lipetz, L. E. (1991) The anatomical basis for colour vision in the vertebrate retina. In *The Perception of Colour*, Gouras, P. (ed.), Vol. 6, *Vision and Visual Dysfunction*. Macmillan: London, pp. 129–145.

König, A. (1929) *Physiologische Optik*. Akademische Verlagsgesellschaft: Leipzig, pp. 129–146.

Krauskopf, J. (1999) A journey in color space. *International Colour Vision Society, XVth Symposium*, Göttingen, Abstract T1.

Krauskopf, J. and Farell, B. (1991) Vernier acuity: effects of chromatic content, blur and contrast. *Vision Research* **31**, 735–749.

Krauskopf, J., Williams, D. R. and Heeley, D. W. (1982) Cardinal directions in color space. *Vision Research* **22**, 1123–1131.

Kremers, J. (2003) The assessment of L- and M-cone specific electroretinographical signals in the normal and abnormal human retina. *Progress in Retinal and Eye Research* **22**, 579–605.

Kremers, J., Lee, B. B. and Kaiser, P. K. (1992) Sensitivity of macaque retinal ganglion cells and human observers to combined luminance and chromatic temporal modulation. *Journal of the Optical Society of America* **A9**, 1477–1485.

Kuffler, S. W. (1953) Discharge patterns and functional organization of mammalian retina. *Journal of Neurophysiology* **16**, 37–68.

von Kries, J. (1905) Die Gesichtsempfindungen. In *Handbuch der Physiologie des Menschen*, Nagel, W. (ed.). Vieweg: Braunschweig, pp. 109–282.

Lambert, J. H. (1760) *Photometria sive de mesura et gradibus luminis colorem et umbrae*. Augustae Vindelicorum. [cited in von Helmholtz, 1911.]

Land, E. H. (1959) Color vision and the natural image. Part I and II. *Proceedings of the National Academy of Sciences of the USA* **45**, 115–129 and 636–644.

Land, E. H. (1983) Recent advances in retinex theory and some implications for cortical computations: color vision and the natural image. *Proceedings of the National Academy of Sciences of the USA* **80**, 5163–5169.

Land, E. H., Hubel, D., Livingstone, M. S., Perry, S. H. and Burns, M. S. (1983) Color-generating interactions across the corpus callosum. *Nature* **303**, 616–618.

Lankheet, M. J. M., Lennie, P. and Krauskopf, J. (1998) Distinctive characteristics of subclasses of red-green P-cells in LGN of macaque. *Visual Neuroscience* **15**, 37–46.

Lee, B. B. (1991a) Die Universität Göttingen und die Entsethung det Farbenlehre. *MPG Spiegel* **3(91)**, pp. 11–15.

Lee, B. B. (1991b) On the relation between cellular sensitivity and psychophysical detection. In *From Pigments to Perception; Advances in Understanding the Visual Process*, Valberg, A. and Lee, B. B. (eds). Plenum: London, pp. 105–116.

Lee, B. B. (1999) Receptor inputs to primate ganglion cells. In *Color Vision. From Genes to Perception*, Gegenfurtner, K. R. and Sharpe, L. T. (eds). Cambridge University Press: Cambridge, pp. 203–217.

Lee, B. B. and Dacey, D. M. (1997) Structure and function in primate retina. In *Color Vision Deficiencies XIII*, Cavonius, C. R. (ed.). Kluwer Academic: Dordrecht, pp. 107–118.

Lee, B. B., Valberg, A., Tigwell, D. A. and Tryti, J. (1987) An account of responses of spectrally opponent neurones in macaque lateral geniculate nucleus to successive contrast. *Proceedings of the Royal Society of London, Series B* **230**, 293–314.

Lee, B. B, Martin, P.R and Valberg, A. (1988) The physiological basis of heterochromatic flicker photometry demonstrated in the ganglion cells of the macaque retina. *Journal of Physiology*, **404**, 323–347.

Lee, B. B., Martin, P. R. and Valberg, A. (1989) Sensitivity of macaque retinal ganglion cells to chromatic and luminance flicker. *Journal of Physiology* **414**, 223–243.

Lee, B. B., Pokorny, J., Smith, V. C., Martin, P. R. and Valberg, A. (1990) Luminance and chromatic modulation sensitivity of macaque ganglion cells and human observers. *Journal of the Optical Society of America* **7**, 2223–2236.

Lee, B. B., Martin, P. R., Valberg, A. and Kremers, J. (1993) Physiological mechanisms underlying psychophysical sensitivity to combined luminance and chromatic modulation. *Journal of the Optical Society of America* **10**, 1403–1412.

Lee, B. B., Wehrhahn, C., Westheimer, G. and Kremers, J. (1995) The spatial precision of macaque ganglion cell responses in relation to vernier acuity of human observers. *Vision Research* **35**, 2743–2758.

Lee, B. B., Smith, V. C., Pokorny, J. and Kremers, J. (1997) Rod inputs to macaque ganglion cells. *Vision Research* **37**, 2813–2828.

Le Grand, Y. (1968) *Light, Colour and Vision*. Chapman and Hall: London.

Lennie, P. (1998) Single units and visual cortical organization. *Perception* **27**, 889–935.

Lennie, P., Krauskopf, J. and Sclar, G. (1990) Chromatic mechanisms in striate cortex of macaque. *Journal of Neuroscience* **10**, 649–669.

Lennie, P., Pokorny, J. and Smith, V. C. (1993) Luminance. *Journal of the Optical Society of America* **10**, 1283–1293.

Lie, I. (1963) Dark adaptation and the photochromatic interval. *Document Ophthalmologica* **17**, 411–510.

Lindsey, D. T. and Teller, D. Y. (1989) Influence of variations in edge blur on minimally distinct border judgements: a theoretical and empirical investigation. *Journal of the Optical Society of America A* **6**, 446–458.

Livingstone, M. S. and Hubel, D. (1984) Specificity of intrinsic connections in primate primary visual cortex. *Journal of Neuroscience* **4**, 2830–2835.

Livingstone, M. and Hubel, D. (1988) Segregation of form, color, movement and depth: anatomy, physiology and perception. *Science* **240**, 740–750.

Logothetis, N. K. (2002) Vision: a window on consciousness. *Scientific American*, **August**: 18–25.

Logvinenko, A. (1999) Lightness induction revisited. *Perception* **28**, 803–816.

Lohne, J. (1959) Thomas Harriot (1560–1621). The Tycho Brahe of Optics. *Centaurus* **6**(2), 113–121.

Luther, R. (1927) Aus dem Gebiet der Farbreizmetrik. *Zeitschrift für technische Physik* **8**, 540.

MacAdam, D. L. (1942) Visual sensitivity to color differences in daylight. *Journal of the Optical Society of America* **32**, 247–274.

MacAdam, D. L. (ed.) (1970) *Sources of Color Science*. MIT Press: Cambridge, MA.

Mach, E. (1965). On the effect of the spatial distribution of the light stimulus on the retina. In *Mach Bands. Quantitative Studies on Neural Networks in the Retina*. Holden-Day: New York, pp. 253–271. [Originally published in *Sitzungsberichte der Mat. Nat. Wiss. Classe der kaiserlichen Akademie der Wissenschaften, Wien*, **52**(2), 303–322 (1865).]

MacLeod, D. and Boynton, R. M. (1979) A chromaticity diagram showing cone excitation by stimuli of equal lumianance. *Journal of the Optical Society of America* **69**, 1183–1185.

Magnussen, S. (1987) Temporal aspects of vision: psychophysical studies of B- and D-channels. In *Problems of Visual Preception*, Lomov, B., Zabrodin, Y., Saugstad, P. and Magnussen, S. (eds). USSR Academy of Sciences: Moscow.

Marr, D. (1982) *Vision. A Computational Investigation into the Human Representation and Processing of Visual Information*. Freeman: San Fransisco, CA.

Martin, P. R., White, A. J. R., Goodchild, A. K., Wilder, H. D. and Sefton, A. E. (1997) Evidence that the blue-on cells are part of the third geniculocortical pathway in primates. *European Journal of Neuroscience* **9**, 1536–1541.

Martinez-Oriegas, E. (1994) Chromatic-achromatic multiplexing in human color vision. In *Visual Science and Engineering. Models and Applications*, Kelly, D. H. (ed.). Marcel Dekker: New York, pp. 117–187.

Maxwell, J. C. (1872) Theory of the perception of colours. *Transactions of the Royal Scottish Society of Arts* **4**, 394–400. [Reprinted in MacAdam, D. L. (ed.) (1970), *Sources of Colour Science*. MIT Press: Cambridge, MA, pp. 75–83.]

McCann, J. J., McKee, S. P. and Tylor, T. H. (1976) Quantitative studies in retinex theory. *Vision Research* **16**, 445–458.

McIlwain, J. T. (1966) Some evidence concerning the physiological basis of the periphery effect in cat's retina. *Experimental Brain Research* **1**, 265–271.

Meyer-Arendt, J. R. (1995) *Introduction to Classical and Modern Optics*. Prentice Hall: Engelwood Cliffs, NJ.

Michael, C. R. (1978a) Color vision mechanisms in monkey cortex: dual opponent cells with concentric receptive fields. *Journal of Neurophysiology* **41**, 572–588.

Michael, C. R. (1978b) Color vision mechanisms in monkey striate cortex: simple cells with dual opponent receptive fields. *Journal of Neurophysiology* **41**, 1233–1249.

Michael, C. R. (1985) Laminar segregation of color cells in the monkey's striate cortex. *Vision Research* **25**, 415–423.

Miescher, K. (1948) Neuermittlung der Urfarben und deren Bedeutung für die Farbordnung. *Helvetica Physiologica Acta* **6**, C12–C13.

Miescher, K., Hofman, K.-D., Weisenhorn, P. and Früh, M. (1961) Über das natürliche Farbsystem. *Die Farbe* **10**, 115–144.

Miescher, K., Richter, K. and Valberg, A. (1982) Farbe und Farbsehen. Beschreibung von Experimenten für die Farbenlehre. *Farbe + Design* **23/24**, 2–23.

Miller S. T. (2000) Retinal imaging and vision at the frontiers of adaptive optics. *Physics Today,* **January**, 31–36.

Mollon, J. D. (1993) Palmer, George. In *The Dictionary of National Biography. Missing Persons.* Oxford University Press: Oxford, pp. 509–510.

Mollon, J. D. (1995) George Palmer (1740–1795): glass-seller, visual theorist and draper. In *The Theory of Colours and Vision.* Drapers Hall: London, abstract.

Mollon, J. D. and Jordan, G. (1997) On the nature of unique hues. In *John Dalton's Colour Vision Legacy,* Dickinson, C., Murray, I. and Carden, D. (eds). Taylor & Francis: London, pp. 381–392.

Moser, E. and Paulsen, O. (2001) New excitement in cognitive space: between place cells and spatial memory. *Current Opinion in Neurobiology* **11**, 745–751.

Mullen, K. T. (1985) The contrast sensitivity of human colour vision to red–green and blue–yellow chromatic gratings. *Journal of Physiology* **359**, 381–400.

Müller, G. E. (1930) *Ueber die Farbempfindungen. Psychophysische Untersuchungen.* Barth: Leipzig.

Nagy, L. A. (1979) Unique hues are not invariant with brief stimulus durations. *Vision Research* **19**, 1427–1432.

Naka, K. I. and Rushton, W. H. A. (1966) 5-potentials from colour units in the retina of fish (*Cyprinidae*). *Journal of Physiology* **185**, 536–555.

Nakano, Y., Ikeda, M. and Kaiser, P. K. (1988) Contributions of the opponent mechanisms to brightness and nonlinear model. *Vision Research* **28**, 799–810.

Nakayama, K. and Mackeben, M. (1982) Steady state visual evoked potentials in the alert primate. *Vision Research* **22**, 1261–1271.

Nathans, J., Thomas, D. and Hogness, D. S. (1986) Molecular genetics of human color vision: the genes encoding blue, green and red pigments. *Science* **232**, 193–202.

Neitz, J., Neitz, M. and Jacobs, G. H. (1993) More than three different cone pigments among people with normal color vision. *Vision Research* **33**, 117–122.

Neitz, J., Neitz, M., He, J. C. and Shevell, S. K. (1999) Trichromatic color vision with only two spectrally distinct photopigments. *Nature Neuroscience* **2**, 884–889.

Neitz, J., Carroll, J. and Neitz, M. (2001) Color vision. *Optics and Photonics News,* **12**, 26–33.

Newton, I. (1979) *Opticks.* Dover: New York. [First published 1704.]

Noell, E. (1995) Issues, problems and opportunities. Appropriate lighting for aging vision and health. In *Lighting for Aging Vision and Health. Proceedings of the 3rd International Symposium, Orlando,* Lighting Research Institute: USA, pp. 149–157.

Noorlander, C. and Koenderink, J. J. (1983) Spatial and temporal discrimination ellipsoids in color space. *Journal of the Optical Society of America* **73**, 1533–1543.

Nothdurft, H.-C. and Lee, B. B. (1982) Responses to coloured patterns in the macaque lateral geniculate nucleus: pattern processing in single neurones. *Experimental Brain Research* **48**, 43–54.

Nyberg, N. D. (1928) Zum Aufbau des Farbenkörpers im Raume aller Lichtempfindungen. *Zeitschrift für Physik* **52**, 406.

Nygård, G. E., Valberg, A. and Rudvin, I. (2002) Comparison of VEP response elicited by achromatic and chromatic stimuli. *ISCEV Symposium,* Leuven.

Optical Society of America (1994) Edwin Land Special Issue. *Optics and Photonic News*, **October**.

Ostwald, W. (1921) *Mathetische Farbenlehre*. Unesma: Leipzig.

Oyster, C. W. (1999) *The Human Eye. Structure and Function*. Sinauer: Boston, MA.

Packer, O. S. and Dacey, D. M. (2002) Receptive field structure of H1 horizontal calls in macaque monkey retina. *Journal of Vision* **2**, 272–292.

Padmos, P. and van Norren, D. (1975) Cone systems interactions in single neurones of the lateral geniculate nucleus of the macaque. *Vision Research* **15**, 617–619.

Palmer, E. S. (1999) Color, consciousness and the isomorphism constraint. *Behavioral and Brain Sciences* **22**, 923–989.

Pardham, S. and Elliott, D. B. (1991) Clinical measurements of binocular summation and inhibition in patients with cataract. *Clinical Vision Science* **8**, 355–359.

Peli, E., Goldstein, R., Young, G., Trempe, C. and Buzney, S. (1991) Image enhancement for the visually impaired: Simulations and experimental results. *Investigative Ophthalmology and Visual Science* **32**, 2337–2350.

Perry, V. H., Oehler, R. and Cowey, A. (1984) Retinal ganglion cells that project to the dorsal lateral geniculate nucleus in the macaque monkey. *Neuroscience* **12**, 1110–1123

Peterhans, E. and Heydt, R. von der (1991) Subjective contours-bridging the gap between psychophysics and physiology. *Trends in Neuroscience* **14**, 112–119.

Pinna, B. and Brelstaff, G. J. (2000) A new visual illusion of relative motion. *Vision Research* **40**, 2091–2096.

Pinna, B., Brelstaff, G and Spillmann, L. (2001) Surface color boundaries: a new 'watercolor' illusion. *Vision Research* **41**, 2669–2676.

Pirenne, M. H. (1967) *Vision and the Eye*. Chapman and Hall: London.

Posner, M. I. and Raichle, M. E. (1994) *Images of the Mind*. Scientific American Library, Freeman Philadelphia, PA.

Purdy, D. M. (1931a) Spectral hue as a function of intensity. *American Journal of Psychology* **43**, 541–559.

Purdy, D. M. (1931b) On the saturations and chromatic thresholds of spectral colours. *British Journal of Psychology* **21**, 283–313.

Purkinje, J. (1823) *Beobachtungen und Versuche zur Physiologie der Sinne*. Clave: Prague.

Purpura, K., Kaplan, E. and Shapley, R. M. (1988) Background light and the contrast gain of primate P and M retinal ganglion cells. *Proceedings of the National Academy of Sciences of the USA* **85**, 4534–4537.

Quigely, H. A., Dunkelberger, G. R. and Green, W. R. (1988) Chronic human glaucoma causes selectively greater loss of large optic nerve fibers. *Ophthalmology* **95**, 357–363.

Raichle, M. E. (1994) Visualizing the mind. *Scientific American* **April**, 36.

Ramachandran, V. S. (1995) 2-D or not 2-D – that is the question. In *The Artful Eye*, Gregory, R., Harris, J., Heard, P. and Rose, D. (eds). Oxford University Press: Oxford.

Ramachandran, V. S., Tyler, R. L., Gregory, R. L., Rogers-Ramachandran, D., Duensing, S., Pillsbury, C. and Ramachandran, C. (1996) Rapid adaptive camouflage in tropical flounders. *Nature* **379**, 815–818.

Ratliff, F. (1965) *Mach Bands. Quantitative Studies on Neural Networks in the Retina*. Holden-Day: New York.

Regan, D. (1986) Form from motion parallax and form from luminance contrast: Vernier discrimination. *Spatial Vision* **1**, 305–318.

Reid, R. C. and Shapley, R. M. (1992) Spatial structure of cone inputs to receptive fields in primate lateral geniculate nucleus. *Nature* **356**, 716–718.

Ribe, N. and Steinle, F. (2002) Exploratory experimentation: Goethe, Land and color theory. *Physics Today* July, 43–49.

Richter, K. D. E. (1969) Anatgonistische Signale beim Farbensehen und ihr Zusammenhang mit der empfindungsgemässen Farbordnung. Thesis, University of Basel.

Richter, K. (1996) *Computergrafik und Farbmetrik.* VDE-Verlag: Berlin.

Rizzolatti, G., Fadiga, L., Gallesi, V. and Fogassi, L. (1996) Premotor cortex and the recognition of motor actions. *Brain Research and Cognitive Brain Research* **3**, 131–141.

Rodieck, R. W. (1991) Which cells code for color? In *From Pigments to Perception*, Valberg, A. and Lee, B. B. (eds). Plenum: New York, pp. 83–93.

Roe, A. W. and Ts'o, D. Y. (1995) Visual topography in primate V2: multiple representation across functional stripes. *Journal of Neuroscience* **15**, 3689–3715.

Rood, O. N. (1899) A photometric method which is independent of color. On the flicker photometer. *American Journal of Science* **8**, 194–198. [First published 1893.]

Roorda, A. and Williams, D. R. (1999) The arrangement of the three classes in the living human eye. *Nature* **397**, 520–522.

Ross, J., Burr, D. and Morrone, C. (1996) Suppression of the magnocellular pathway during saccades. *Behavioural Brain Research* **80**, 1–8.

Rudvin, I., Valberg, A. and Kilavik, B. E. (2000) Visual evoked potentials and magnocellular and parvocellular segregation. *Visual Neuroscience* **17**, 579–590.

Sacks, O. (1995) *An Anthropologist on Mars.* Macmillan: London.

Sacks, O. (1997) *The Island of the Colorblind and Cycad Island.* Alfred A. Knopf: New York.

Saunders, B. A. C. and van Brakel, J. (1997) Are there nontrivial constraints on color categorization? With open peer commentary. *Behavioral and Brain Sciences* **20**, 167–228.

Savoie, R. E. (1973) Bezold–Brucke effect and visuel non-linearity. *Journal of the Optical Society of America* **63**, 1253–1261.

Schanda, J. D. (1998) Future trends in photometry. In *Handbook of Applied Photometry*, DeCusatis, C. (ed). Springer: New York.

Scheibner, H. M. O. and Boynton, R. M. (1968) Residual red–green discrimination in dichromats. *Journal of the Optical Society of America* **58**, 1151–1158.

Schiller, P. H. (1992) The ON and OFF channels of the visual system. *Trends in Neuroscience* **15**, 86–92.

Schober, H. (1957) *Das Sehen*, Vol. I. Fachbuchverlag: Leipzig.

Schober, H. (1958) *Das Sehen*, Vol. II. Fachbuchverlag: Leipzig.

Schober, H. and Rentschler, I. (1979) *Das Bild als Schein der Wirklichkeit.* Moos: München.

Schrödinger, E. (1920a) Theorie der Pigmente von grösster Leuchtkraft. *Annalen Physik (IV)* **62**, 603–622.

Schrödinger, E. (1920b) Grundlinien einer Theorie der Farbenmetrik im Tagessehen. *Annalen Physik (IV)* **63**, 397 and 481.

Schrödinger, E. (1925) Ueber das Verhältnis der Vierfarben- zur Dreifarbentheorie. *Sitzungsberichte der Akademie der Wissenschaften, Wien* **IIa**(134), 471–490.

Schultze, M. (1866) *Über den gelben Fleck der Retina, seinen Einfluss auf normales Sehen und auf Farbenblindheit.* Cohen: Bonn.

Schumann, J., Sivak, M., Flannagan, M. J., Traube, E. C., Hashimoto, H. and Kojima, S. (1996). Brightness of colored retroreflective materials. University of Michigan, Report no. UMTRI-96–33.

Seim, T. and Valberg, A. (1980) Physiological response and the scaling of color differences. *Experimental Brain Research* **41**, A39.

Seim, T. and Valberg, A. (1986) Towards a uniform color space: A better formula to describe the Munsell and OSA color scales. *Color Research and Application* **11**, 11–24.

Seim, T. and Valberg, A. (1988) Hvordan bruk av kulørte farger kan bedre synsforholdene for svaksynte. Rapport 88–02. Fysisk institutt, Universitetet i Oslo.

Seim, T. and Valberg, A. (1993) Image diffusion in cataracts affects chromatic and achromatic contrast perception differently. In *Colour Vision Deficiencies XI*, Drum, B. (ed.). Kluwer Academic: Dordrecht, pp. 153–161.

von Senden, M. (1960) *Space and Sight:* The Perception of Space and Shape in the Congenitally Blind Before and after Operation. Methuen: London. [First published 1932.]

Shapley, R. and Victor, J. D. (1986) Hyperacuity in cat retinal ganglion cells. *Science* **231**, 999–1002.

Sharpe, L. T., Stockman, A., Jägle, H. and Nathans, J. (1999) Opsin genes, cone photopigments, color vision and color blindness. In *Color Vision. From Genes to Perception*, Gegenfurtner, K. R. and Sharpe, L. T. (eds). Cambridge University Press: Cambridge, pp. 3–51.

Sheppard, J. J. (1968) *Human Color Perception. A Critical Study of the Experimental Foundation*. Elesevier: New York.

Shinomori, K., Schefrin, B. E. and Werner, J. S. (1997) Spectral mechanisms of spatially induced blackness: data and quantitative model. *Journal of the Optical Society of America* **A14**, 372–387.

Singer, W. (1993) Synchronization of cortical activity and its putative role in information processing and learning. *Annual Review of Physiology* **55**, 349–374.

Singer, C. and Hughes, R. (1995) Clinical use of bright light in geriatric neuropsychiatry. In *Lighting for Aging Vision and Health. Proceedings of the 3rd Interanional Symposium, Orlando*. Lighting Research Institute: USA, pp. 143–147.

Smith, V. C. and Pokorny, J. (1975) Spectral sensitivity of the foveal cone photopigments between 400 and 500 nm. *Vision Research* **15**, 161–171.

Smith, V. C., Lee, B. B., Pokorny, J., Martin, P. R. and Valberg, A. (1992) Responses of macaque ganglion cells to the relative phase of heterochromatically modulated lights. *Journal of Physiology* **458**, 191–221.

Solomon, S. G., White, A. J. R. and Martin, P. R. (2002) Extraclassical receptive field properties of parvocellular, magnocellular and koniocellular cells in the primate lateral geniculate nucleus. *Journal of Neuroscience* **22**, 338–349.

Spillmann, L. (1971) Foveal receptive fields in the human visual system measured with simultaneous contrast in grids and bars. *Pflügers Archiv für die gesamte Physiologie* **326**, 281–299.

Spillmann, L. and Dresp, B. (1995) Phenomena of illusory form: can we bridge the gap between levels of explanation? *Perception* **24**, 1333–1364.

Spillmann, L. and Ehrenstein, W. H. (2003) Gestalt factors in the visual neurosciences. In *The Visual Neurosciences*, Chalupa, L. and Werner, J. H. (eds). MIT Press: Boston, MA.

Stabell, U. and Stabell, B. (1982a) Color vision in the peripheral retina under photopic conditions. *Vision Research* **22**, 839–844.

Stabell, B. and Stabell, U. (1982b) Bezold–Brucke phenomenon of the far peripheral retina. *Vision Research* **22**, 845–849.

Stevens, S. S. (1961) To honor Fechner and repeal his law. *Science* **133**, 80–86.

Stettler, D. D., Das, A., Bennett, J. and Gilbert, C. D. (2002) Lateral connectivity and contextual interactions in macaque primary visual cortex. *Neuron* **36**, 739–750.

Stiles, W. S. and Burch, J. M. (1955) Interim report to the Commission Internationale de l'Eclairage, 1955, on the National Physical Laboratory's investigation of colour matching. *Optica Acta* **2**, 168–181.

Stiles, W. S. and Burch, J. M. (1959) NPL colour-matching investigation: final report (1958) *Optica Acta* **6**, 1–26.

Stöcklin, S. (2002) Ein tiefer Blick in die Evolution der Augen. *Basler Nachrichten* 8 March.

Stockman, A. and Sharpe, L. T. (1999) Cone spectral sensitivities and color coding. In *Color Vision. From Genes to Perception*, Gegenfurtner, K. R. and Sharpe, L. T. (eds). Cambridge University Press: Cambridge, pp. 53–87.

Stockman, A. and Sharpe, L. T. (2000) The spectral sensitivities of the middle- and long-wavelength sensitive cones derived from measurements in observers of known genotype. *Vision Research* **40**, 1711–1737. Available at: http://cvision.ucsd.edu/database/data/cones/linss2_10e_1.txt

Stockman, A., MacLeod, D. I. A. and Johnson, N. E. (1993) Spectral sensitivities of human cones. *Journal of the Optical Society of America* **A10**, 2491–2521.

Streri, A. (1993) *Seeing, Reaching, Touching, The relations between vision and touch in infancy.* Harvester Wheatsheaf: New York.

Svaetichin, G. (1956) Spectral response curves from single cones. *Acta Physiologica Scandinavica* **39** (Suppl. 134), 17–46.

Swanson, W. H., Ueno, T., Smith, V. C. and Pokorny, J. (1987) Temporal modulation sensitivity and pulse detection thresholds for chromatic and luminance perturbations. *Journal of the Optical Society of America* **4**, 1992–2005.

Teller, D. Y. (1990) *Teller Acuity Cards (TAC). Instruction Manual.* Vistech Consultants: Dayton, OH.

Thompson, E. (1995) *Colour Vision.* Routledge: London.

Thorell, L. G., DeValois, R. and Albrecht, D. G. (1984) Spatial mapping of monkey V1 cells with pure color and luminance stimuli. *Vision Research* **24**, 751–769.

Trauzettel-Klosinski, S., MacKeben, M., Reinhard, J., Feucht, A., Dürrwächter, U. and Klosinski, G. (2002) Pictogram naming in dyslexic and normal children assessed by SLO. *Vision Research* **42**, 789–799.

Trendelenburg, W. (1943) *Der Gesichtsinn.* Springer: Berlin, pp. 80–84.

Tryti, J. (1985) Modellutvikling og kvantitativ analyse av celleresponser i synssystemet. (Model and quantitative analysis of cell responses in the visual system.) Master Thesis in Physics. University of Oslo. (In Norwegian.)

Ts'o, D. Y. and Gilbert, C. D. (1988) The organization of chromatic and spatial interactions in the primate striate cortex. *Journal of Neurosience* **8**(5), 1712–1727.

Tucson, I. I. (1996) *Toward a Science of Consciousness*, 8–13 April 1996. Abstract and comments on the Internet: www.imprint-academic.demon.co.uk/SPECIAL/tucson.html

Uchikawa, K. and Sato, M. (1995) Saccadic suppression of achromatic and chromatic responses measured by increment-threshold spectral sensitivity. *Journal of the Optical Society of America* **A12**, 661–666.

Ungerleider, L. G. and Mishkin, M. (1982) Two cortical visual systems. In *Analysis of Visual Behavior*, Engle, D. J., Goodale, M. A. and Mansfield, R. J. (eds). MIT Press: Cambridge, MA, pp. 549–586.

Valberg, A. (1971) A method for the precise determination of achromatic colors including white. *Vision Research* **11**, 157–160.

Valberg, A. (1974) Color induction: dependence on luminance, purity and dominant or compementary wavelength of inducing stimuli. *Journal of the Optical Society of America* **64**, 1531–1540.

Valberg, A. (1981) Advantages of an opponent colour metrics and the opponent purity concept. *Die Farbe* **29**, 127–144.

Valberg, A. (2001) Unique hues: an old problem for a new generation. *Vision Research* **41**, 1645–1657.

Valberg, A. and Fosse, P. (1997) Vision with age-related macular degeneration. *Perception* **26** (Suppl.), 37.

Valberg, A. and Fosse, P. (2002) Binocular contrast inhibition in subjects with age-related macular degeneration. *Journal of the Optical Society of America* **19**, 223–228.

Valberg, A. and Lange Malecki, B. (1990) 'Colour constancy' in Mondrian patterns: a partial cancelation of physical chromaticity shifts by simultaneous contrast. *Vision Research* **30**, 371–380.

Valberg, A. and Lee, B. B. (1989) Detection and discrimination of colour, a comparison of physiological and psychophysical data. *Physica Scripta* **39**, 178–186.

Valberg, A. and Lee, B. B. (1992) Main cell systems in primate visual pathways. *Current Opinions in Ophthalmology* **3**, 813–823.

Valberg, A. and Rudvin, I. (1997) Possible contributions of magnocellular- and parvocellular-pathway cells to transient VEPs. *Visual Neuroscience* **14**, 1–11.

Valberg, A. and Seim, T. (1983) Chromatic induction: responses of neurophysiological double opponent units? *Biological Cybernetics* **46**, 149–158.

Valberg, A. and Tansley, B. (1977) Tritanopic purity-difference function to describe the properties of minimally distinct borders. *Journal of the Optical Society of America* **67**, 1330–1336.

Valberg, A., Seim, T. and Sällström, P. (1979) Colour rendering and the three-band fluorescent lamp. *Proceedings of the 19th Session of the CIE*, Kyoto, pp. 218–223.

Valberg, A., Lee, B. B., Creutzfeldt, O. D. and Tigwell, D. A. (1983) Luminance ratio and spectral responsiveness of cells in the macaque lateral geniculate nucleus. In *Colour Vision*, Mollon, J. D. and Sharpe, L. T. (eds). Academic Press: London, pp. 235–243.

Valberg, A., Lee, B.B., Tigwell, D. A. and Creutzfeldt, O. D. (1985a) A simultaneous contrast effect of steady remote surrounds on responses of cells in macaque lateral geniculate nucleus. *Experimental Brain Research* **58**, 604–608.

Valberg, A., Tryti, J. and Lee, B. (1985b) Computation of responses of opponent-cells in the macaque lateral geniculate nucleus to light stimuli varying in luminance, wavelength and purity. Institute of Physics Report Series 85-29, University of Oslo, pp. 1–36.

Valberg, A., Seim, T., Lee, B. B. and Tryti, J. (1986a) Reconstruction of equidistant color space from responses of visual neurones of macaques. *Journal of the Optical Society of America* **A:3**, 1726–1734.

Valberg, A., Lee, B. B. and Tigwell, D. A. (1986b) Neurones with strong inhibitory S-cone inputs in the macaque lateral geniculate nucleus. *Vision Research* **26**, 1061–1064.

Valberg, A., Lee, B. B. and Tryti, J. (1987) Simulation of responses of spectrally-opponent neurones in the macaque lateral geniculate nucleus to chromatic and achromatic light stimuli. *Vision Research* **27**, 867–882.

Valberg, A. Lee, B. B. and Creutzfeldt, O. (1991a). Remote surrounds and the sensitivity of primate P-cells. In *From Pigments to Perception*, Valberg, A. and Lee, B. B. (eds). Plenum: London, pp. 177–180.

Valberg, A., Lange-Malecki, B. and Seim, T. (1991b). Colour changes as a function of luminance contrast. *Perception* **20**, 655–668.

Valberg, A., Lee, B. B., Kaiser, P. K. and Kremers, J. (1992) Responses of macaque ganglion cells to movement of chromatic borders. *Journal of Physiology* **458**, 579–602.

Valberg, A., Seim, T., Zhang, W. and Fosse, P. (1994) Vigra-C. *Optikeren* **4**, 3–5. (In Norwegian.)

Valberg, A., Fosse, P., and Gjerde, T. (1997) Chromatic contrast sensitivity without correction for chromatic aberration. *Investigative Ophthalmology and Visual Science* **38/4**, S893.

Valeton, J. M. and Van Norren, D. (1983) Light adaptation of primate cones: an analysis based on extracellular data. *Vision Research* **23**, 1539–1547.

Venable, W. H. and Hale, W. N. (1996) Color and nighttime pedestrian safety markings. *Color Research and Application* **21**, 305–309.

Vidyasagar, T. R., Kulikowski, J. J., Lipnicki, D. M. and Dreher, B. (2002) Convergence of parvocellular and magnocellular information channels in the primary visual cortex of the macaque. *European Journal of Neuroscience* **16**, 945–956.

Vienot, F. (2001) Report on a fundamental chromaticity diagram with physiologically significant axes. Association Internationale de la Coulour, Rochester, NY.

Vienot, F., Brettel, H., Ott, L., Ben M'Barek, A. and Mollon, J. D. (1995) What do colour-blind people see? *Nature* **376**, 127–128.

Vimal, R. L. P., Pokorny, J. and Smith, V. C. (1987) Appearance of steadily viewed lights. *Vision Research* **27**, 1309–1318.

Virsu, V. and Rovamo, J. (1979) Visual resolution, contrast sensitivity and cortical magnification factor. *Experimental Brain Research* **37**, 475–494.

Volbrecht, V. J. and Kliegl, R. (1998) The perception of blackness: An historical and contemporary review. In *Color Vision*, Backhaus, W. G. K., Kliegl, R. and Werner, J. S. (eds). De Gruyter: Berlin.

Von Bekesy, G. (1968) Mach- and Hering type lateral inhibition in vision. *Vision Research* **8**, 1483–1499.

Vos, J. J. (1978) Colorimetric and photometric properties of a 2° fundamental observer. *Color Research and Application* **3**, 125–128.

Vos, J. J. and Walraven, P. L. (1972) An analytical description the line element in the zone fluctuation model of colour vision. II. The derivation of the line element. *Vision Research* **12**, 1345–1365.

Waaler, G. H. M. (1969) Studies in colour vision. Norwegian Academy of Sciencies (Nor. Vidensk. Akad.) Oslo. Mat-naturvit. Kl. N. S. no. 12.

Wachtler, T., Sejnowski, T. J. and Albright, T. D. (2003) Representation of color stimuli in awake macaque primary visual cortex. *Neuron* **37**, 681–691.

Wachtler, T., Dohrmann, U., and Hertel, R. (2004) Modeling color percepts of dichromats. *Vision Research* **44**, 2843–2855.

Wade, N. J. (1996) Descriptions of visual phenomena from Aristotle to Wheatstone. *Perception* **25**, 1137–1175.

Wade, N. J. (1998) *A Natural History of Vision.* MIT Press, Cambridge, MA.

Walls, G. L. (1956) The G. Palmer story (or what it's like, sometimes, to be a scientist). *Journal of the History of Medicine and Allied Sciences*, **11**, 66–96.

Walraven, P. L. (1961) On the Bezold-Brucke phenomenon. *Journal of the Optical Society of America* **S1**, 1113–1116.

Ware, C. V. and Cowan, W. (1983) *Specification of Heterochromatic Brightness Matches: a Conversion Factor for Calculating Luminances of Stimuli that are Equal in Brightness.* NRC publication no. 26055. NRC: Ottawa.

Watson, A., Barlow, H. B. and Robson, J. G. (1983) What does the eye see best? *Nature* **302**, 419–422.

Watt, R. J. and Morgan, M. J. (1984) Spatial filters and the localization of luminance changes in human vision. *Vision Research* **24**, 1387–1397.

Werner, J. S., Cicerone, C. M., Kliegl, R and Dellarosa, D. (1984) Spectral efficiency of blackness induction. *Journal of the Optical Society of America* **A1**, 981–986.

WHO (1973) *The Prevention of Blindness.* World Health Organization Technical Report Series no. 518. WHO: Geneva.

WHO (2001) *International Classification of Functioning, Disability and Health (ICF).* WHO: Geneva. Available at: http://www3.who.int/icf/icftemplate.cfm.

Wiesel, T. N. and Hubel, D. H. (1966) Spatial and chromatic interactions in the lateral geniculate body of the rhesus monkey. *Journal of Neurophysiology* **29**, 1115–1156.

Williams, D. R., Brainard, D. H, McMahon M. J. and Navarro, R. (1994) Double-pass and interferometric measures of the optical quality of the eye. *Journal of the Optical Society of America* **11**, 3123–3135.

Wilson, H. R. and Gelb, D. J. (1984) Modified line element theory for spatial frequency and width discrimination. *Journal of the Optical Society of America* **A1**, 124–131.

Wilson, H. R., Levi, D., Maffei, L., Rovamo, J. and DeValois, R. (1990) The perception of form. Retina to striate cortex. In *Visual Perception. The Neurophysiological Foundations*, Spillmann, L. and Werner, J. S. (eds). Academic Press: London.

Wittgenstein, L. (1977) *Remarks on Colour.* Blackwell: Oxford. Bemerkungen über die Farben. Suhrkamp: Frankfurt (1979)

Wold, J. H. (1992) Opponente gangliecellers reseptoriske felter. (The receptive fields of opponent ganglion cells.) *Norwegian. Annual Biophysics Meeting*, Kongsvoll.

Wold, J. H. (1998) Demonstrations of Mach bands and area fill-in contrast. In *Color between Art and Science. Proceedings of the Oslo International Colour Conference*, Oslo, pp. 177–179.

Wold, J. H. and Valberg, A. (2001) The derivation of XYZ tristimulus spaces: a comparison of two alternative methods. *Color Research and Application* **26**. S222–S224.

Wold, J. H. and Valberg, A. (1999) General method for deriving an *XYZ* tristimulus space exemplified by use of the Stiles–Burch (1955) 2° color matching data. *Journal of the Optical Society of America* **16**, 2845–2858.

Wright, W. D. (1928/1929) A redetermination of the trichromatic coefficients of the spectral colours. *Transactions of the Optical Society* **30**, 141–164.

Wyszecki, G. and Stiles, W. S. (1982) *Color Science. Concepts and Methods. Quantitative Data and Formulae.* Wiley: New York. [First published 1967.]

Xiao, Y., Wang, Y.I and Fellman, J. (2003) A spatial organized representation of colour in macaque cortical area V2. *Nature* **421**, 535–539.

Yaguchi, H and Ikeda, M. (1983) Contribution of opponent-colour channels to brightness. In *Colour Vision: Physiology and Psychophysics*, Mollon, J. D. and Sharpe, L. T. (eds). Academic Press: London, pp. 353–360.

Yaguchi, H., Kawada, A., Shioiri, S. and Miyake, Y. (1993) Individual differences of the contribution of chromatic channels to brightness. *Journal of the Optical Society of America* **10**, 1373–1379.

Zajonc, A. (1993) *Catching the Light. The Entwined History of Light and Mind.* Oxford University Press: Oxford.

Zeki, S. (1980) The representation of colours in the cerebral cortex. *Nature* **284**, 412–418.

Zeki, S. (1983a) Color coding in the cerebral cortex: the reaction of cells in monkey visual cortex to wavelengths and colors. *Neuroscience* **9**, 741–765.

Zeki, S. (1983b) Color coding in the celebral cortex: the responses of wavelength-selective and color-coded cells in monkey visual cortex to changes in wavelength composition. *Neuroscience* **9**, 767–781.

Zeki, S. (1993) *A Vision of the Brain.* Blackwell: Oxford.

Zeki, S. (2003) Improbable areas in the visual brain. *Trends in Neurosciences* **26**, 23–26.

Zhang, W. (1994) Analysis of waveforms in transient visual evoked potentials. Master Thesis in Physics. University of Oslo.

Zihl, J., von Cramon, D. and Mai, N. (1983) Selective disturbance of movement vision after bilateral brain damage. *Brain* **106**, 313–340.

Zrenner, E. (1983) Neurophysiological aspects of color mechanisms in the primate retina. In *Color Vision: Physiology and Psychophysics*, Mollon, J. and Sharpe, L. T. (eds). Academic Press: New York, pp. 195–210.

Zrenner, E., Abramov, I., Akita, M., Cowey, A. Livingstone, M. and Valberg, A. (1990) Color perception. Retina to cortex. In *Visual Perception. The Neurophysiological Foundations*, Spillmann, L. and Werner, J. (eds). Academic Press: New York, pp. 163–204.

Some interesting web sites

http://webvision.med.utah.edu/

http://www.hhmi.org/lectures/

http://www.richardgregory.org/

http://cvision.ucsd.edu/database/data/cones/linss2_10e_1.txt

Eye and camera: http://hyperphysics.phy-astr.gsu.edu/hbase/vision/rfreye.html

Light refraction:www.phy.ntnu.edu.tw/java/light/flashLight.html

Optics of thin lenses: www.phy.ntnu.edu.tw/java/Lens/lens_e.html

WHO (2001) International Classification of Functioning, Disability and Health (ICF), Geneva; www3.who.int/icf/icftemplate.cfm

www.viperlib.com

http://www.phys.ntnu.no/~arneval/illusjoner/Mach.doc

Index

Light Vision Color. Arne Valberg
© 2005 John Wiley & Sons Ltd